RAS Signaling Pathway in Cancer Therapy

RAS Signaling Pathway in Cancer Therapy

Guest Editor

Matthias Drosten

MDPI

Basel • Beijing • Wuhan • Barcelona • Belgrade • Novi Sad • Cluj • Manchester

Guest Editor
Matthias Drosten
Centro de Investigación
del Cáncer
Salamanca
Spain

Editorial Office
MDPI AG
Grosspeteranlage 5
4052 Basel, Switzerland

This is a reprint of the Special Issue, published open access by the journal *Cancers* (ISSN 2072-6694), freely accessible at: https://www.mdpi.com/journal/cancers/special_issues/PD34O9UUSY.

For citation purposes, cite each article independently as indicated on the article page online and as indicated below:

Lastname, A.A.; Lastname, B.B. Article Title. *Journal Name* **Year**, *Volume Number*, Page Range.

ISBN 978-3-7258-6942-8 (Hbk)
ISBN 978-3-7258-6943-5 (PDF)
https://doi.org/10.3390/books978-3-7258-6943-5

Contents

About the Editor

Matthias Drosten

Matthias Drosten is a tenured scientist at the Spanish National Research Council (CSIC) and a principal investigator at the Centro de Investigación del Cáncer (CIC) in Salamanca, Spain. His research focuses on the molecular mechanisms underlying RAS-driven lung cancer, with particular emphasis on dissecting RAS signaling pathways to identify novel and actionable therapeutic targets. Dr. Drosten has developed and characterized a range of genetically engineered mouse models that have been instrumental in advancing the understanding of RAS biology in lung tumor initiation, progression, and therapeutic resistance. His work has provided key insights into how oncogenic RAS signaling shapes tumor behavior and response to treatment, contributing to the identification of new vulnerabilities and treatment strategies. Through an integrative approach combining in vivo models, molecular biology, and translational research, his laboratory aims to improve therapeutic options for patients with RAS-driven lung cancer.

cancers MDPI

Article

K-Ras Binds Calmodulin-Related Centrin1 with Potential Implications for K-Ras Driven Cancer Cell Stemness

Ganesh babu Manoharan, Christina Laurini, Sara Bottone, Nesrine Ben Fredj and Daniel Kwaku Abankwa *

Cancer Cell Biology and Drug Discovery Group, Department of Life Sciences and Medicine, University of Luxembourg, L-4362 Esch-sur-Alzette, Luxembourg
* Correspondence: daniel.abankwa@uni.lu

Simple Summary: Trafficking chaperones facilitate the spatio-temporal distribution pattern of proteins inside cells. In the case of the membrane-anchored protein Ras, trafficking chaperones typically bind to the C-terminal farnesyl-moiety. Thus shielded from the aqueous environment, Ras can diffuse more efficiently through the cytoplasm. The calcium-binding protein calmodulin (CaM) was proposed as a K-Ras trafficking chaperone. However, CaM has many different functions inside the cell. Centrin proteins are highly related to calmodulin, and we find that they also bind to K-Ras. Unexpectedly, this interaction depends on the activation state and the effector binding site of K-Ras, not on the farnesyl-anchor. Overall, CaM and centrin1 appear to enable only a fraction of K-Ras membrane anchorage. Given that CaM inhibitors also affect the K-Ras/centrin1 interaction and the very similar distribution of centrin1 and CaM throughout the cell cycle, the dependence of K-Ras on either protein may be difficult to determine.

Abstract: Recent data suggest that K-Ras4B (hereafter K-Ras) can drive cancer cell stemness via calmodulin (CaM)-dependent, non-canonical Wnt-signalling. Here we examined whether another Ca^{2+}-binding protein, the CaM-related centrin1, binds to K-Ras and could mediate some K-Ras functions that were previously ascribed to CaM. While CaM and centrin1 appear to distinguish between peptides that were derived from their classical targets, they both bind to K-Ras in cells. Cellular BRET- and immunoprecipitation data suggest that CaM engages more with K-Ras than centrin1 and that the interaction with the C-terminal membrane anchor of K-Ras is sufficient for this. Surprisingly, binding of neither K-Ras nor its membrane anchor alone to CaM or centrin1 is sensitive to inhibition of prenylation. In support of an involvement of the G-domain of K-Ras in cellular complexes with these Ca^{2+}-binding proteins, we find that oncogenic K-RasG12V displays increased engagement with both CaM and centrin1. This is abrogated by addition of the D38A effector-site mutation, suggesting that K-RasG12V is held together with CaM or centrin1 in complexes with effectors. When treated with CaM inhibitors, the BRET-interaction of K-RasG12V with centrin1 was also disrupted in the low micromolar range, comparable to that with CaM. While CaM predominates in regulating functional membrane anchorage of K-Ras, it has a very similar co-distribution with centrin1 on mitotic organelles. Given these results, a significant overlap of the CaM- and centrin1-dependent functions of K-Ras is suggested.

Keywords: K-Ras; centrin; calmodulin; mitosis; centrosome; BRET

1. Introduction

KRAS is the most frequently mutated oncogene and in addition mutated in congenital disorders, called RASopathies [1,2]. It is not fully understood why *KRAS* is more frequently mutated in cancer than the other *RAS* genes, *NRAS* and *HRAS*. Several facets of Ras biology may contribute to the higher exploitation of *KRAS*, such as its higher expression level, its specific intracellular trafficking and distribution, or its distinct nanoscale organization in the plasma membrane that imposes differential effector usage [3–5]. Another less characterized difference is the ability of Ras proteins to drive stemness properties in cells [6,7]. Notably, the most common *KRAS* splice variant, K-Ras4B (hereafter K-Ras), but not H-Ras, mediates stemness properties via calmodulin (CaM)-dependent non-canonical Wnt-signalling [6]. In line with this, CaM inhibitors block the stemness properties of *K-RAS*-mutant cancer cells [8,9]. However, the exact mechanism of how CaM mediates K-Ras-driven stemness is not resolved.

Previous cellular data showed that K-Ras/CaM complexes are disrupted by phosphomimetic mutations of Ser181 at the C-terminus of K-Ras. Conversely, CaM binding blocked phosphorylation at that site [10]. Intriguingly, the phosphomimetic mutation of K-RasG12V on Ser181 reduces its ability to drive stemness [6]. Mutations at this site also modulate the interaction with another trafficking chaperone PDE6D [11], which traffics several prenylated proteins to stemness mediating organelles [12]. Hence, CaM may not be alone in mediating the K-Ras-stemness activity.

CaM possesses two Ca^{2+}-binding lobes, which can each encase 15–20 residue long peptide stretches of classical target proteins in their hydrophobic surfaces [13]. Classical target peptides are typically helical, positively charged, and contain hydrophobic anchor residues. Very similar biochemical characteristics are found in singly lipidated, polybasic termini of prenylated or myristoylated proteins, which have emerged as non-canonical targets of CaM [14]. CaM facilitates the Ca^{2+}-dependent cytoplasmic solubilization of K-Ras by sequestering its farnesyl-tail from the aqueous environment [15]. This contrasts to the GTP-Arl2/3 triggered release of PDE6D cargo [16]. PDE6D and CaM share the preference for K-Ras amongst the Ras isoforms as palmitoylation obstructs access to the hydrophobic pockets, making K-Ras4A, N-Ras, and H-Ras clients only in their non-palmitoylated states [17,18]. Both trafficking chaperones are found in the cyto- and nucleoplasm and on centriolar structures, such as the primary cilium and the centrosomes [16,19,20]. Hence, it is plausible to assume that these two chaperones have overlapping, yet distinct roles in coordinating trafficking of prenylated proteins spatio-temporally.

In cell lysates, CaM engages more with GTP-loaded K-Ras than with its inactive counterpart [17,21]. Furthermore, complexes between K-Ras, CaM, and PI3K p110 subunits have been proposed as being relevant for Akt activation during platelet-derived growth factor receptor (PDGFR)-mediated cell migration [22,23]. The fact that the activation state of Ras matters for its interaction with CaM contrasts with in vitro and structural data. Only weak transient contacts of CaM with non-farnesylated K-Ras were observed in NMR-experiments, while the farnesylated poly-lysine stretch of K-Ras comprising residues 180–185 was sufficient for CaM binding [15,24]. In vitro data further suggest that the polybasic and farnesylated C-terminus of K-Ras binds to either of the Ca^{2+}-bound lobes of CaM, but without involvement of the G-domain [25]. Thus, it appears that the farnesylated C-terminus of K-Ras is sufficient for micromolar binding to CaM. However, in cells, there may be CaM/ K-Ras complexes that depend on the activation state of K-Ras.

Inhibitors of CaM alter its conformation, thus preventing binding of canonical target peptides and non-canonical targets [9,13,26,27]. The covalent CaM inhibitor ophiobolin A disrupts binding of K-Ras to CaM and K-Ras membrane anchorage by irreversibly modifying Lys75, 77, and 148 of CaM [8,9,28]. We recently developed an alternative, less toxic covalent inhibitor of CaM, called Calmirasone1, which is much more suitable for cell biological applications [9].

Centrin (or caltractin) proteins are highly related to CaM with the same bi-lobal structure, however, only the C-terminal lobe binds and senses Ca^{2+} with high affinity [29]. This leaves the centrin-specific N-terminus free for mediating self-assembled extended structures of centrins, which are Ca^{2+}-dependent due to allosteric coupling with the C-terminus [30]. In humans, three centrin paralogs (centrin1-3, *CETN1-3*) are known [31]. While centrin2 and centrin3 are ubiquitously expressed, centrin1 expression is limited to male germ cells, neurons, and ciliated cells [32]. Centrin2 is probably best known for binding and stabilizing XPC (xeroderma pigmentosum group C), which is involved in DNA repair [33]. In addition, centrins have been implicated in nuclear pore functions and proteasomal activities [32]. Like CaM, centrins appear to recognize a hydrophobic motif of 15–20 residues in such classical target proteins [34].

The activity of centrins can be regulated by several phosphorylation and SUMOylation events [34]. Nuclear localization of centrin2 is enhanced by its SUMOylation [35]. Phosphorylation of T118 in the third EF-hand of the centrin2 C-terminal lobe is required for Ca^{2+}-binding and its centrosomal localisation [32]. Centrins localise to distal and intermediate regions preferentially from the mother centrioles and are part of a set of 14 ancient and highly conserved centriolar proteins [36,37]. Hence, loss of centrins broadly affects centriolar functions, including organisation of the microtubule network or overall biogenesis of centrioles [32]. Based on the essential roles of centrins in uni-cellular organisms that depend on cilia formation, it is plausible to assume that an important role also exists for centrins in vertebrate/mammalian ciliogenesis [32]. In line with this, ciliogenesis is reduced upon depletion of centrin2 in hTERT-RPE1 cells [38].

Given the highly similar bi-lobal structure with hydrophobic binding pockets, we hypothesized that centrins also bind to non-classical targets of CaM, such as K-Ras. Here we show that K-Ras binds to centrin1 in cells in a similar manner to CaM. Our results suggest that binding of K-Ras to these Ca^{2+}-binding proteins in cells is largely independent of the prenylation of K-Ras and involves the G-domain. Given that CaM inhibitors also affect the K-Ras/centrin1 interaction and the very similar distribution of centrin1 and CaM throughout the cell cycle, the dependence of K-Ras on either protein may be difficult to determine.

2. Experimental Procedures

2.1. Plasmids, siRNAs and Inhibitors

All construct names contain the tag at a position corresponding to its location in the protein sequence, e.g., GFP2-CaM, contains the GFP2-tag at the N-terminus of CaM. All plasmids employed in the study were produced by multi-site gateway cloning [39]. The human CaM (*CALM1*) entry clone with L1–L2 recombination sites was obtained from the NCI RAS Initiative. The K-Ras4b entry clone was from RAS mutant clone collection (Kit #1000000089) procured from Addgene (Watertown, MA, USA). Custom-synthesised entry clones encoding human centrin1 (*CETN1*) or the CTK fragment with L1–L2 recombination sites in pDONR221 vector were commercially obtained from Genecust, Boynes, France. An LR recombination reaction comprising three entry clones encoding the CMV promoter, a tag (Rluc8 or GFP2) and the protein of interest (CTK, K-Ras wt, CaM and centrin1); a destination vector, pDest-305 vector, was performed to obtain the recombinant plasmids. In a single-site LR recombination reaction, CaM or centrin1 entry clones were combined with the destination vector, pDest-527, to produce bacterial expression plasmids encoding N-terminally His6-tagged CaM and centrin1. The positive clones were selected using ampicillin in *E. coli* DH10B. The pmCherry-CaM, pEGFP-centrin1 plasmids and plasmids encoding N-terminal Rluc8 or GFP2-tagged K-RasG12V and H-RasG12V were previously described [9,26]. siRNA for *CALM1* (Hs_CALM1_6, SI02224222), and *FNTA* (Hs_FNTA_6, SI02661995) were obtained from Qiagen (Venlo, The Netherlands). The siRNA for *CETN1* (ON-TARGETplus SMARTpool siRNA, L-011831-00-0005) and negative control siRNA (ON-TARGETplus Non-targeting pool, D-001810-10-05) were obtained from Dharmacon (Cambridge, UK). Mevastatin (J61357, Alfa Aesar, Leuven, Belgium), calmidazolium chloride (sc-201494, Santa Cruz, Heidelberg, Germany), and ophiobolin A (sc-202266, Santa Cruz, Heidelberg, Germany) were commercially acquired from the sources given in parenthesis. Calmirasone1 was synthesized as previously described by us [9].

2.2. Protein Sequence Analyses

The protein sequences encoded by *CALM1-3* and *CETN1-3* genes were collected from the uniprot database (http://unirprot.org/; last accessed 2 April 2023) and a multiple sequence alignment was performed using Clustal Omega (https://www.ebi.ac.uk/Tools/

msa/clustalo/; last accessed 2 April 2023). For paralog number analysis, the protein coding genes of calmodulin and centrin were searched for each species in the NCBI protein database (https://www.ncbi.nlm.nih.gov/; last accessed 2 April 2023). The *CALM1* or *CETN1* genes were given as search query and orthologs were identified from the annotation pipeline. A process flow was then generated using RefSeq to identify a set of comparable proteins including orthologs and similar proteins. Note that only protein encoding genes were considered and pseudogenes were discarded.

2.3. Protein Purification

The His6-tagged human CaM and centrin1 proteins were purified as described previously [9]. Briefly, the pDest527-His6-CaM or pDest527-His6-centrin1 plasmid transformed *E. coli* BL21 (DE3) cells were grown in LB medium supplemented with 100 μg/mL of ampicillin. At 0.4–0.6 OD, 0.5 mM IPTG was used to induce the culture with subsequent overnight incubation at 25 °C with shaking. After centrifugation of the culture, its pellet was suspended in a lysis buffer composed of 20 mM HEPES, pH 7.6, 150 mM NaCl, 5 mM $MgCl_2$, 0.5 mg/mL lysozyme, and 700 units DNase I. For the pellet from 1 l of cell culture 20 mL of lysis buffer was used. After cell lysis by sonication, the His-tagged proteins were purified using HisTrapTM HP Prepacked Columns (GE Healthcare, Leuven, Belgium) on the ÄKTAprime plus chromatography system (GE Healthcare). A buffer composed of 50 mM Tris HCl, pH 7.5, 150 mM NaCl, and 35 mM imidazole was used to equilibrate the column, and His-tagged proteins were eluted using 250 mM imidazole elution buffer. Afterwards, the eluted fractions were dialyzed for 16 h at 4 °C in dialysis buffer (50 mM Tris HCl, pH 7.5, 150 mM NaCl, and 2 mM $CaCl_2$). Using a NanoDrop 2000c Spectrophotometer (Thermo Fisher Scientific, Merelbeke, Belgium), the protein concentration was determined by absorbance.

2.4. Fluorescence Polarisation Binding Assay

Fluorescence polarisation assays were performed as established previously by us [9,26]. The fluorescein-labelled PMCA- and CaMKII-peptides were custom synthesized by Genscript (Piscataway, NJ, USA) and Pepmic (Suzhou, China), respectively. The PMCA peptide was derived from 1086-LRRGQ-ILWFR-GLNRI-QTQIK-1105 of human PMCA and fluorescein was attached to the C-terminal native Lys. The CaMKII peptide sequence was derived from 294-NARRK-LKGAI-LTTML-ATRN-312 of human CaMKII and fluorescein was attached to a non-native cysteine added to the N-terminus. The N-terminal His6-tagged CaM or centrin1 proteins were 2-fold diluted in a buffer composed of 20 mM Tris Cl pH 7.5, 50 mM NaCl, 1 mM $CaCl_2$ and 0.005% *v*/*v* Tween 20 in a black, low volume, round bottom 384-well plate (cat. no. 4514, Corning, Amsterdam, The Netherlands). Then, 10 nM of fluorescein-labelled peptide was added to the protein dilution series. The reaction mix was incubated for 20 min at RT before anisotropy measurements.

The Sfi1 peptide was derived from 670-REVAA-RESQH-NRQLL-RGALR-RWK-692 of human Sfi1 and the fluorescein was attached to the native C-terminal Lys. Sfi1 peptide titration was performed in a buffer composed of 10 mM HEPES pH 7.4, 100 mM $CaCl_2$, and 0.005% *v*/*v* Tween 20. For binding, to a 2-fold dilution series of centrin1, 100 nM of Sfi1 peptide was added, and the reaction mix was incubated for 45 min at RT before anisotropy measurements. For measuring the IC_{50} of inhibitors to centrin1, to the 3-fold dilution series of inhibitors in the assay buffer, a complex of 100 nM fluorescein labelled Sfi1 peptide and 250 nM His-centrin1 was added in 20 μL volume in a 384-well plate. The fluorescence anisotropy was measured after overnight incubation at RT.

The fluorescence anisotropy was measured on a Clariostar (BMG Labtech, Ortenberg, Germany) plate reader using the fluorescence intensity signal recorded from vertical (I_v)- and horizontal (I_h)-polarised light using a fluorescence polarisation module ($\lambda_{excitation}$ 482 ± 8 nm and $\lambda_{emission}$ 530 ± 20 nm). Fluorescence anisotropy was calculated from the measured fluorescence intensities according to $r = \frac{I_v - G(\lambda)I_h}{I_v + 2G(\lambda)I_h}$, where r is the fluorescence anisotropy value and I_v and I_h are the fluorescence emission intensities detected with vertical and horizontal polarisation, respectively. The instrument specific correction factor $G(\lambda)$ was set to 1 and not determined further. A quadratic equation as described [40,41] by others was defined in Prism (GraphPad, version 9.5.1, La Jolla, CA, USA) and was used to determine the K_D value of the fluorescein tagged peptides to target protein.

$$y = \frac{Af + (Ab - Af) * (Lt + K_D + x - \sqrt{(Lt + K_D + x)^2 - 4 * Lt * x}}{2Lt}$$

Here, Af is the anisotropy value of the free fluorescent probe, Ab is the anisotropy value of the fluorescent probe/protein complex, Lt is the total concentration of the fluorescent probe, K_D is the equilibrium dissociation constant, x is total concentration of protein, and y is measured anisotropy value. K_D is measured in the same unit of x. Note that variations in the active fraction of the home-made proteins and different methods used to determine the protein concentrations of the obtained K_D values can vary from those reported.

The IC_{50} value of inhibitors was determined by plotting the log concentration of inhibitor against fluorescence anisotropy values and fitting the data to log inhibitor vs. response—variable slope (four parameters) equation in Prism (GraphPad). The IC_{50} of the inhibitor was converted into K_d as described earlier using the equation [42],

$$K_d = \frac{[I]_{50}}{1 + \frac{[P]_{50}}{K_{D,probe}} + \frac{[E]_0}{K_{D,probe}}}$$

where $[I]_{50} = IC_{50} - [EI]_{50}$, in which $[EI]_{50}$ is the concentration of the centrin1:inhibitor complex at 50% displacement, $[I]_{50}$ is the free inhibitor concentration at 50% displacement, $[P]_{50}$ is the concentration of the free probe, F-Sfi1 at 50% displacement, $[E]_0$ is concentration of free centrin1 at 0% displacement, and $K_{D,probe}$ is the dissociation constant of the complex of centrin1 and Sfi1.

2.5. Co-Immunoprecipitation Experiments

About 800,000 HEK293-ebna cells were seeded in 60 mm dishes and cultured in Dulbecco's Modified Eagle's Medium (DMEM) supplemented with 10% v/v Foetal Bovine Serum (FBS), 2 mM L-glutamine (cat. no. 25030-024, Gibco, Thermo Fisher Scientific), and 1% v/v penicillin/ streptomycin (cat. no. 15140122, Gibco, Thermo Fisher Scientific) overnight. The next day, cells were transiently transfected with 4 μg plasmids encoding the indicated combinations of constructs using jetPRIME (cat. no. 114-75, Polyplus, Leuven, Belgium) according to the manufacturer's instructions. At 48 h post-transfection, the cells were lysed using 200 μL of Lysis buffer (10 mM Tris Cl pH 7.5, 150 mM NaCl, 2 mM $CaCl_2$, 0.2% v/v NP40) supplemented with protease inhibitor cocktail (cat. no. A32955, Pierce, Thermo Fisher Scientific). After 30 min incubation on ice, the lysate was cleared by centrifugation for 10 min at 4 °C and 17,000× g. The cleared lysate was transferred to a clean tube and 15 μL sample was withdrawn (as "Input" for Western blot analysis). The lysate was diluted with 300 μL of dilution buffer (10 mM Tris Cl pH 7.5, 150 mM NaCl, 2 mM $CaCl_2$) supplemented with protease inhibitor cocktail. Then 25 μL of GFP-trap Beads Slurry (ChromoTek GFP-Trap Agarose, cat. no. gta, Proteintech Europe, Manchester, UK) were added to the diluted lysate and rotated end-over-end for 1 h at 4 °C. Then, the beads were washed 3 times with Wash buffer (10 mM Tris Cl pH 7.5, 150 mM NaCl, 2 mM $CaCl_2$, 0.02% v/v NP40). Bound proteins were eluted by the addition of 2 × Laemlli buffer and boiling for 10 min at 95 °C. The eluted proteins were subsequently analysed by SDS-PAGE on 10% acrylamide gels. Using the Trans-Blot Turbo Transfer system (Bio-Rad, Temse, Belgium), proteins were transferred onto a 0.2 μm nitrocellulose membrane (Bio-Rad) and incubated with a primary antibody. The following primary antibodies were used: anti-GFP (SAB4301138, Sigma-Aldrich, Overijse, Belgium, at dilution ratio 1:5000), anti-Renilla Luciferase (ab187338, Abcam, Cambridge, UK, at dilution ratio 1:3000), and anti β-actin (A5441, Sigma-Aldrich, Overijse, Belgium, at dilution ratio 1:5000). Anti-rabbit IRDye 680RD or anti-mouse IRDye 800CW secondary antibodies (LI-COR Biosciences, Bad Homburg vor der Höhe, Germany) were used to visualise the proteins on an Odyssey CLx system (LI-COR). The relative expression level of proteins was densitometrically quantified from images of membranes analysed using Image Studio software (LI-COR, version 5.2). For the quantitative analysis of the pull-down proteins, the signal of the Rluc8-tagged prey proteins was normalized with the signal from the GFP-tagged bait protein. Next, the signal intensity of the GFP2-K-RasG12V + Rluc8-CaM transfected sample was used to normalize the other samples.

2.6. BRET Donor Saturation Titration Assays

The detailed method of our BRET assay can be found in [9,43]. Briefly, ~200,000 HEK293-ebna cells were seeded per well of a 12-well plate (cat. no. 665180, Greiner Bio-One, Vivoorde, Belgium) and grown in 1 mL of complete DMEM. The next day, ~1 µg of BRET sensor plasmids was transfected using 2.5 µL of jetPRIME. The concentration of donor plasmid was 25 ng, and that of the acceptor plasmid had increased from 25 ng to 1000 ng for titration curves. Cells were treated with inhibitors or vehicle control (DMSO at 0.2% *v*/*v*) 24 h after transfection. Cells were collected the following day in PBS and plated in white, flat bottom 96-well plates (cat. no. 236108, Nunc, Thermo Fisher Scientific). BRET measurements were performed on a Clariostar plate reader (BMG Labtech). Three channels were read. The first channel was First, GFP2-fluorescence ($\lambda_{excitation}$ 405 ± 10 nm and $\lambda_{emission}$ 515 ± 10 nm), which is directly proportional to the acceptor concentration (RFU). Second, the channel was read after adding coelenterazine 400a (cat. no. C-320, GoldBio, Saint Louis, MO, USA; at 10 µM final concentration) BRET-channel (515 ± 15 nm) readings in well-mode, and then again (third) with the luminescence channel (410 ± 40 nm), and recordings were made. Signals corresponded to the BRET signal and donor (RLU) signals. The ratio of BRET signal/RLU gave the raw BRET ratio. The final BRET ratio (BRET in plots) was obtained by subtracting the raw BRET ratio from the background raw BRET ratio of cells expressing only the donor. The relative expression is calculated as the ratio of RFU/RLU and denoted as [Acceptor]/[Donor]. The BRET ratio vs [Acceptor]/[Donor] ratio data from biological repeats (typically three) were plotted together, and the data were fitted by a hyperbolic equation in Prism. The BRETtop value represents the top asymptote of the BRET ratio reached within the defined [Acceptor]/[Donor] ratio. The one phase association equation of Prism 9 (GraphPad) was used to predict the top asymptote Ymax-value, which was taken as the BRETtop. Statistical analysis between the BRETtop values was performed using the Extra sum-of-squares F test.

2.7. Dose Response Analysis of Inhibitors and siRNA Knockdown in BRET Assays

For dose response analysis of inhibitors, on day one, ~200,000 HEK293-ebna cells were seeded per well of a 12-well plate (cat. No. 665180, Greiner Bio-One, Vivoorde, Belgium) and grown in complete DMEM. On day two, ~1 µg of BRET sensor plasmids were transfected at the indicated donor/acceptor plasmid ratio using jetPRIME, as mentioned in the corresponding figure legends. On day three, the medium was exchanged with fresh medium containing various doses of inhibitors. After 24 h incubation, on day four, the cells were collected in PBS, and the BRET assay was performed. The log inhibitor vs BRET ratio was plotted, and the data were fitted by a log (inhibitor) vs. response variable slope (four parameters) equation of Prism, and the IC_{50} values were calculated.

For studying the effect of siRNA-mediated knockdown, on day one, the HEK293-ebna cells were seeded in 12-well plates in 1 mL of growth medium. On day two, cells were transfected using 3.5 µL Lipofectamine RNAiMAX (cat. no. 13778, Thermo Fisher Scientific) and Opti-MEM medium (cat. no. 31985062, Gibco, Thermo Fisher Scientific) as the vehicle with 100 nM of siRNA per well. On the next day, the medium was exchanged, and the cells were transfected with ~1 µg of BRET sensor plasmids using 3 µL jetPRIME reagent and expressed for 48 h. The transfected donor/acceptor plasmid ratio is indicated in corresponding figure legends. On day five, the BRET assay was performed as indicated above.

2.8. siRNA-Mediated Knockdown and Western Blotting

About 300,000 HEK293-ebna cells were seeded per well of a 6-well plate (cat. no. 657160, Cellstar, Greiner Bio-One) and grown in 2 mL of complete DMEM for 24 h. The next day, cells were transfected with 100 nM of siRNA using Lipofectamine RNAiMAX, followed by a medium exchange after 4 h. After 48 h, cells were lysed in RIPA buffer (10 mM Tris, 150 mM NaCl, 0.5 mM EDTA, 0.2% *v/v* NP40) supplemented with protease inhibitor cocktail (cat. no. A32955, Pierce, Thermo Fisher Scientific). The protein amount in cell lysates were quantified using Bio-Rad protein assay kit (cat. no. 5000006). Cell lysate containing 50 µg of protein per lane was resolved in Mini-PROTEAN precast 4–20% acrylamide gels. Proteins were subsequently transferred onto a nitrocellulose membrane 0.2 µm (Bio-Rad) using the Trans-Blot Turbo Transfer system (Bio-Rad) and probed with the mix of primary antibodies against the protein of interest and the loading control. The primary antibodies employed were anti-FNTA (cat. no. ab109738-1001, Abcam, at 1:1000), anti-centrin1 (cat. no. 12794-1-AP, Proteintech, Manchester, UK, at dilution ratio 1:500) and anti-GAPDH (cat. no. G8796 mouse and G9545 rabbit, Sigma-Aldrich, at 1:10,000). Anti-mouse or anti-rabbit IRDye 800CW or 680RD secondary antibodies (LI-COR) were used subsequently to develop the membrane, and the proteins were detected using an Odyssey CLx system (LI-COR).

2.9. Three-Dimensional Spheroid Assay

MDA-MB-231 and MCF-7 cells were seeded in 12-well plates (cat. No. 665180, Greiner Bio-One) and transfected with either 100 nM negative control siRNA or siRNA targeting *CALM1* or *CETN1* using Lipofectamine RNAiMAX. A day later, cells were harvested and plated into low-attachment, suspension cell culture 96-well plates (cat. no. 655185, Cellstar, Greiner Bio-One) for 3D spheroid suspension culture. About 1000 MDA-MB-231 or 2500 MCF-7 cell were seeded per well of the 96-well plate in 50 µL of RPMI medium (cat. no. 52400-025, Gibco, Thermo Fisher Scientific) or DMEM, respectively, containing 0.5% *v/v* MethoCult (cat. no. SFH4636, Stemcell technologies, Grenoble, France), 1x B27 (cat. no. 17504044, Gibco, Thermo Fisher Scientific), 25 ng/mL EGF (cat. no. E9644, Sigma-Aldrich), and 25 ng/mL FGF (cat. no. RP-8628, Thermo Fisher Scientific). Cells were incubated in a cell culture incubator for 6 days, and fresh growth medium was supplemented on the third day. After six days of incubation, the alamarBlue reagent (cat. No. DAL1025, Invitrogen, Thermo Fisher Scientific) was added to each well (10% final volume) for 4 h at 37 °C. Using a Clariostar plate reader, the fluorescence signal ($\lambda_{excitation}$ 560 ± 5 nm and $\lambda_{emission}$ 590 ± 5 nm) was recorded. Fluorescence signals were normalized to negative control siRNA, which was set to 100% sphere formation.

2.10. Confocal Microscopy

HeLa cells were seeded on glass coverslips 1.5H (cat. no. LH22.1, Carl Roth, Karlsruhe, Germany) in 6-well plates (cat. no. 657160, Cellstar, Greiner Bio-One) and grown in complete DMEM for 24 h. The next day, the cells were transiently co-transfected with pmCherry-CaM and pmGFP-K-RasG12V using jetPRIME. At 48 h after transfection, cells were fixed using 4% *v/v* formaldehyde (cat. no. 43368, Alfa Aesar) in PBS for 10 min at room temperature. The fixation solution was then replaced with PBS-Tween 0.05% *v/v* (cat. no. 9127.1, CarlRoth). After permeabilization in PBS-Triton X100 0.5% *v/v* (cat. no. T8787, Merck, Overijse, Belgium) for 10 min and blocking for 30 min in 2% *v/v* solution of BSA (A6588, Applichem, Darmstadt, Belgium) in PBS, the cells were incubated for 1 h at room temperature with primary antibody against centrin1 (rabbit polyclonal, cat no.12794-1-AP, Proteintech). After washing with PBS-Tween 0.05% *v/v*, the secondary antibody AlexaFluor 667 goat anti-rabbit (cat no. A21244, Life Technologies, Thermo Fisher Scientific) was applied for 1 h at room temperature. A 1 mg/mL solution of DAPI (cat.

no. D1306, Thermo Fisher Scientific) in PBS for 10 min was used for DNA-staining. Using Vectashield (cat. no. H-1000, Vector Laboratories, Brussels, Belgium) coverslips were mounted onto glass slides. Images were captured on a spinning disk confocal microscope (Andor, Oxford Instruments, Belfast, UK) fitted with a Zyla 5.5 sCMOS camera (Andor, Oxford Instruments) and using a plan APO 60×/1.40 Ph3 DM oil immersion objective (Nikon, Brussels, Belgium) and NIS-Elements Imaging Software (Nikon, Version 5.42.02).

2.11. Data and Statistical Analysis

Prism 9 (GraphPad) was used for the preparation of plots, data, and statistical analysis. The number of independent biological repeats (n) and the type of statistical analysis used are indicated in the corresponding figure legends. A p-value < 0.05 is considered statistically significant, and the statistical significance levels are annotated as follows: * $p < 0.05$; ** $p < 0.01$; *** $p < 0.001$; **** $p < 0.0001$, or ns = not significant.

3. Results and Discussion

3.1. Binding Studies Support Specific Canonical Target Peptides for CaM or Centrin1

Calmodulin (CaM) and centrin proteins are highly related, both at the sequence level (Figure 1A) with 54% sequence identity between CaM and centrin1, and structurally (Figure 1B), with the most obvious difference being the N-terminal extension of centrins. While three CaM genes encode proteins with the exact same sequence, the three centrin paralogs are more divergent. Centrin1 and −2 are ~84% identical in sequence, while centrin3 differs significantly from centrin1 with only 58% sequence similarity. In several vertebrates, at least two paralog genes from each family of Ca^{2+}-binding proteins are found, supporting their cell biological significance (Figure 1C).

Current evidence suggests that both CaM and centrins have distinct target protein selectivities [34]. We therefore examined whether centrin1 could also bind to classical CaM target proteins, such as the plasma membrane calcium transporting ATPase isoform 4b (PMCA) and CaM-dependent kinase II (CaMKII). PMCA removes intracellular calcium and, a 20-residue stretch mediates its regulation by CaM to which it binds with low nanomolar affinity [44]. An even higher picomolar affinity has been reported for the 19-residues of CaMKII [45].

We employed fluorescence polarization experiments to measure the binding of fluorescein-labelled peptides of these target proteins, F-PMCA and F-CaMKII, to His-tagged CaM and centrin1, respectively. Similar to previous observations with bovine CaM [26], we found that both peptides bound to human CaM with low nanomolar affinity (F-PMCA, $K_D = 36 \pm 5$ nM; F-CaMKII, $K_D = 6.6 \pm 0.2$ nM) (Figure 1D). By contrast, no binding of either peptide to human centrin1 was observed, even at 2 μM centrin1 concentration (Figure 1E). However, when testing a fluorescently labelled 18-residue long centrin1-specific target peptide, F-Sfi1, derived from the mitotic spindle regulator Sfi1, we observed a nanomolar affinity ($K_D = 30 \pm 12$ nM) (Figure 1F), which was higher than the reported micromolar affinity [46]. This deviation could be partially explained by the applied methods, as in the latter case, isothermal titration calorimetry was used.

Overall, these results suggest that the sequence divergence between CaM and centrin1 is sufficient to define specific binding to their classical targets that contain a distinct peptide recognition sequence.

3.2. Cellular BRET Data Suggest That the K-Ras G-Domain Participates in Complexes with Either CaM or Centrin1

Given the high sequence similarity between CaM and centrin1 (Figure 1A), we investigated whether farnesylated K-Ras could bind to centrin1 as a non-canonical target. Centrin1 was chosen due to its expression in ciliated cells, notably stem cells [32,47]. We therefore established a cellular Bioluminescence Resonance Energy Transfer (BRET)-assay to test binding of wild-type K-Ras or oncogenic K-RasG12V to centrin1 as compared to CaM.

We genetically fused the donor emission enabling Renilla Luciferase-derivative Rluc8 to the N-terminus of the K-Ras protein and the acceptor GFP2 to the N-terminus of centrin1 or CaM. If donor- and acceptor-tagged proteins interact, the BRET signal increases with increasing acceptor-to-donor ratio and may reach a saturation value. Commonly, the BRETmax value describes an absolute saturation value [48], which is typically not reached in most BRET titration experiments, and is therefore associated with significant extrapolation.

We here introduce the BRETtop value that characterizes the highest BRET value reached within a defined acceptor-to-donor ratio titration range. By keeping the titration range constant, we can compare different BRETtop values with each other. As with FRET, the BRET-values depend on the distance between the luminophores in the complex of interacting proteins. Therefore, only if the binding modes, i.e., the structure of the complexes are comparable, such as can be reasonably assumed for point mutants or paralogs of a protein, higher BRETtop values indicate a higher interaction probability and strength of examined BRET-pairs in cells. Like BRETmax, BRETtop would then correlate with the relative number of binding-sites and the relative affinities.

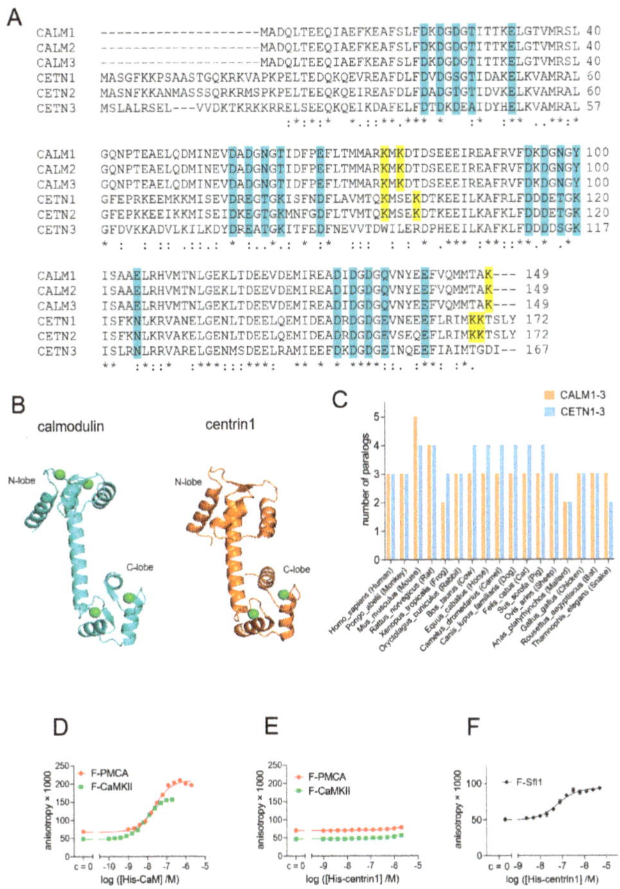

Figure 1. Despite its high similarity to CaM, centrin1 does not recognize CaM-target peptides. (**A**) Multiple sequence alignment of human CaM (*CALM1-3*) and centrin (*CETN1-3*) protein paralogs,

designated by the encoding gene names. The Ca^{2+}-binding residues are highlighted in cyan. Lysines 75, 77, and 148 of CaM, which become covalently modified by CaM inhibitor ophiobolin A, are highlighted in yellow. The same highlight was used for lysine residues at similar positions in centrin1 and centrin2, while no such lysine residues could be identified for centrin3. Note that the CaM protein numbering starts at Ala, as the N-terminal, and native Met is removed in most organisms [49]. (**B**) Structures of human CaM (PDB ID 1CLL) and human centrin1 (PDB ID 2GGM). Calcium ions are marked as green spheres. Structures were generated using the PyMOL Molecular Graphics System, Version 2.4.0, Schrödinger, LLC. (**C**) Analysis of the number of paralog coding genes of *CALM1-3* and *CETN1-3* in different species. Data were curated from the NCBI protein database. (**D–F**) Binding of 10 nM fluorescein-labelled F-CaMKII and F-PMCA (**D**,**E**) or 100 nM F-Sfi1 (**F**) peptides to His-tagged human CaM or centrin1 was detected using fluorescence anisotropy measurements. * $p < 0.05$; ** $p < 0.01$; *** $p < 0.001$; **** $p < 0.0001$.

We previously observed a higher interaction BRET-signal of oncogenic K-Ras as compared to its wild-type (wt) counterpart with CaM [9]. In line with these data, both CaM and centrin1 were significantly more co-immunoprecipitated with oncogenic GFP2-tagged K-RasG12V than with wt K-Ras (Figure 2A,B; Figure S1).

Consistent with our previous BRET-data, we also found that K-RasG12V had a significantly higher BRETtop with CaM than wt K-Ras (Figure 2C). Similarly, the BRETtop of K-RasG12V with centrin1 was significantly higher than that of wt K-Ras with centrin1 (Figure 2D). As expected, a control BRET-pair showed significantly lower BRET-values than the weakest BRET-interaction pair studied (Figure S2). The higher BRET of K-RasG12V with the Ca^{2+}-binding proteins was surprising given the afore-mentioned in vitro binding data [15,25]. Ras binding to some effectors can be reduced by the D38A-mutation, which abolishes major contacts preserved in several effector complexes [50,51]. Addition of the D38A mutation reduced the BRET to the level of wt K-Ras for both CaM and centrin1 (Figure 2C,D). This may suggest a dependence on some effectors or other effector lobe binders; however, more comprehensive studies are required to demonstrate this.

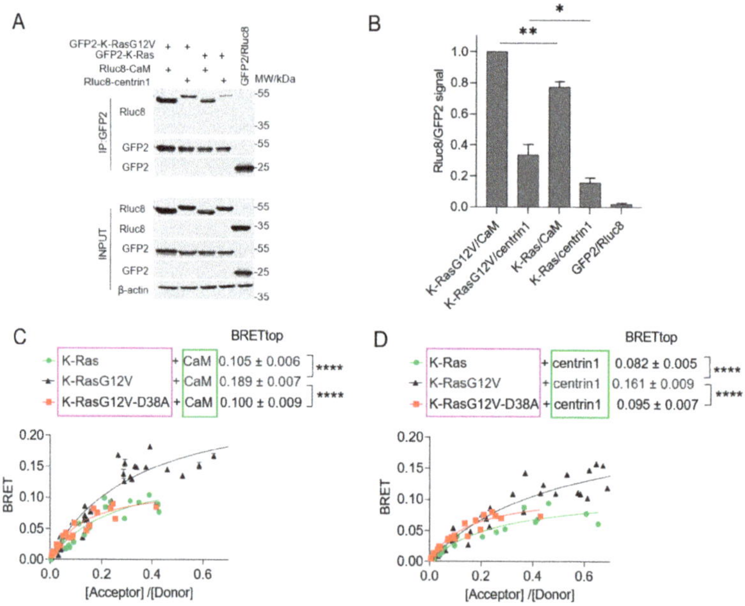

Figure 2. The interaction of CaM or centrin is increased with oncogenic K-Ras. (**A**) Co-immunoprecipitation of Rluc8-CaM or Rluc8-centrin1 with GFP2-K-RasG12V or GFP2-K-Ras wt. Pull-down was performed

using lysates of HEK293-ebna cells transfected with combinations of GFP2-K-RasG12V/Rluc8-CaM, GFP2-K-RasG12V/Rluc8-centrin1, GFP2-K-Ras/Rluc8-CaM, GFP2-K-Ras/Rluc8-centrin1 and GFP2/Rluc8 and expressed for 48 h. The GFP2-tagged protein was bound using GFP-trap beads, and the samples were analysed using anti-Rluc8 and anti-GFP antibodies. See Figure S1 for the original images of Western blots. (**B**) Immunoprecipitated Rluc8-tagged protein signals were normalized to GFP-tagged protein signals. The signal intensity of the GFP2-K-RasG12V/Rluc8-CaM transfected sample was set to 1 in each experiment and was used to normalize the other samples. The plot shows mean ± SEM and the statistical analysis was performed using one-way ANOVA test. (**C**,**D**) Interaction of Rluc8-K-Ras wt, Rluc8-K-RasG12V, and Rluc8-K-RasG12V-D38A with GFP2-CaM (**C**) or GFP2-centrin1 (**D**). All samples were treated with 0.2% *v*/*v* DMSO for 24 h, n = 3. Statistics of BRETtop values were analysed using the F-test. BRET donor protein is boxed purple, acceptor protein is boxed green. * $p < 0.05$; ** $p < 0.01$; **** $p < 0.0001$.

3.3. CaM Inhibitors Bind to Centrin

Given that centrin1 possesses lysines on positions 96, 100, 167, and 168 that are homologous to those targeted by covalent CaM inhibitors (Figure 1A), we tested whether covalent CaM inhibitors ophiobolin A and calmirasone1 or the potent non-covalent CaM inhibitor calmidazolium would disrupt binding of K-Ras to centrin1 in cells. Indeed, treatment with any of these CaM inhibitors lowered the BRETtop of K-RasG12V/centrin1 (Figure 3A). The inhibition of this interaction occurred at IC_{50} (calmidazolium) = 10.44 ± 0.05 μM and IC_{50} (calmirasone1) = 41.6 ± 0.3 μM (Figure 3B), the latter of which was comparable to what was previously observed with CaM [9]. For centrin1, fluorescence anisotropy data revealed that CaM inhibitors can displace fluorescently labelled Sfi1 from it, indicating their direct binding to centrin1 (Figure 3C, Table 1).

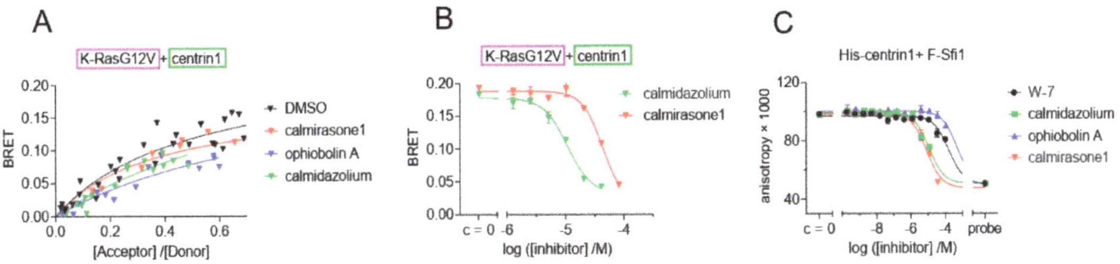

Figure 3. The interaction of K-Ras with centrin1 is modulated by direct binding of CaM inhibitors to centrin1. (**A**) HEK293-ebna cells were transfected with Rluc8-K-RasG12V/GFP2-centrin1 BRET sensor plasmids for 24 h followed by treatment with ophiobolin A (2.5 μM), calmidazolium (10 μM), calmirsone1 (20 μM) or equal volume of DMSO (0.2% *v*/*v*) for another 24 h, n = 3. (**B**) HEK293-ebna cells were transfected with BRET sensor plasmids Rluc8-K-RasG12V/ GFP2-centrin1 at a ratio of 1/19, respectively, for 24 h followed by a 24 h treatment with 2-fold dilution series of calmidazolium or calmirasone1 ranging from 80 μM to 0.1 μM. Data represent mean ± SEM, n = 2. BRET donor protein is boxed purple, acceptor protein is boxed green. (**C**) Displacement of fluorescent F-Sfi1 from centrin1 by CaM inhibitors. The inhibitors were 3-fold diluted in assay buffer, followed by addition of the complex of 100 nM F-Sfi1 and 250 nM His-centrin1. The fluorescence anisotropy was measured after overnight incubation at RT.

Table 1. Comparison of K_d values of CaM inhibitors with centrin1 and CaM determined by fluorescence anisotropy measurements. The competition assay derived K_d values of inhibitors to CaM were previously reported by us, using F-PMCA peptide as probe.

Inhibitor	Centrin1	CaM
	Mean K_d (Repeat Values)	K_d (References)
calmidazolium	1.6 (1.4; 1.8) μM	13.5 nM [26]
W-7	18.2 (17.8; 18.5) μM	1.47 μM [26]
ophiobolin A	49 (58; 39) μM	3.5 μM [9]
calmirasone1	0.9 (1.0; 0.8) μM	0.87 μM [9]

The sensitivity of the K-Ras/centrin1 interaction to CaM inhibitors suggests conserved inhibitor binding sites and a similar mode of interaction between K-Ras and the Ca^{2+}-binding proteins. Importantly, treatment with several CaM inhibitors may therefore also affect centrin1 biology, making it potentially difficult to interpret inhibitor-dependent phenotypic observations.

3.4. Inhibition of Prenylation Does Not Disrupt the BRET-Interaction of K-Ras with CaM or Centrin1 in Cells

Agamasu et al. have previously reported that the K-Ras-derived farnesylated and carboxymethylated KSKTKC-peptide is sufficient to bind to CaM in vitro [25]. To test whether non-prenylated K-Ras can still bind to the Ca^{2+}-binding proteins in cells, we tested the effects of the prenylation inhibitor mevastatin in our BRET-assays. Statins such as mevastatin inhibit the HMG-CoA pathway, and thus, provision of prenylpyrophosphate substrates for protein prenylation [52]. We therefore expected that treatment of cells with high concentrations of mevastatin would abrogate farnesyl-mediated K-Ras/CaM interaction. Surprisingly, mevastatin treatment did not significantly affect the BRET-levels of K-Ras with either CaM or centrin1 (Figure 4A,B). The higher BRETtop of K-Ras with CaM (Figure 4A) than with centrin1 (Figure 4B) may relate to the fact that only one Ca^{2+}-binding lobe is found in centrins [32], which may allow for the binding of only one K-Ras per centrin1 protein.

We next examined whether the C-terminal membrane targeting sequence of K-Ras alone (residues 166–188), CTK, was sufficient to mediate binding to CaM, as suggested by in vitro data, and whether the same would apply for binding to centrin1. In agreement with in vitro data, the BRET between CTK and CaM indicated binding; it had a lower BRETtop than full length K-Ras (Figure 4C), but the BRET-values were still above background (Figure S2). The CTK interaction with centrin1 was comparable to that with CaM (Figure 4C,D). As with full-length K-Ras, mevastatin treatment did not decrease the BRET of CTK with either of the Ca^{2+}-binding proteins (Figure 4C,D).

This mevastatin insensitivity was overall unexpected, given the strong contribution of the farnesyl-moiety to CaM-binding in vitro [15,24,25], but it was in line with data showing binding of non-farnesylated K-RasG12V to CaM in vitro [53].

Taken together with the activation-state dependent complexation of K-Ras with CaM or centrin1, this may suggest that these proteins exist in cellular complexes that are largely prenylation independent yet involve the C-terminal poly-lysine stretch of K-Ras and depend on the activation state of K-Ras. Alternatively, similarly sized, distinct pools of K-Ras in complex with the Ca^{2+}-binding proteins exist, and they require a subset of the aforementioned features.

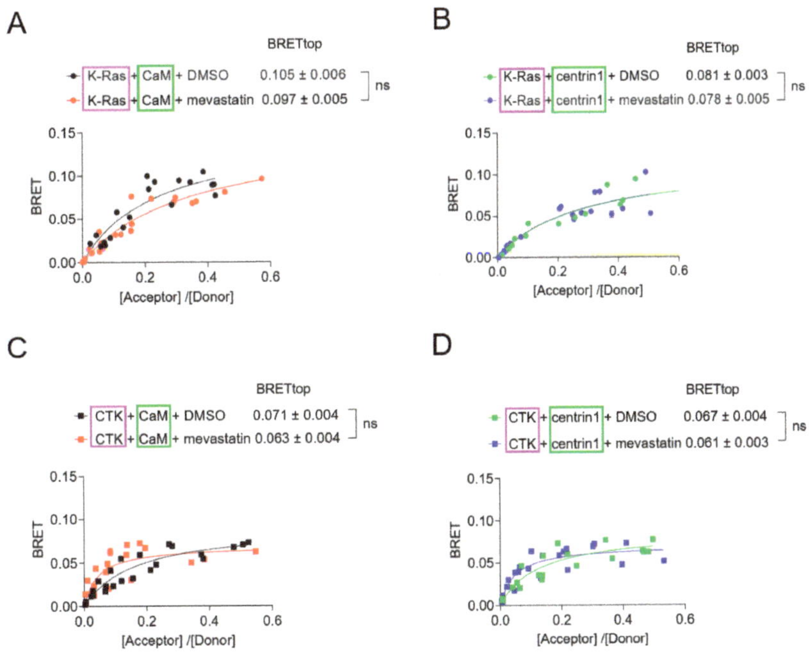

Figure 4. Inhibition of prenylation by mevastatin does not disrupt BRET between K-Ras and CaM or centrin1. (**A–D**) BRET-sensors Rluc8-K-Ras/GFP2-CaM (**A**), Rluc8-K-Ras/GFP2-centrin1 (**B**), Rluc8-CTK/GFP2-CaM (**C**) and Rluc8-CTK/GFP2-centrin1 (**D**) were transfected into HEK293-ebna cells, and cells were treated with 10 μM mevastatin or the vehicle control, DMSO 0.2% v/v for 24 h, n = 3. Statistics of BRETtop values was analysed using the F-test. BRET donor protein is boxed purple, acceptor protein is boxed green. ns = not significant.

3.5. Membrane Targeting and Anchorage of K-Ras Depends More on CaM Than on Centrin1

Prenyl-binding chaperone proteins can effectively facilitate diffusion of their target proteins in cells, as they shield the hydrophobic prenyl-moiety and thus allow for a longer residence in the aqueous cytoplasm [4]. Others suggested that CaM can extract and solubilize K-Ras and act as a trafficking chaperone [15].

We previously showed that inhibition of CaM selectively reduces K-RasG12V- as compared to H-RasG12V-BRET signals that originate from nanoclustering of active Ras on the plasma membrane [9]. This nanoclustering-dependent BRET-signal is sensitive to disruption, not only of Ras nanoclustering, but of any process upstream that interferes with functional membrane anchorage, such as disrupted trafficking or inhibition of the Ras lipid modification [54]. Similar to CaM inhibition, knockdown of another trafficking chaperone, PDE6D, which also binds to prenylated proteins and facilitates K-Ras localization at the plasma membrane, reduces K-Ras membrane anchorage associated FRET [55].

In line with the mevastatin data (Figure 4), binding of K-Ras to CaM or centrin1 was essentially insensitive to inhibition of prenylation by the knockdown of the shared α-subunit of farnesyl- and geranylgeranyl-transferases (*FNTA*) (Figure 5A–C). However, the same treatment significantly abrogated the membrane anchorage-BRET signal of both K-RasG12V or H-RasG12V (Figure 5E,F), consistent with the significance of prenylation for Ras membrane anchorage [56].

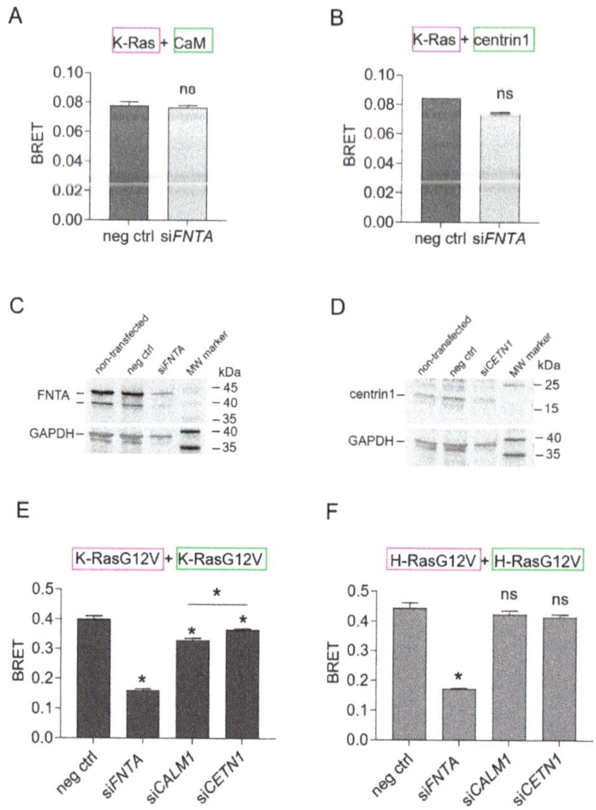

Figure 5. K-Ras membrane anchorage is selectively affected by CaM-, but less so by centrin1-depletion. (**A**,**B**) Rluc8-K-Ras was transfected with GFP2-CaM (**A**) or GFP2-centrin1 (**B**) plasmids at a donor/acceptor plasmid ratio of 1/5 into HEK293-ebna cells. BRET donor protein is boxed purple, acceptor protein is boxed green. Data represent mean ± SEM, n = 2 to 4. Statistical significance between negative control siRNA and sample siRNA was analysed using Mann–Whitney test. (**C**,**D**) HEK293-ebna cells were transfected with 100 nM of negative control siRNA or si*FNTA* or si*CETN1* for 48 h and cell lysates were immunoblotted as indicated. See Figures S3 and S4 for the original images of Western blots. (**E**,**F**) HEK293-ebna cells were transfected with 100 nM siRNA for 24 h, followed by BRET sensor transfection. Rluc8-/GFP2-tagged K-RasG12V (**E**) or H-RasG12V (**F**) nanoclustering-BRET sensor plasmids were transfected at a donor/acceptor plasmid ratio of 1/15. BRET donor protein is boxed purple, acceptor protein is boxed green. Data represent mean ± SEM, n = 4. Statistical significance between negative control siRNA and sample siRNA was analysed using Mann–Whitney test. * $p < 0.05$; ns = not significant.

As observed previously, knockdown of CaM selectively reduced the membrane anchorage-BRET signal of K-RasG12V (Figure 5E) but not H-RasG12V (Figure 5F). By contrast, knockdown of centrin1 decreased the BRET-signal K-Ras-selectively and to a significantly lesser extent than knockdown of CaM (Figure 5D–F). Immunoblotting confirmed the significant knockdown of *FNTA* and centrin1 (*CETN1)* expression in HEK293-ebna cells (Figure 5C,D; Figures S3 and S4), while that of CaM (*CALM1)* was previously validated by us using RT-qPCR [9].

These data suggest that CaM is more important to facilitate membrane trafficking of K-Ras in cells than centrin1.

3.6. Centrin1 Co-Distributes with CaM during the Cell Cycle

We previously observed that CaM inhibitors decrease stemness properties of KRAS-mutant cancer cell lines [8,9]. The clonogenic growth of cancer cell spheroids is employed as a surrogate measure for cancer cell stemness [57]. We therefore tested the effect of the knockdown of CaM (*CALM1*) and centrin1 (*CETN1*) on MDA-MB-231 and MCF-7 derived spheroids (Figure 6A,B). Both *CALM1* and *CETN1* (Figures S5 and S6) knockdown decreased the formation of spheroids derived from these cell lines. However, the effect was more pronounced in the KRAS-mutant MDA-MB-231 cell line (Figure 6A). Moreover, the knockdown of *CALM1* decreased spheroid growth significantly more in this cell line, which correlated with the overall stronger effect of this knockdown treatment on K-RasG12V membrane anchorage BRET (Figure 5E).

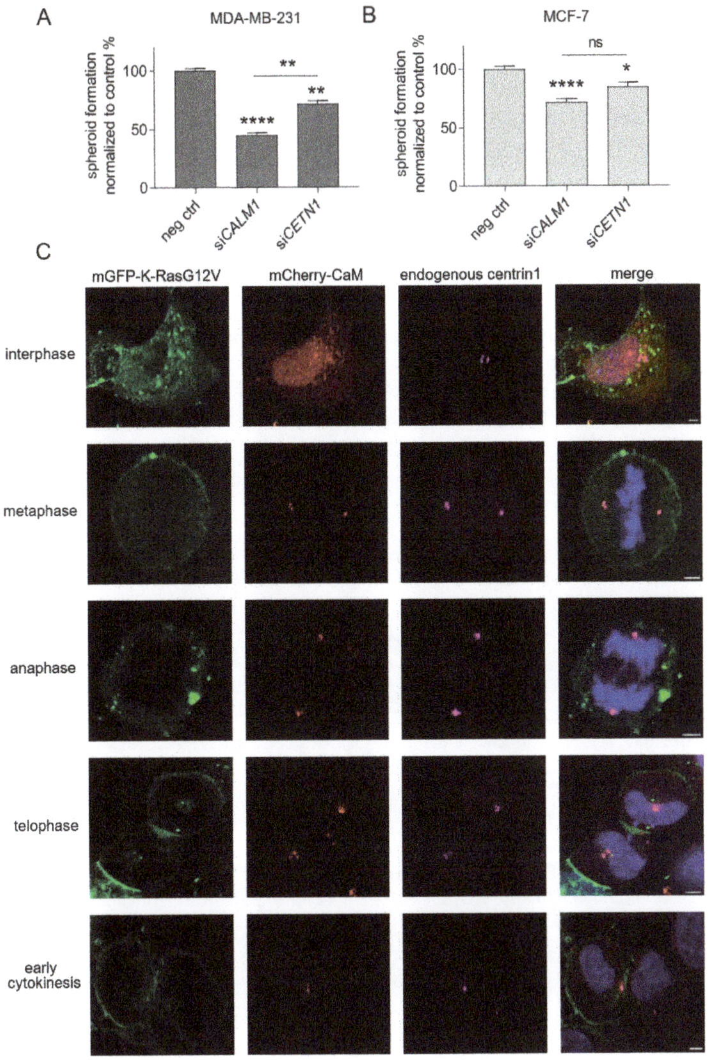

Figure 6. CaM and centrin1 distribute to the centrosomes, and their loss more potently affects spheroid formation of *KRAS* mutant MDA-MB-231 cells. (**A**,**B**) Effect of knockdown of CaM (*CALM1*)

and centrin1 (*CETN1*) on spheroids derived from MDA-MB-231 (**A**) and MCF-7 (**B**) cells. The knockdown efficiency was compared to spheroids grown from negative control siRNA transfected cells. Data represent mean ± SEM of four biological repeats. Statistical analysis was performed using Mann–Whitney test. (**C**) Representative images of HeLa cells that were co-transfected with mGFP-K-RasG12V (green) and mCherry-CaM (red). Endogenous centrin1 was immunostained (purple), and DNA was stained using DAPI (blue). Cell-cycle stages are indicated on the left. Scale bar is 5 µm. * $p < 0.05$; ** $p < 0.01$; **** $p < 0.0001$; ns = not significant.

Only limited conclusions in regard to cancer cell stemness can be derived from these experiments, which are essentially assaying the ability of cancer cells to evade anoikis and which bear some similarity to culture conditions employed for stem/progenitor cells.

The fate of stem and progenitor cells is decided during the cell cycle, which can proceed to symmetric or asymmetric cell divisions [58]. Oncogenes are suggested to shift the mode of cell division that has more symmetric divisions and produce more stem cells [59]. Stemness can be mediated by centriolar organelles, such as the centrosomes, specifically the mother centrosome [60]. Interestingly, mCherry-tagged CaM localises during different cell cycle phases to the centrosomes and the midbody in HeLa cells (Figure 6C), as observed previously by others [19,27]. The same is essentially seen for endogenous centrin1, which also localises to these structures (Figure 6C). However, during interphase, CaM has a more pronounced cyto/nucleoplasmic distribution, while centrin1 discretely localises to the centrosomes.

4. Conclusions

Our data show that K-Ras does not only interact with CaM, but also with the highly related protein centrin1. While both Ca^{2+}-binding proteins distribute to similar mitotic structures, notably the centrosomes, CaM appears to have a stronger impact on K-Ras functional membrane organisation at the plasma membrane. Centrin1 may instead function to localize K-Ras to certain structures, such as the centrosomes, while it appears to have only a minor role in K-Ras trafficking. These distinct functions of CaM and centrin proteins are difficult to tell apart using pharmacological inhibitors against CaM, which we found affect binding of K-Ras to centrin1 as well.

While previous in vitro data demonstrated that the farnesylated C-terminus of K-Ras was sufficient for binding to CaM, our data here suggest that farnesylation is essentially dispensable for most of the interaction of K-Ras with either CaM or centrin1 in cells. Given that oncogenic K-Ras engages more with either of these proteins, we propose that most of the K-Ras/CaM and K-Ras/centrin1 pairs are found in complexes that can recognize the activation state of K-Ras. This recognition is typically afforded by effectors, hence it is plausible to assume that most of the K-Ras binding to these Ca^{2+}-binding proteins happens in higher order complexes that contain effectors. Others have previously proposed PI3Kα-containing complexes with K-Ras and CaM [61]. Our data encourage further investigation of these potential complexes and their function inside cells.

Supplementary Materials: The following supporting information can be downloaded at: https://www.mdpi.com/article/10.3390/cancers15123087/s1. Figure S1: (Supporting data for Figure 2A,B). Figure S2: (Supporting data for Figure 2C,D). Figure S3: (Supporting data for Figure 5C). Figure S4: (Supporting data for Figure 5D). Figure S5: (Supporting data for Figure 6A). Figure S6: (Supporting data for Figure 6B).

Author Contributions: Conceptualization, G.b.M. and D.K.A.; Methodology, G.b.M., C.L., S.B. and N.B.F.; Software, G.b.M., C.L., S.B. and N.B.F.; Validation, G.b.M. and C.L.; Formal analysis, G.b.M., C.L. and S.B.; Investigation, G.b.M.; Writing—original draft, G.b.M.; Writing—review & editing, D.K.A.; Supervision, D.K.A.; Funding acquisition, D.K.A. All authors have read and agreed to the published version of the manuscript.

Funding: The study was supported by internal funds of the University of Luxembourg and the Luxembourg National Research Fund (FNR) grant C19/BM/13673303-PolaRAS2 to D.K.A.

Institutional Review Board Statement: Not applicable.

Informed Consent Statement: Not applicable.

Data Availability Statement: Any data that support the findings of this study are included within the article and Supplementary Materials.

Conflicts of Interest: The authors declare no conflict of interest.

References

1. Prior, I.A.; Hood, F.E.; Hartley, J.L. The Frequency of Ras Mutations in Cancer. *Cancer Res.* **2020**, *80*, 2969–2974. [CrossRef] [PubMed]
2. Castel, P.; Rauen, K.A.; McCormick, F. The duality of human oncoproteins: Drivers of cancer and congenital disorders. *Nat. Rev. Cancer* **2020**, *20*, 383–397. [CrossRef] [PubMed]
3. Abankwa, D.; Gorfe, A.A. Mechanisms of Ras Membrane Organization and Signaling: Ras Rocks Again. *Biomolecules* **2020**, *10*, 1522. [CrossRef] [PubMed]
4. Schmick, M.; Kraemer, A.; Bastiaens, P.I. Ras moves to stay in place. *Trends Cell Biol.* **2015**, *25*, 190–197. [CrossRef]
5. Newlaczyl, A.U.; Coulson, J.M.; Prior, I.A. Quantification of spatiotemporal patterns of Ras isoform expression during development. *Sci. Rep.* **2017**, *7*, 41297. [CrossRef]
6. Wang, M.T.; Holderfield, M.; Galeas, J.; Delrosario, R.; To, M.D.; Balmain, A.; McCormick, F. K-Ras Promotes Tumorigenicity through Suppression of Non-canonical Wnt Signaling. *Cell* **2015**, *163*, 1237–1251. [CrossRef]
7. Quinlan, M.P.; Quatela, S.E.; Philips, M.R.; Settleman, J. Activated Kras, but not Hras or Nras, may initiate tumors of endodermal origin via stem cell expansion. *Mol. Cell. Biol.* **2008**, *28*, 2659–2674. [CrossRef]
8. Najumudeen, A.K.; Jaiswal, A.; Lectez, B.; Oetken-Lindholm, C.; Guzman, C.; Siljamaki, E.; Posada, I.M.; Lacey, E.; Aittokallio, T.; Abankwa, D. Cancer stem cell drugs target K-ras signaling in a stemness context. *Oncogene* **2016**, *35*, 5248–5262. [CrossRef]
9. Okutachi, S.; Manoharan, G.B.; Kiriazis, A.; Laurini, C.; Catillon, M.; McCormick, F.; Yli-Kauhaluoma, J.; Abankwa, D. A Covalent Calmodulin Inhibitor as a Tool to Study Cellular Mechanisms of K-Ras-Driven Stemness. *Front. Cell Dev. Biol.* **2021**, *9*, 665673. [CrossRef]
10. Alvarez-Moya, B.; Lopez-Alcala, C.; Drosten, M.; Bachs, O.; Agell, N. K-Ras4B phosphorylation at Ser181 is inhibited by calmodulin and modulates K-Ras activity and function. *Oncogene* **2010**, *29*, 5911–5922. [CrossRef]
11. Dharmaiah, S.; Bindu, L.; Tran, T.H.; Gillette, W.K.; Frank, P.H.; Ghirlando, R.; Nissley, D.V.; Esposito, D.; McCormick, F.; Stephen, A.G.; et al. Structural basis of recognition of farnesylated and methylated KRAS4b by PDEdelta. *Proc. Natl. Acad. Sci. USA* **2016**, *113*, E6766–E6775. [CrossRef]
12. Chippalkatti, R.; Abankwa, D. Promotion of cancer cell stemness by Ras. *Biochem. Soc. Trans.* **2021**, *49*, 467–476. [CrossRef]
13. Tidow, H.; Nissen, P. Structural diversity of calmodulin binding to its target sites. *FEBS J.* **2013**, *280*, 5551–5565. [CrossRef]
14. Grant, B.M.M.; Enomoto, M.; Ikura, M.; Marshall, C.B. A Non-Canonical Calmodulin Target Motif Comprising a Polybasic Region and Lipidated Terminal Residue Regulates Localization. *Int. J. Mol. Sci.* **2020**, *21*, 2751. [CrossRef]
15. Grant, B.M.M.; Enomoto, M.; Back, S.I.; Lee, K.Y.; Gebregiworgis, T.; Ishiyama, N.; Ikura, M.; Marshall, C.B. Calmodulin disrupts plasma membrane localization of farnesylated KRAS4b by sequestering its lipid moiety. *Sci. Signal.* **2020**, *13*, eaaz0344. [CrossRef]
16. Ismail, S.A.; Chen, Y.X.; Rusinova, A.; Chandra, A.; Bierbaum, M.; Gremer, L.; Triola, G.; Waldmann, H.; Bastiaens, P.I.; Wittinghofer, A. Arl2-GTP and Arl3-GTP regulate a GDI-like transport system for farnesylated cargo. *Nat. Chem. Biol.* **2011**, *7*, 942–949. [CrossRef]
17. Villalonga, P.; Lopez-Alcala, C.; Bosch, M.; Chiloeches, A.; Rocamora, N.; Gil, J.; Marais, R.; Marshall, C.J.; Bachs, O.; Agell, N. Calmodulin binds to K-Ras, but not to H- or N-Ras, and modulates its downstream signaling. *Mol. Cell. Biol.* **2001**, *21*, 7345–7354. [CrossRef]
18. Chandra, A.; Grecco, H.E.; Pisupati, V.; Perera, D.; Cassidy, L.; Skoulidis, F.; Ismail, S.A.; Hedberg, C.; Hanzal-Bayer, M.; Venkitaraman, A.R.; et al. The GDI-like solubilizing factor PDEdelta sustains the spatial organization and signalling of Ras family proteins. *Nat. Cell Biol.* **2011**, *14*, 148–158. [CrossRef]
19. Li, C.J.; Heim, R.; Lu, P.; Pu, Y.; Tsien, R.Y.; Chang, D.C. Dynamic redistribution of calmodulin in HeLa cells during cell division as revealed by a GFP-calmodulin fusion protein technique. *J. Cell Sci.* **1999**, *112*, 1567–1577. [CrossRef]
20. Yu, Y.Y.; Chen, Y.; Dai, G.; Chen, J.; Sun, X.M.; Wen, C.J.; Zhao, D.H.; Chang, D.C.; Li, C.J. The association of calmodulin with central spindle regulates the initiation of cytokinesis in HeLa cells. *Int. J. Biochem. Cell Biol.* **2004**, *36*, 1562–1572. [CrossRef]
21. Abraham, S.J.; Nolet, R.P.; Calvert, R.J.; Anderson, L.M.; Gaponenko, V. The hypervariable region of K-Ras4B is responsible for its specific interactions with calmodulin. *Biochemistry* **2009**, *48*, 7575–7583. [CrossRef] [PubMed]
22. Liao, J.; Planchon, S.M.; Wolfman, J.C.; Wolfman, A. Growth factor-dependent AKT activation and cell migration requires the function of c-K(B)-Ras versus other cellular ras isoforms. *J. Biol. Chem.* **2006**, *281*, 29730–29738. [CrossRef] [PubMed]
23. Fischer, R.; Julsgart, J.; Berchtold, M.W. High affinity calmodulin target sequence in the signalling molecule PI 3-kinase. *FEBS Lett.* **1998**, *425*, 175–177. [CrossRef] [PubMed]
24. Wu, L.J.; Xu, L.R.; Liao, J.M.; Chen, J.; Liang, Y. Both the C-terminal polylysine region and the farnesylation of K-RasB are important for its specific interaction with calmodulin. *PLoS ONE* **2011**, *6*, e21929. [CrossRef] [PubMed]

25. Agamasu, C.; Ghirlando, R.; Taylor, T.; Messing, S.; Tran, T.H.; Bindu, L.; Tonelli, M.; Nissley, D.V.; McCormick, F.; Stephen, A.G. KRAS Prenylation Is Required for Bivalent Binding with Calmodulin in a Nucleotide-Independent Manner. *Biophys. J.* **2019**, *116*, 1049–1063. [CrossRef] [PubMed]
26. Manoharan, G.B.; Kopra, K.; Eskonen, V.; Härmä, H.; Abankwa, D. High-throughput amenable fluorescence-assays to screen for calmodulin-inhibitors. *Anal. Biochem.* **2019**, *572*, 25–32. [CrossRef]
27. Berchtold, M.W.; Villalobo, A. The many faces of calmodulin in cell proliferation, programmed cell death, autophagy, and cancer. *Biochim. Biophys. Acta* **2014**, *1843*, 398–435. [CrossRef]
28. Faust, F.M.; Slisz, M.; Jarrett, H.W. Calmodulin is labeled at lysine 148 by a chemically reactive phenothiazine. *J. Biol. Chem.* **1987**, *262*, 1938–1941. [CrossRef]
29. Bhattacharya, S.; Bunick, C.G.; Chazin, W.J. Target selectivity in EF-hand calcium binding proteins. *Biochim. Biophys. Acta (BBA) Mol. Cell Res.* **2004**, *1742*, 69–79. [CrossRef]
30. Yang, A.; Miron, S.; Mouawad, L.; Duchambon, P.; Blouquit, Y.; Craescu, C.T. Flexibility and plasticity of human centrin 2 binding to the xeroderma pigmentosum group C protein (XPC) from nuclear excision repair. *Biochemistry* **2006**, *45*, 3653–3663. [CrossRef]
31. Friedberg, F. Centrin isoforms in mammals. Relation to calmodulin. *Mol. Biol. Rep.* **2006**, *33*, 243–252. [CrossRef]
32. Dantas, T.J.; Daly, O.M.; Morrison, C.G. Such small hands: The roles of centrins/caltractins in the centriole and in genome maintenance. *Cell. Mol. Life Sci.* **2012**, *69*, 2979–2997. [CrossRef]
33. Puumalainen, M.R.; Ruthemann, P.; Min, J.H.; Naegeli, H. Xeroderma pigmentosum group C sensor: Unprecedented recognition strategy and tight spatiotemporal regulation. *Cell. Mol. Life Sci.* **2016**, *73*, 547–566. [CrossRef]
34. Sanz, J.M.; Grecu, D.; Assairi, L. Ca^{2+} signaling and Target Binding Regulations: Calmodulin and Centrin In Vitro and In Vivo. *Bioenergetics* **2016**, *5*, 1000144. [CrossRef]
35. Klein, U.R.; Nigg, E.A. SUMO-dependent regulation of centrin-2. *J. Cell Sci.* **2009**, *122*, 3312–3321. [CrossRef]
36. Paoletti, A.; Moudjou, M.; Paintrand, M.; Salisbury, J.L.; Bornens, M. Most of centrin in animal cells is not centrosome-associated and centrosomal centrin is confined to the distal lumen of centrioles. *J. Cell Sci.* **1996**, *109*, 3089–3102. [CrossRef]
37. Hodges, M.E.; Scheumann, N.; Wickstead, B.; Langdale, J.A.; Gull, K. Reconstructing the evolutionary history of the centriole from protein components. *J. Cell Sci.* **2010**, *123*, 1407–1413. [CrossRef]
38. Graser, S.; Stierhof, Y.D.; Lavoie, S.B.; Gassner, O.S.; Lamla, S.; Le Clech, M.; Nigg, E.A. Cep164, a novel centriole appendage protein required for primary cilium formation. *J. Cell Biol.* **2007**, *179*, 321–330. [CrossRef]
39. Wall, V.E.; Garvey, L.A.; Mehalko, J.L.; Procter, L.V.; Esposito, D. Combinatorial assembly of clone libraries using site-specific recombination. *Methods Mol. Biol.* **2014**, *1116*, 193–208. [CrossRef]
40. Zimmermann, G.; Papke, B.; Ismail, S.; Vartak, N.; Chandra, A.; Hoffmann, M.; Hahn, S.A.; Triola, G.; Wittinghofer, A.; Bastiaens, P.I.; et al. Small molecule inhibition of the KRAS-PDEdelta interaction impairs oncogenic KRAS signalling. *Nature* **2013**, *497*, 638–642. [CrossRef]
41. Vaasa, A.; Viil, I.; Enkvist, E.; Viht, K.; Raidaru, G.; Lavogina, D.; Uri, A. High-affinity bisubstrate probe for fluorescence anisotropy binding/displacement assays with protein kinases PKA and ROCK. *Anal. Biochem.* **2009**, *385*, 85–93. [CrossRef]
42. Sinijarv, H.; Wu, S.; Ivan, T.; Laasfeld, T.; Viht, K.; Uri, A. Binding assay for characterization of protein kinase inhibitors possessing sub-picomolar to sub-millimolar affinity. *Anal. Biochem.* **2017**, *531*, 67–77. [CrossRef] [PubMed]
43. Manoharan, G.B.; Okutachi, S.; Abankwa, D. Potential of phenothiazines to synergistically block calmodulin and reactivate PP2A in cancer cells. *PLoS ONE* **2022**, *17*, e0268635. [CrossRef] [PubMed]
44. Liyanage, M.R.; Zaidi, A.; Johnson, C.K. Fluorescence polarization assay for calmodulin binding to plasma membrane Ca^{2+}-ATPase: Dependence on enzyme and Ca^{2+} concentrations. *Anal. Biochem.* **2009**, *385*, 1–6. [CrossRef]
45. Waxham, M.N.; Tsai, A.L.; Putkey, J.A. A mechanism for calmodulin (CaM) trapping by CaM-kinase II defined by a family of CaM-binding peptides. *J. Biol. Chem.* **1998**, *273*, 17579–17584. [CrossRef] [PubMed]
46. Zhao, Y.; Guo, X.; Yang, B. Calcium-induced human centrin 1 self-assembly and double-regulating the binding with peptide R18-Sfi1p. *Int. J. Biol. Macromol.* **2019**, *128*, 314–323. [CrossRef]
47. Bodle, J.C.; Loboa, E.G. Concise Review: Primary Cilia: Control Centers for Stem Cell Lineage Specification and Potential Targets for Cell-Based Therapies. *Stem Cells* **2016**, *34*, 1445–1454. [CrossRef]
48. Pfleger, K.D.; Eidne, K.A. Illuminating insights into protein-protein interactions using bioluminescence resonance energy transfer (BRET). *Nat. Methods* **2006**, *3*, 165–174. [CrossRef]
49. Halling, D.B.; Liebeskind, B.J.; Hall, A.W.; Aldrich, R.W. Conserved properties of individual Ca^{2+}-binding sites in calmodulin. *Proc. Natl. Acad. Sci. USA* **2016**, *113*, E1216–E1225. [CrossRef]
50. Herrmann, C.; Horn, G.; Spaargaren, M.; Wittinghofer, A. Differential interaction of the ras family GTP-binding proteins H-Ras, Rap1A, and R-Ras with the putative effector molecules Raf kinase and Ral-guanine nucleotide exchange factor. *J. Biol. Chem.* **1996**, *271*, 6794–6800. [CrossRef]
51. Vetter, I.R.; Linnemann, T.; Wohlgemuth, S.; Geyer, M.; Kalbitzer, H.R.; Herrmann, C.; Wittinghofer, A. Structural and biochemical analysis of Ras-effector signaling via RalGDS. *FEBS Lett.* **1999**, *451*, 175–180. [CrossRef]
52. Thurnher, M.; Gruenbacher, G.; Nussbaumer, O. Regulation of mevalonate metabolism in cancer and immune cells. *Biochim. Biophys. Acta (BBA) Mol. Cell Res.* **2013**, *1831*, 1009–1015. [CrossRef]
53. Abdelkarim, H.; Leschinsky, N.; Jang, H.; Banerjee, A.; Nussinov, R.; Gaponenko, V. The dynamic nature of the K-Ras/calmodulin complex can be altered by oncogenic mutations. *Curr. Opin. Struct. Biol.* **2021**, *71*, 164–170. [CrossRef]

54. Parkkola, H.; Siddiqui, F.A.; Oetken-Lindholm, C.; Abankwa, D. FLIM-FRET Analysis of Ras Nanoclustering and Membrane-Anchorage. *Methods Mol. Biol.* **2021**, *2262*, 233–250. [CrossRef]
55. Siddiqui, F.A.; Alam, C.; Rosenqvist, P.; Ora, M.; Sabt, A.; Manoharan, G.B.; Bindu, L.; Okutachi, S.; Catillon, M.; Taylor, T.; et al. PDE6D Inhibitors with a New Design Principle Selectively Block K-Ras Activity. *ACS Omega* **2020**, *5*, 832–842. [CrossRef]
56. Kohnke, M.; Schmitt, S.; Ariotti, N.; Piggott, A.M.; Parton, R.G.; Lacey, E.; Capon, R.J.; Alexandrov, K.; Abankwa, D. Design and application of in vivo FRET biosensors to identify protein prenylation and nanoclustering inhibitors. *Chem. Biol.* **2012**, *19*, 866–874. [CrossRef]
57. Dontu, G.; Abdallah, W.M.; Foley, J.M.; Jackson, K.W.; Clarke, M.F.; Kawamura, M.J.; Wicha, M.S. In vitro propagation and transcriptional profiling of human mammary stem/progenitor cells. *Genes Dev.* **2003**, *17*, 1253–1270. [CrossRef]
58. Santoro, A.; Vlachou, T.; Carminati, M.; Pelicci, P.G.; Mapelli, M. Molecular mechanisms of asymmetric divisions in mammary stem cells. *EMBO Rep.* **2016**, *17*, 1700–1720. [CrossRef]
59. Cicalese, A.; Bonizzi, G.; Pasi, C.E.; Faretta, M.; Ronzoni, S.; Giulini, B.; Brisken, C.; Minucci, S.; Di Fiore, P.P.; Pelicci, P.G. The tumor suppressor p53 regulates polarity of self-renewing divisions in mammary stem cells. *Cell* **2009**, *138*, 1083–1095. [CrossRef]
60. Chen, C.; Yamashita, Y.M. Centrosome-centric view of asymmetric stem cell division. *Open Biol.* **2021**, *11*, 200314. [CrossRef]
61. Nussinov, R.; Zhang, M.; Tsai, C.J.; Jang, H. Calmodulin and IQGAP1 activation of PI3Kalpha and Akt in KRAS, HRAS and NRAS-driven cancers. *Biochim. Biophys. Acta (BBA) Mol. Basis Dis.* **2018**, *1864*, 2304–2314. [CrossRef] [PubMed]

cancers

MDPI

Article

Proteomic Mapping of the Interactome of KRAS Mutants Identifies New Features of RAS Signalling Networks and the Mechanism of Action of Sotorasib

Aoife Nolan [1], Cinzia Raso [1], Walter Kolch [1,2], Alex von Kriegsheim [1,3], Kieran Wynne [1] and David Matallanas [1,*]

[1] Systems Biology Ireland, School of Medicine, University College Dublin, Belfield, D04 V1W8 Dublin, Ireland; aoife.a.nolan@ucdconnect.ie (A.N.); cinzia.raso@gmail.com (C.R.); walter.kolch@ucd.ie (W.K.); alex.vonkriegsheim@igmm.ed.ac.uk (A.v.K.); kieran.wynne1@ucd.ie (K.W.)

[2] Conway Institute of Biomolecular and Biomedical Research, University College Dublin, D04 V1W8 Dublin, Ireland

[3] Edinburgh Cancer Research UK Centre, MRC Institute of Genetics and Molecular Medicine, University of Edinburgh, Edinburgh EH4 2XU, UK

* Correspondence: david.gomez@ucd.ie

Simple Summary: Cancer is caused by changes in DNA called mutations that alter the way proteins work. We know that RAS proteins are one of the most commonly mutated proteins in cancer cells. These proteins work like a light switch and mutant RAS remain in the ON position sending signals to the cell to keep dividing when they should not. Until recently, it was thought that all RAS mutations were equal, but differences among these mutants have been identified. Here we used a technique called proteomics to decipher the differences among RAS mutants. We find that each mutant binds a different set of proteins and can regulate different signals. We also find that a clinically approved drug that inhibits one RAS mutant regulates the interaction of RAS proteins with other proteins. Our findings extend our knowledge of how the RAS mutants work, which can potentially be used to improve cancer treatments.

Abstract: RAS proteins are key regulators of cell signalling and control different cell functions including cell proliferation, differentiation, and cell death. Point mutations in the genes of this family are common, particularly in *KRAS*. These mutations were thought to cause the constitutive activation of KRAS, but recent findings showed that some mutants can cycle between active and inactive states. This observation, together with the development of covalent KRASG12C inhibitors, has led to the arrival of KRAS inhibitors in the clinic. However, most patients develop resistance to these targeted therapies, and we lack effective treatments for other KRAS mutants. To accelerate the development of RAS targeting therapies, we need to fully characterise the molecular mechanisms governing KRAS signalling networks and determine what differentiates the signalling downstream of the KRAS mutants. Here we have used affinity purification mass-spectrometry proteomics to characterise the interactome of KRAS wild-type and three KRAS mutants. Bioinformatic analysis associated with experimental validation allows us to map the signalling network mediated by the different KRAS proteins. Using this approach, we characterised how the interactome of KRAS wild-type and mutants is regulated by the clinically approved KRASG12C inhibitor Sotorasib. In addition, we identified novel crosstalks between KRAS and its effector pathways including the AKT and JAK-STAT signalling modules.

Keywords: KRAS; proteomics; JAK1; RADIL; Sotorasib; SOS1

1. Introduction

The three genes of the RAS family, *HRAS*, *KRAS*, and *NRAS*, code for four proteins and were described 40 years ago [1]. These proteins are members of the small GTPase superfamily, and they cycle between an inactive state when they are bound to GDP and an active state when they exchange GDP for GTP [2]. Inactivation occurs when GTP is hydrolysed to GDP in a reaction that can be catalysed by RAS proteins intrinsic GTPase activity. However, the activation switch of RAS proteins is tightly regulated by several proteins including the guanine nucleotide exchange factors (GEFs) that promote the exchange of GDP for GTP and the RAS GTPase activating proteins (GAPs) that accelerate the hydrolysis of GTP. Activated RAS-GTP binds to effector proteins that are ultimately responsible for the (patho)physiological functions mediated by RAS proteins including proliferation, migration, and differentiation [3]. Extensive work has identified more than 20 bona fide and putative RAS effectors [2]. However, despite intensive research, we lack a clear picture on how these proteins regulate their complex signalling networks.

Mutations of the RAS family of genes are common in cancer and they are driving oncogenes in some of the cancer types with higher prevalence [4]. In cancer cells, the three genes are mutated at different rates. KRAS is the most commonly mutated gene (85% of mutations), followed by NRAS (14%), while HRAS mutations are less frequent (~1%) [4]. Traditionally, hyperactivating mutations of RAS proteins were thought to lock RAS in the active GTP-bound state, due to the conformational changes in protein structure caused by mutations in RAS [1]. However, recent evidence has disproved this dogma as it has been shown that many KRAS mutants retain varying levels of GTPase activities [5–9]. The current view is that the amino acid substitution involved is the determining factor of KRAS mutant GTPase activity, and subsequent activation of downstream pathways [7]. Biochemical studies have determined that the KRASG13D mutant has the highest GTPase activity with a 14 times faster GDP–GTP exchange rate than wild-type KRAS (hence forth KRAS WT) and nine times slower GAP-mediated GTP hydrolysis rate [6,9]. In clear contrast, KRASG12V shows a 1.8-fold reduced GDP–GTP exchange rate but it is insensitive to GAP-mediated GTP hydrolysis [6]. However, the determination of the GTPase activity of the different KRAS mutants is not completed and conflicting results in a study performed in MCF10A breast cancer cells found that KRASG12D and KRASG13D mutants had similar GTPase activity to KRAS WT, but that KRASG12C and KRASG12V had higher GTP binding [7]. Importantly, the binding of many KRAS effectors is GTP-dependent, including the best characterised KRAS effectors RAF1, PI3K, RALGDS, and PLCε [10,11]. Thus, the fact that the different mutants are regulated by GTP/GDP exchange can indicate that different KRAS mutants can have different affinities for interacting proteins and some mutations can result in the formation of oncogenic neo-interactions with proteins that do not bind to KRAS WT. This hypothesis is supported by recent data showing that KRAS interaction affinity with the RAF RAS-binding domain (RBD) decreases, dependent on the mutation present at KRAS codons 12, 13, and 61 [9]. In this case, KRASG13D, KRASQ61L, KRASG12A, and KRASG12C mutants had only slightly lower interaction affinity with RAF-RBD compared with KRAS WT. Conversely, KRASG12V, KRASG12D, and KRASG12R mutants had significantly lower interaction affinity with RAF-RBD. Thus, different types of KRAS mutations can profoundly change the interaction with effectors and downstream signalling.

Importantly, in the last decade there has been a paradigm change in the way we approach the development of targeted therapies against RAS-driven tumours. For decades, RAS proteins were considered undruggable. The development of a series of covalent inhibitors specific for KRASG12C and the FDA approval of Sotorasib (AMG510) and Adagrasib demonstrate that we can develop effective RAS inhibitors, at least for some of the most common mutants [12]. Unfortunately, while the initial response rates to Sotorasib and Adagrasib are promising, ranging from 32% to 46%, respectively, in non-small cell lung cancer (NSCLC), resistance inevitably develops in these patients [13]. These drugs can also lead to severe side effects in the patients and more information is necessary to fully characterise their mechanism of action [14].

Traditionally, all KRAS point mutations observed in human tumours were considered to mediate their transforming potential through the same effector pathways. Importantly, increasing evidence shows that specific mutations are associated with different transforming potential and differential regulation of metastasis in various cell and animal models [7,15]. The existence of pathophysiological differences among the different mutant KRAS is further supported by clinical observations indicating that patients with mutant KRAS colorectal cancer may respond to targeted therapy depending on the type of point mutation they express [16]. In particular, KRASG13D but not KRASG12V tumours were shown to be sensitive to EGFR inhibition. These differences are probably related to the different GTPase activity of these mutants which likely results in different interactomes associated with specific KRAS mutants. For this reason, in the past few years, several groups have performed affinity purification mass-spectrometry (AP-MS) studies to characterise the specific interactome of several mutant KRAS [10]. Collectively, these studies clearly support the existence of specific mutant KRAS interactomes.

Here, we extend these studies using AP-MS to compare the interactome of KRAS WT, KRASG12V, and KRASG13D in non-tumoral and tumoral cell lines. Using AP-MS we identify specific transient interactors and novel RAS interactors. We experimentally validated our findings, extending our studies to other KRAS mutants, demonstrating that KRAS mutants have different affinities for some of their interactors including SOS1. In light of these findings, we applied our AP-MS proteomic approach to characterise the effect of Sotorasib on the interactome of KRASG12C and KRAS WT. Our findings show a regulation of the KRASG12C interactome by this drug. Surprisingly, we also see a Sotorasib-mediated regulation of KRAS WT interacting proteins and an activation of AKT. Our analysis identified a novel interaction of KRASG12C with JAK1, with an unexpected regulation of KRAS protein expression by the JAK pathway. We also identified proteins of the AKT pathway which may be involved in the activation of AKT. Altogether, our results expand our understanding of KRAS signalling networks and provide mechanistic insights into why not all KRAS mutants are created equal.

2. Materials and Methods

2.1. Cell Culture and Transfection

HEK293 (ATCC), RAS-less mouse embryonic fibroblasts (MEF) isogenic cell lines [17] (from the RAS initiative at Frederick National Laboratory), and HKe-3 (Shirasawa et al. 1993) cells were maintained in T75 flasks at 37 °C in 95% humidified air containing 5% CO_2 in HEPA class 100 steri-cycle CO_2 incubator (Thermo Electron Corporation, Vantaa Finland) and cultured in Dulbecco's Modified Eagle Medium (DMEM) containing 10% Foetal Bovine Serum and 2 mM L glutamine (all from Gibco, Waltham, MA, USA). NIH3T3 (ATCC) cells were maintained at similar conditions and cultured in filtered DMEM containing 10% Calf Serum (Sigma-Aldrich, Burlington MA, USA) and 2 mM L-glutamine. Cells were transfected with Lipofectamine 2000 (Invitrogen, Waltham, MA, USA), according to the manufacturer's instructions. All cells were validated by genome sequencing in the last 12 months. For stimulation with EGF (Roche, Basel, Switzerland), following 24 h transfection of HEK293 cells, cells were starved for 16 h prior to stimulation with 3 µM EGF for 5 and 10 min, and cells were then lysed. Cells were treated with 5 µM Sotorasib (MedChemExpress, Monmouth Junction, NJ, USA) and 5 µM Ruxolitinib (SelleckChem, Berlin, Germany) for 24 h in normal growth conditions prior to protein extraction by cell lysis.

2.2. Immunoblotting, Antibodies and Reagents

Cell extract protein concentrations were measured by Pierce BCA assay (Invitrogen) as outlined by the manufacturer's instruction. Extracts and immunoprecipitates were denaturalised and analysed by SDS-polyacrylamide gel electrophoresis and transferred to PVDF membranes (Millipore, Darmstadt, Germany), and blotted with commercial antibodies. Rabbit polyclonal RADIL (Proteintech, Manchester, UK), monoclonal anti-FLAG M2-HRP (MERK, Darmstadt, Germany), mouse monoclonal α-Tubulin antibody (TU-02; Santa Cruz, Dallas, TX, USA), rabbit polyclonal GAPDH (Cell Signaling, Danvers, MA, USA), goat polyclonal RIN1 (N-19; Santa Cruz), mouse monoclonal p-ERK (E-4; Santa

Cruz), rabbit polyclonal ERK1 (C-16; Santa Cruz), rabbit polyclonal p-S473-AKT (Cell Signaling), rabbit polyclonal AKT1 (N-19, Santa Cruz), rabbit polyclonal Vinculin (Cell Signaling), mouse monoclonal BRAF (F-7; Santa Cruz), mouse monoclonal KRAS (F234; Santa Cruz), rabbit polyclonal MEK (12-B; Santa Cruz), rabbit polycloncal p-S217/221-MEK (Cell Signaling), mouse monoclonal panRAS (Ab-3; EMD Millipore, Germany), mouse monoclonal RAF1 (E-10; Santa Cruz), rabbit polyclonal p-S289/296/301-RAF1 (Cell Signaling), rabbit polyclonal SAPK/JNK (Cell Signaling), rabbit polyclonal p-T183/Y185-SAPK/JNK (Cell Signaling), rabbit polyclonal STAT5 (C-17; Santa Cruz), rabbit polyclonal p-Y705-STAT3 (Cell Signaling), rabbit polyclonal JAK1 (HR-785; Santa Cruz), rabbit polyclonal p-Y1034/1035-JAK1 (Cell Signaling), rabbit polyclonal Cleaved Caspase 3 (Cell Signaling), and mouse monoclonal AU5 (Covance, Burlington, NC, USA).

2.3. Constructs and siRNA

Constructs encoding pCEFL-FLAG-KRAS WT, pCEFL-FLAG-KRASG13D, pCEFL-FLAG-KRASG12V, pCEFL-FLAG-KRASG12C, pCEFL-FLAG-KRASG12D, pCEFL-HA-NRASG12V, pCEFL-HA-KRAS WT, pCEFL-HA-KRASG13D, pCEFL-HA-KRASG12V, pCEFL-HA-KRASG12C, pCEFL-HA-KRASG12D, and pCEFL-AU5-SOS1 have been described before [18–21]. Small interfering RNAs (siRNAs) against RADIL human (SO-2776281G) and mouse (SO-2858456G) and non-targeted were from Dharmacon/Horizon (Lafayette, CO, USA).

2.4. Immunoprecipitation (IP) and in Beads Tryptic Digestion

Cells were transfected with pCEFL-FLAG-KRAS-WT, -KRASG13D, -KRASG12V, -KRASG12C, -KRASG12D, and lysed after 48 h in 25 mM HEPES (pH7.5), 150 mM NaCl, 10 mM $MgCl_2$, 1 mM EDTA (all Sigma-Aldrich), 1% NP-40 (Calbiochem, San Diego, CA, USA), protease, and phosphatase inhibitors (Roche). Cellular debris was removed by centrifugation at 20,000× *g* at 4 °C for 10 min. Baits were immunoprecipitated with anti-FLAG M2 conjugated agarose beads (Sigma-Aldrich). After 2 h incubation at 4 °C, immunoprecipitates were washed twice with lysis buffer containing 1% NP-40 and separated by SDS-PAGE. Blots were quantified using ImageJ version 1.50. For immunoprecipitation of endogenous RADIL the lysates were incubated with 10 µL Protein-G Sepharose beads (Sigma-Aldrich) with 2.6 µg/mL RADIL antibody at 4 °C for 2 h. Samples used in the MS analyses were immunoprecipitated with anti-FLAG M2 conjugated agarose beads, performed as above but after 2 h incubation at 4 °C the immunoprecipitates were digested with proteases as described before [22]. Briefly, immunoprecipitates were washed three times with lysis buffer containing 1% NP-40 and twice with lysis buffer that does not contain NP-40. Proteins were eluted from beads with 2 M Urea, 50 mM Tris-HCl (pH7.5) containing 5 µg/mL modified sequencing-grade trypsin (Promega, Madison, WI, USA) for 30 min. Proteins were digested in 2 M Urea, 50 mM Tris-HCl pH7.5, and 1 mM DTT (all from Sigma-Aldrich) containing 5 µg/mL modified sequencing-grade trypsin overnight at room temperature. Peptides were alkylated using iodoacetamide (5 mg/mL) and incubated in the dark for 30 min at room temperature. C18 Stage Tips were used to desalt samples, as described by [23] and analysed by mass spectrometry. For double digestion, proteins were denatured in 8 M Urea and reduced in 1 mM DTT for 30 min at room temperature. Proteins were alkylated in the dark for 30 min at room temperature with iodoacetamide (3 mM). Proteins were digested in 8 M Urea containing 3 ng/µL Lysyl Endopeptidase (Lys-C, Osaka, Wako, Japan) for 4 h at 37 °C. Samples were diluted to a final concentration of 2 M Urea and digested again overnight with 3 ng/µL modified sequencing grade trypsin. C18 Stage tips were used to desalt samples, as before, and analysed by mass spectrometry.

2.5. Mass Spectrometry Analysis of Affinity Purification-MS (AP-MS)

Experiment IP KRAS WT, KRASG12V, and KRASG13D: the samples were run on a ThermoScientific Q-Exactive mass spectrometer (ThermoFisher Scientific, Waltham, MA, USA) connected to Dionex Ultimate 3000 (RSLCnano, ThermoFisher, Waltham, MA, USA) chromatography system. Tryptic peptides were resuspended in 0.5% acetic acid, and 2% acetonitrile (both Sigma-Aldrich). Each sample was loaded onto a fused silica emitter, 75 μm ID, pulled using a laser puller (Sutter Instruments P2000, Novato, CA, USA), packed with Reprocil Pur C18 (1.9 μm, Germany) reverse phase media, and was separated by an increasing acetonitrile gradient over 120 min at a flow rate of 200 nL/min. The mass spectrometer was operated in positive ion mode with a capillary temperature of 320 °C, and with a potential of 2300 V applied to the frit. All data were acquired with the mass spectrometer operating in automatic data-dependent switching mode. A high resolution (70,000) MS scan (350–1600 m/z) was performed using the Q-Exactive to select the 12 most intense ions prior to MS/MS analysis using HCD.

Experiment IP Sotorasib treatment KRAS WT and KRASG12C: Samples were analysed on a Bruker timsTof Pro mass spectrometer (Bruker, Billerica, MA, USA) connected to an Evosep One liquid chromatography system (Eveosep, Odense, Denmark) or a Bruker nanoElute nanoflow chromatography system. In the case of the Evosep platform tryptic peptides were resuspended in 0.1% formic acid (Sigma-Aldrich) and each sample was loaded onto an Evosep tip. The Evosep tips were placed in position on the Evosep One, in a 96-tip box. The autosampler was configured to pick up each tip, and elute and separate the peptides using a set chromatography method (30 samples a day) [24]. For the Bruker nanoElute nano-LC chromatography system, each sample was loaded onto Acclaim PepMap C18 trap cartridge (0.3 mm inside diameter, 5 mm length, ThermoFisher Scientific) and then separated on an Aurora UHPLC column (25 cm × 75 μm ID, C18, 1.6 μm, Ionopticks, Fitzroy Australia) with an increasing acetonitrile gradient over 30 min at a flow rate of 250 nL/min [25]. The chromatography buffers used for both systems were Buffer A: 99.9% water, and 0.1% formic acid; and Buffer B: 99.9% acetonitrile, and 0.1% formic acid. All solvents used were LCMS grade. The mass spectrometer was operated in positive ion mode with a capillary voltage of between 1600 and 1800 V, dry gas flow of 3 L/min and a dry temperature of 180 °C. All data were acquired with the instrument operating in trapped ion mobility spectrometry (TIMS) mode. Trapped ions were selected for ms/ms using parallel accumulation serial fragmentation (PASEF). A scan range of (100–1700 m/z) was performed at a rate of 5 PASEF MS/MS frames to 1 MS scan with a cycle time of 1.03 s [26].

2.6. Analysis of AP-MS Data

The resulting mass spectra were analysed with MaxQuant [27] software, containing a built-in Andromeda search engine to identify the proteins from the UniProt HUMAN database [28] (release 2014_02). Protein interactions were quantified by LFQ Intensity. This consisted of 3 biological replicates and 2 technical replicates with 3 baits and empty vector controls in 2 cell lines. The raw data were searched against the Homo sapiens subset of the Uniprot Swissprot database (reviewed) using the search engine Maxquant (release 2.0.1.0) using specific parameters for trapped ion mobility spectra data-dependent acquisition (TIMS DDA). Each peptide used for protein identification met specific Maxquant parameters, i.e., only peptide scores that corresponded to a false discovery rate (FDR) of 0.01 were accepted from the Maxquant database search. The normalized protein intensity of each identified protein was used for label-free quantitation (LFQ) [29].

Contaminants and reverse peptides were removed. The statistical analysis was performed as previously reported [23,30]. In essence, the average protein interaction for each bait was measured by LFQ values. As the first step in the analysis, significant proteins were identified in comparison to the empty vector control. The ratio of protein abundance and Student's t-test were used to select proteins significantly interacting with the bait proteins compared with the empty vector control (ratio > 2, p-value < 0.01). Specific proteins were pooled. For the experiment performed with HKe-3 cells described in Section 3.2 bait protein levels (KRAS) were assessed and normalised to remove potential protein interaction bias for one bait over the others. The ratio between conditions were calculated pairwise. To select proteins significantly interacting compared with other baits, the ratio of each protein, combined with Student's *t*-test were used to identify proteins significantly interacting (ratio > 2 OR < 0.5, *p*-value < 0.01).

Cytoscape (version 3.7.2) was used to construct the interactomes of each KRAS isoform in both cell lines. String DB [31], REACTOME [32], and PANTHER databases [33] were used for gene ontology analysis, network reconstruction, and K-means cluster identification.

2.7. Transformation Assay

Transformation assays were performed as described by Aaronson et al. [34] and further explained in Herrero et al. [35]. Briefly, NIH3T3 cells were transfected with pCEFL-HA-KRAS WT, KRASG13D, KRASG12V, KRASG12C, KRASG12D, and/or 20 nM siRNA-RADIL. Cells were grown in culture conditions for at least two weeks with the media changed every 2–3 days. When macroscopic foci were visible, the cells were fixed with methanol for 5 min and stained with 5% Giemsa (in PBS) and the transformed foci were counted. The number of foci counted were quantified relative to 1 μL of DNA. Three biological replicates were quantified and averaged.

2.8. Migration Assay

HKe-3 cells were transfected with pCEFL-FLAG-KRAS-WT, -KRASG13D, -KRASG12V, -KRASG12C, -KRASG12D, and +/− 4nM siRNA-RADIL; 24 h post-transfection, cells were counted and seeded into migration culture inserts (Ibidi, Grafelding, Germany). Culture inserts were removed 24 h post-seeding of cells. Cells were supplemented with DMEM containing 1% FBS and 1% L-glutamine. Pictures of the cell gap were taken at 0 h, 6 h and 24 h. The area of the gap was measured using ImageJ. Results were normalised to the 0 h time point for each condition, as outlined by Cappiello et al. [36] and percentage gap closure was quantified.

2.9. Survival Curves

To measure survival in HKe-3 we used MTS assays. To perform this, cells were transfected with pCEFL-FLAG-KRAS-WT, -KRASG13D, -KRASG12V, -KRASG12C, -KRASG12D, and/or 10 nM siRNA-RADIL. Cells were collected at the indicated times and MTS assay was performed as outlined by the manufacturer (Promega).

2.10. Statistics

Transformation assay: Number of foci were counted and quantified relative to 1μg DNA. Positive (upper) error bars are Max-Q3 (maximum [highest data point in the data set]—Q3 [third quartile, the median of the upper half of the dataset]), minus (lower) error bars are the error amount set to 100%. Upper (dark grey) boxes are Q3-Med (Q3 [third quartile, the median of the upper half of the dataset]—Med [median, the middle value in the dataset]). Lower light grey boxes are Med-Q1 (Med [median, the middle value in the dataset]—Q1 [first quartile, the median of the lower half of the dataset]). *p*-values were determined by two-tail, two-sample equal variance Student's *t*-test; * $p < 0.05$, ** $p < 0.01$, *** $p < 0.001$; differences without *p*-values indicated were not significant.

Migration assay: Percentage cell gap was quantified using ImageJ and normalised to the 0 h time point gap as described by Cappiello et al. [36]. Error bars are +/− standard error of the mean (SEM) of three biological replicates. *p*-values were determined by two-tail, two-sample equal variance Student's *t*-test; * $p < 0.05$, ** $p < 0.01$, *** $p < 0.001$; differences without *p*-values indicated were not significant.

Western blot quantification: Protein band intensity was quantified by ImageJ. Immunoprecipitates were normalised to the band intensity of bait proteins, phospho-proteins were normalised to band intensity of the corresponding total-protein levels. All samples were normalised to the control (KRAS WT) levels. *p*-values were determined by one-tail, two-sample equal variance Student's *t*-test; * $p < 0.05$, ** $p < 0.01$, *** $p < 0.001$; differences without *p*-values indicated were not significant.

MTS cell survival assay: Three biological replicates with three technical replicates each. Percentage cell viability was normalised to the control (live cells). Error bars are +/− standard error of the mean (SEM) of three biological replicates. *p*-values were determined by two-tail, two-sample equal variance Student's *t*-test; * $p < 0.05$; differences without *p*-values indicated were not significant.

3. Results

3.1. KRAS Mutants Differentially Regulate the RAF and AKT Effector Pathways in HEK293 Cells

Different lines of evidence have shown that the specific point mutations of the KRAS oncogene are not equal [37–41]. To obtain a better understanding of the biochemical and functional differences among these proteins, we first tested the effects of the expression of KRAS WT and four of the most common oncogenic KRAS mutants, KRASG13D, KRASG12V, KRASG12C, and KRASG12D, on the regulation of ERK and AKT pathways in HEK293 cells. As expected, overexpression of KRAS WT caused a modest increase on ERK phosphorylation, while overexpression of the four KRAS mutants increased ERK phosphorylation (Figure 1A,B). KRASG12D induced phospho-ERK levels 12-fold compared with WT, followed by KRASG12C and KRASG13D with a 7.3- and 5.6-fold increase, respectively. KRASG12V was consistently less potent and induced ERK phosphorylation only 3-fold even though it was expressed at higher levels than KRASG13D. The effect on the levels of phospho-AKT were less striking, but KRASG12C and KRASG12D significantly induced phosphorylation of this kinase while expression of KRASG13D and KRASG12V had no effect (Figure 1A,C). Altogether, the results show that KRASG12C and KRASG12D have similar effects in the activation of ERK and AKT. KRASG13D does not activate the AKT pathway while KRASG12V has a lower effect in the activation of the ERK pathway than the other mutants and no effect on the activation of AKT.

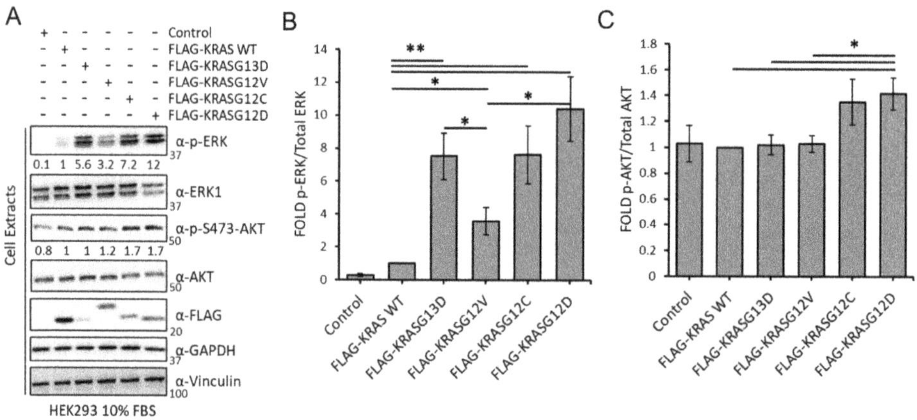

Figure 1. (**A**) HEK293 cells transfected with the indicated constructs were lysed after 48 h. Lysates were Western blotted with indicated antibodies. The whole Western blot figure can be found in Supplementary Materials Original Blots and quantification for Figure 1A. Numbers indicate fold of phosphorylation of the indicated proteins normalised to the total protein. (**B**) graph shows the fold of phospho-ERK normalised to ERK expression ($n = 3$). Error bars show standard deviation. * *p* value < 0.05; ** *p* value < 0.01. (**C**) graph shows the fold of phospho-AKT normalized to AKT expression ($n = 3$). Error bars show standard error of the mean (SEM). * *p* value < 0.05; ** *p* value < 0.01.

3.2. The KRAS Proteins Interactome Differs between Cell Types

We hypothesised that the results of the previous section might be explained, at least in part, by differences in the interactome of these mutants. To test this, we performed a proteomics screen of KRAS mutants interactomes. In order to facilitate the experimental analysis, we selected only two of these mutants, KRASG12V and KRASG13D. The rationale for this selection was that these two KRAS isoforms have the greater variation in the GTPase activity and differential response to EGFR inhibitors [7,16], which we reasoned may be due to the presence of very different interactors in their proteomes. We selected the human HEK-293 cells which are an immortalized cell line commonly used for interaction proteomics and has been used to explore RAS proteins interactomes [38,42,43]. Also, we were interested in determining if there are differences in the interactome of these mutants in different cell lines and, in particular, in a cancer type where KRAS is commonly mutated. In our experience, overexpression of KRAS mutants in cells that have mutations in KRAS or BRAF genes is extremely difficult as this seems to result in oncogene-induced stress. For this reason, in addition to HEK293 cells, we performed a proteomic screen in HKe-3 cells, a colorectal cancer cell line derived from HCT116 cells where the oncogenic KRASG13D allele was knocked out that tolerates overexpression of KRAS mutants and we used before to perform proteomics screens [44,45]. We transfected FLAG-tagged constructs of KRAS WT, KRASG12V, and KRASG13D into these cell types (Figure 2A). Following IP, we determined the specific interactomes associated with the different KRAS proteins by AP-MS. Using label-free quantitation (LFQ), we performed statistical analysis to check if there were significant changes in the interactome of the three KRAS proteins expressed in the experiments. A first comparison using a Venn diagram representation showed clear differences in the interactomes associated with KRAS proteins in both cell lines with only 12 proteins out the 450 specific interactors being present in KRAS IPs performed in both cell lines (Figure 2B and Tables S1 and S2). These results showed that there are very pronounced cell specific differences in the interactome of KRAS proteins. To continue the systematic analysis, we focused on the differences in the interactome for the three KRAS baits in each cell line.

In HEK293 cells, 1530 proteins were identified in the screening, of which 400 were specifically binding to at least one of the KRAS proteins (Figures 2B,C and S1 and Table S1). Using the statistical analysis explained in the material and methods section, we observed remarkable variations on the number of proteins binding to each of the KRAS proteins. As expected, there were common interactors to KRAS WT, KRASG13D, and KRASG12V (193), and other proteins were interacting with two of them (Figure 2C). Conversely, there were 46 proteins exclusively interacting with KRAS WT. In the case of the KRAS mutants, we also identified proteins exclusively binding to one of them with 48 proteins shown as unique KRASG13D interactors and only 16 shown as specific for KRASG12V. Similar results were observed in HKe-3 cells, although with clear differences in the distribution of the interactome (Figures 2D and S2 and Table S2). In HKe-3 cells, we identified significantly fewer proteins (713 proteins), of which 62 proteins were shown as specific KRAS interactors. As in HEK293 cells the three proteins expressed shared part of the interactome, but only 11 proteins were common interactors. The most remarkable observation in these cells was that KRASG13D interacted with 22 proteins, a much lower number than the 40 binders of KRAS WT and the 41 interactors of KRASG12V. Despite these differences, all KRAS isoforms have unique interactors, with KRAS WT having the highest number (13), followed by KRASG12V and KRASG13D (13 and 7, respectively). Altogether, these results show that there are clear differences in the interactomes of the three KRAS isoforms and that they show cell specific characteristics.

Next, we evaluated the quality of our experiments by focusing on the proteins identified in our experiments. Several known RAS interactors were found in the analysis of the proteomic interaction data including proteins that have been identified in previous proteomics screenings performed with other methods such as BioID and AP-MS, among them the RAS effectors RAF1, BRAF, NRAS, HRAS, RADIL, or mTOR [37,39,40,46,47] (Figures 2E–N and S2). The high enrichment of known KRAS interactors in our datasets demonstrate the validity of our method to investigate the specific interactomes of the different KRAS isoforms. Moreover, LFQ allowed us to find changes of affinity between the different KRAS baits and specific proteins. For instance, RAF1, BRAF, and ARAF have higher affinity for

KRAS mutants than for KRAS WT in HEK293 cells (Figures 2E,F,J,K and S3A). However, while BRAF and RAF1 have similar affinity for both KRAS mutants, ARAF shows a higher affinity for KRASG12V (Figure S3A). Importantly, in HKe-3 cells we saw further differences in the interaction of KRAS proteins with RAF family proteins. ARAF was not identified, and we saw that both RAF1 and BRAF have much higher affinity for KRASG13D than for KRASG12V (Figure 2J,K) indicating that there are cell specific factors that regulate the interaction of the KRAS mutants with RAF family effectors.

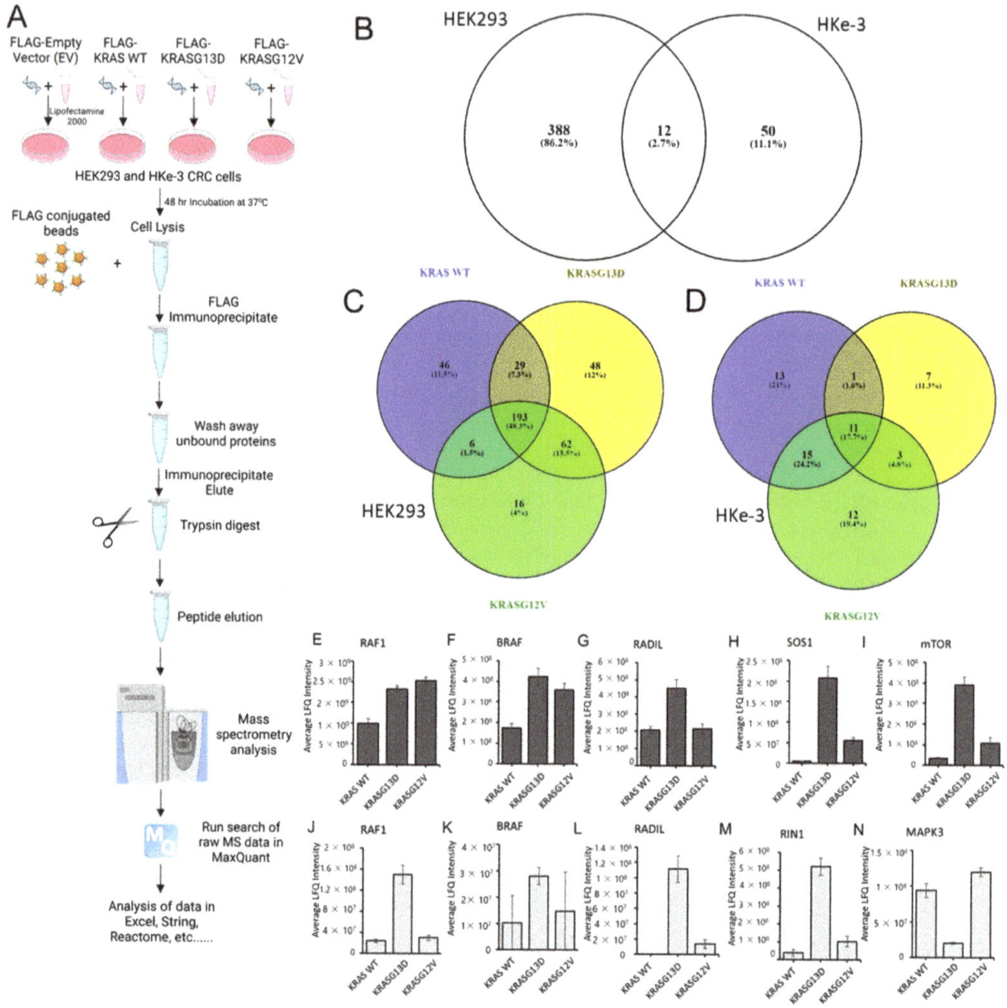

Figure 2. (**A**) Graphical representation of the mass spectrometry affinity purification protocol for KRAS WT, KRASG13D, and KRASG12V in HEK293 and HKe-3 cell lines. Cells were transfected with FLAG-tagged KRAS WT, G13D, and G12V DNA constructs. Cells were lysed after 48 h and incubated with FLAG-M2 conjugated to agarose. Following washes of the beads with lysis buffer, samples were prepared for trypsin digestion. Mass spectrometer protein searches of the raw data were conducted using Max Quant to quantify protein abundance associated with each bait. LFQ intensity of the protein for each condition was used to identify specific KRAS interactors in both cell lines. Created with BioRender.com. (**B**) Venn diagram representing the number of specific protein

interactions binding to KRAS proteins (sorted by gene name) in HEK293 and HKe-3 cell lines (**C**) Venn diagram representing the specific interactors of the indicated KRAS proteins identified in AP-MS proteomics in HEK293 and (**D**) HKe-3 cell lines. (**E–I**) graphs show the average LFQ in HEK293 cells of the indicated proteins in immunoprecipitates of KRAS WT, KRASG13D, or KRASG12V as indicated ($n = 3$). Error bars show SEM. (**J–N**) graphs show the average LFQ in HKe-3 cells of the indicated proteins in immunoprecipitates of KRAS WT, KRASG13D or KRASG12V as indicated ($n = 3$). Error bars show SEM.

Other known interactors of KRAS observed in the analysis of the HEK293 cells included MLLT4, SOS1, RHOA, RHOC, HRAS, and NRAS. Interactors of KRAS observed in the HKe-3 cell line included RIN1, ROCK1, HRAS, and NRAS. Interestingly, but perhaps not surprisingly, we saw that similar to ARAF some of these known interactors were only detected in one of the cell lines. For instance, MLLT4 and SOS1 were specific to the KRAS interactome in HEK293 cells (Figures 2H and S3B). On the other hand, the RAS effector RIN1 was only identified in HKe-3 cell IPs and showed higher affinity for KRASG13D (Figure 2M). This disparity in the number of interactors may be due to the expression levels of these proteins in both cell lines or can be specific to certain cell types, tissues, or the mutational status of oncogenic cell lines. In the case of the KRAS-RIN1 interaction, RIN1 is not expressed in HEK293 cells according to the Human Protein Atlas [48], explaining why we do not see this interaction in HEK293. CAT, MAPK3, NRAS, and DOCK7, which were shown to interact with KRAS baits only in HKe-3 cells (Figures 2N, S2 and S3F and Table S2), are expressed in both cell lines suggesting that these interactions are cell type-specific.

3.3. Bioinformatic Characterisation of KRAS Protein Signalling Networks Show Correlation between KRAS Protein Interactomes and Function

We used gene ontology (GO) and pathway reconstruction to broadly determine the physiological and pathophysiological relevance of our findings. Panther GO enrichment analysis [49] showed some cell-type-specific variations in biological processes mediated by the KRAS proteins as a whole. For example, transporter activity GO is overrepresented in HEK293 but not in HKe-3 while the developmental process is enriched in HKe-3 cells (Figure 3A). Differences are also shown within each cell type for the GO enriched for specific mutants. For instance, in HEK293 cells the KRASG12V and KRASG13D interactomes are significantly enriched in proteins involved in signalling processes and protein binding versus KRAS WT (Figure S4A,C). However, little differences between KRASG13D and KRASG12V were observed. Clearer differences between KRASG12V and KRASG13D interactomes were shown in the HKe-3 cell line (Figure S4B,D). In these cells, KRAS WT and KRASG12V showed similar percentages of proteins involved in cellular and metabolic processes compared with the KRASG13D mutant which had more proteins involved in adhesion and in catalytic activity (Figure S4B,D).

Next, to obtain a network view of the interactomes associated with the three KRAS proteins in the two cell lines we used StringDB [31]. In HEK293 cells, the network showed a high number of connections between KRAS (Figure S4E, red halo), which is to be expected as the dataset includes some of the best-known interactors such as RAF1, BRAF, and FNTA. Representation of a subnetwork containing highly enriched functional clusters identified an overrepresentation of proteins that are involved in metabolic processes, mTOR signalling, and Ras GTPase binding and cell cycle (Figure 3B and Table S3). In the case of HKe-3, the signalling network of specific KRAS interactors showed an enrichment of proteins involved in different aspects of RNA binding and metabolism (Figure 3C). Altogether, these results confirm that the interactomes of KRAS are cell-type-specific and different KRAS proteins regulate specific effector pathways and functions which is in line with findings in recently published interactome studies [10].

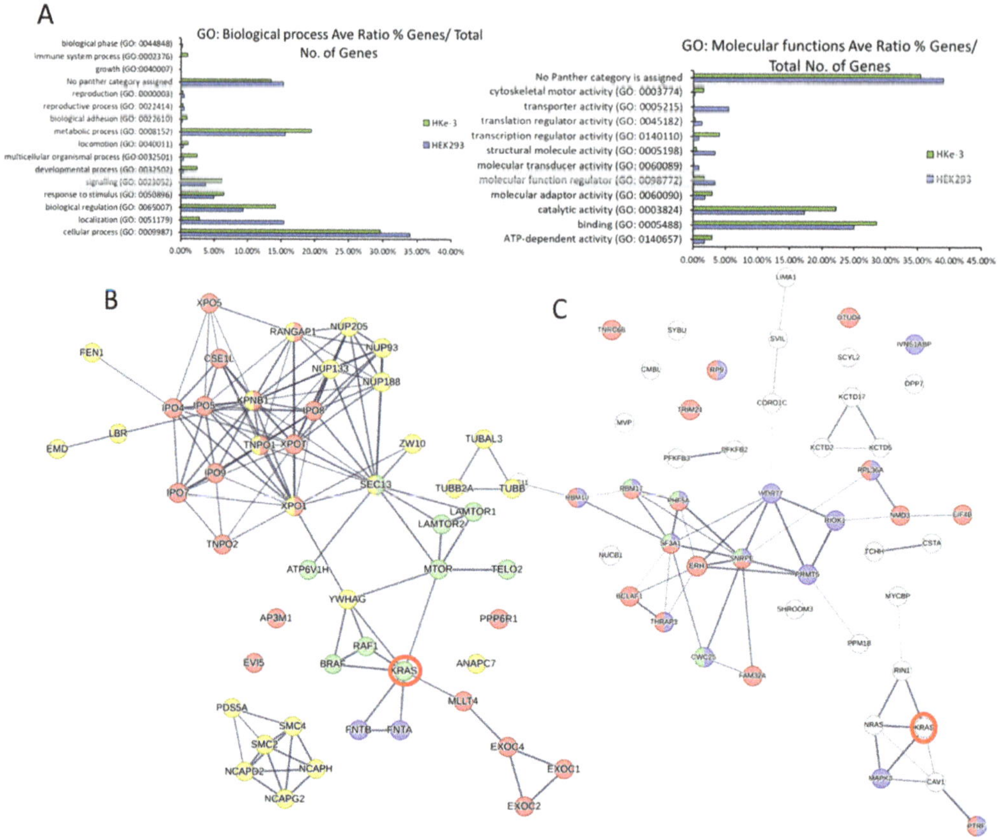

Figure 3. (**A**) The specific KRAS interactomes in HEK293 and HKe-3 cells identified by AP-MS were input into the Panther analysis tool. Graphs show representation of GO terms enriched for biological processes (left) and molecular functions (right). (**B**) Pathway reconstruction and cluster analysis performed with StringDB show HEK293 specific protein interactors of KRAS that are exclusive to the HEK293 cell line that are part of functional clusters. Red halo indicates the KRAS node. Edges represent confidence, line thickness indicates the strength of data support. Proteins coloured red are involved in RAS GTPase binding. Proteins coloured green are involved in mTOR signalling pathway. Yellow proteins are involved in Cell cycle-mitotic. Blue are proteins of the farnesyltransferase complex. (**C**) Specific protein interactors of KRAS that are exclusive to the HKe-3 cell line were reconstructed using StringDB. Red halo indicates the KRAS node. Edges represent confidence, line thickness indicates the strength of data support. Proteins coloured red are involved in RNA binding; proteins coloured blue are involved in RNA metabolic processing; green proteins are involved with the spliceosome.

3.4. Experimental Validation Confirms That KRAS Proteins Interactomes Are Differentially Regulated by KRAS Point Mutation and Shows That RADIL Mediates KRAS-Dependent Migration in a Mutant Specific Manner

Several of the findings described above can be of special functional relevance. In particular, we were intrigued by the observation that SOS1 might have different affinities for KRASG12V and KRASG13D in HEK293 cells (Figure 2H, of note this interactor was not identified in HKe-3). To validate this finding, we performed IP experiments in HEK293 cells co-transfected with AU5-SOS1 and FLAG-KRAS constructs. Of note, SOS1 expression levels were consistently lower when KRAS WT was expressed in these cells than when we expressed KRAS mutant constructs indicating that KRAS WT overexpression can regulate SOS1 levels by yet uncharacterised mechanisms. Nevertheless, as shown in the proteomics experiment, SOS1 had a higher affinity for KRASG13D than for KRASG12V (Figure 4A). Interestingly, KRAS WT had the highest affinity for SOS1 than any of the KRAS mutants. This disparity might be explained by the fact that we immunoprecipitated SOS1 while in the screen we pulled down KRAS complexes. Additionally, we could see that KRASG12C has similar affinity for SOS1 as KRASG13D, while KRASG12D has a higher affinity for SOS1 and KRASG12V has the lowest affinity. These differential affinities of KRAS mutants for SOS1 might have important implications in tumours, as KRAS binds to SOS1 in the enzymatic pocket but can also regulate its function by binding to an allosteric regulatory pocket [50]. Unfortunately, our attempts to experimentally distinguish the contribution of the two binding sites were not conclusive.

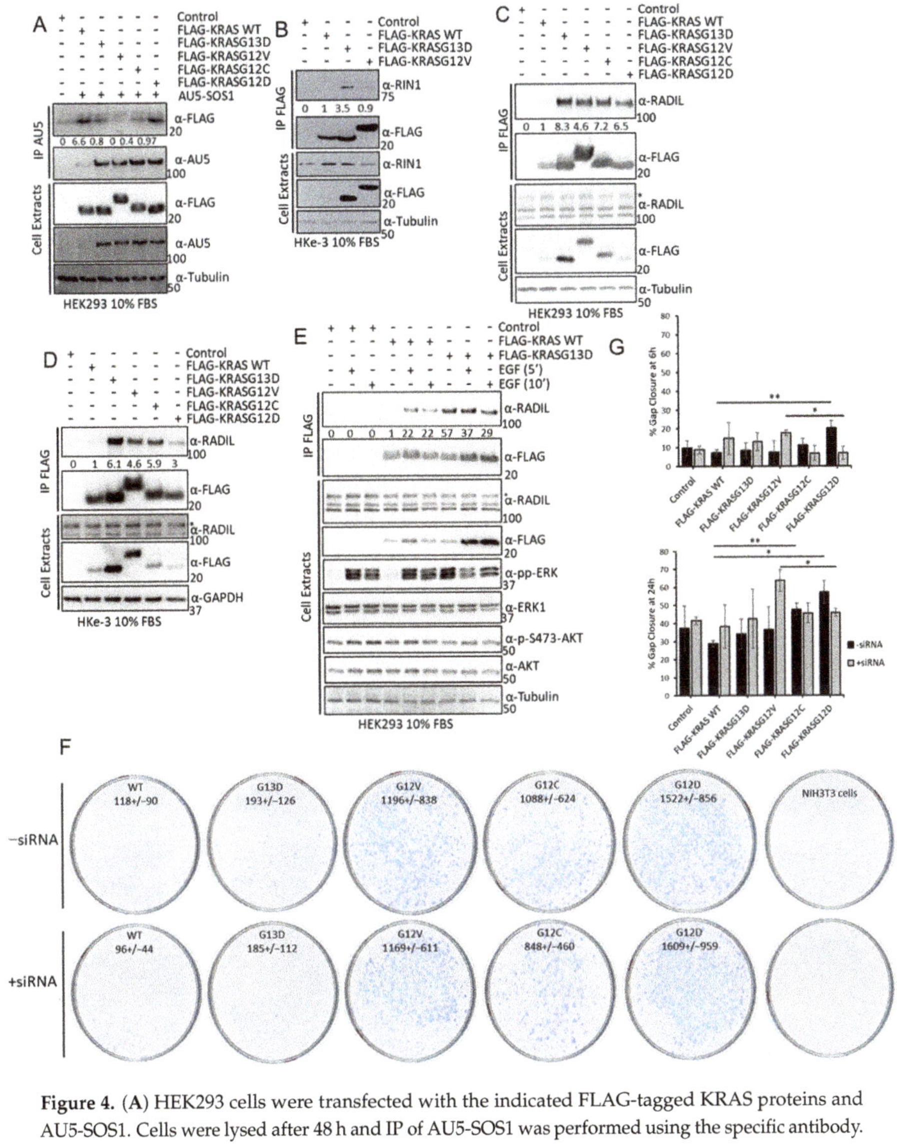

Figure 4. (**A**) HEK293 cells were transfected with the indicated FLAG-tagged KRAS proteins and AU5-SOS1. Cells were lysed after 48 h and IP of AU5-SOS1 was performed using the specific antibody.

IPs and cell extracts were blotted with the indicated antibodies. (**B**) HKe-3 cells were transfected with empty vector, FLAG-KRAS WT, -KRASG12V, or -KRASG13D. Cell lysates were incubated with anti-FLAG antibody and immunoprecipitates were blotted with the indicated antibodies. Band intensities were measured using ImageJ and numbers show fold changes of the ratio RIN1/FLAG intensity in the IPs. (**C**) HEK293 cells or HKe-3 cells (**D**) were transfected with empty vector or the indicated KRAS plasmids and lysates were IP using FLAG antibody and blotted with the indicated antibodies. Numbers show fold changes of RADIL/FLAG intensities as measured using ImageJ. * indicates specific RADIL band in the cell extracts. (**E**) HEK293 cells were transfected with empty vector, FLAG-KRAS WT or –KRASG13D. Following 16 h serum deprivation cells were treated with EGF (3 μM) for the indicated times and lysed 48 h after transfection. FLAG IP was performed with the cell extracts followed by Western blotting with the indicated antibodies. Numbers show fold change in RADIL interaction with FLAG-KRAS proteins. Band intensity was measured using ImageJ. The whole Western blot figures can be found in Supplementary Materials Original Blots and quantification for Figure 4. (**F**) NIH3T3 were transfected with 200 ng of FLAG-KRAS WT (WT), -KRASG13D (G13D), -KRASG12V (G12V), -KRASG12C (G12C), -KRASG12D (G12D), or empty vector and/or RADIL siRNA (20 nM); 14 days after, transfection plates were fixed and stained with Giemsa and macroscopic foci were counted. Numbers show average number of foci per 1 ug DNA +/− standard deviation (SD) (n = 3). (**G**) Cells transfected as in A were grown to confluence in a plate with a rubber stopper. Stoppers were removed and images were taken at the indicated times. Percentage gap closure was quantified using ImageJ and results were normalised to 0 h time point measurements [36]. Error bars are SEM. ** p value < 0.01, * p value < 0.05.

We also confirmed that RIN1 expression is not detected in HEK293 cells, but it is expressed in HKe-3 cells where it binds better to KRASG13D than KRASG12V or KRAS WT (Figure 4B). Finally, we validated the differential interaction of RADIL with the three KRAS proteins used in the proteomics experiments in both cell lines. In this case, we did see a very weak interaction with KRAS WT while the two mutants had higher affinity, with KRASG13D having the highest affinity for RADIL. Moreover, our results showed that the differential affinity of RADIL for the KRAS mutant isoforms also extends to the mutants KRASG12C and KRASG12D, that bind this protein with lower affinity than KRASG13D with KRASG12D, showing the lowest affinity of all the mutants tested (Figure 4C,D). In order to test if activation of KRAS WT increases its interaction with RADIL, we treated HEK293 cells with EGF, which caused a clear increase in KRAS WT-RADIL interaction (Figure 4E). This is similar to recent observations showing that RADIL interaction is increased upon KRAS WT activation [43]. Importantly, when we performed these experiments, RADIL was not considered a KRAS interactor but since then, data from other interaction proteomic studies identified this protein as one of the main interactors in AP-MS experiments using different RAS proteins [37].

Our results confirm that RADIL is an important KRAS interactor protein that is differentially regulated by specific KRAS point mutations. Choi et al. showed that this protein mediates KRAS-dependent proliferation and focal adhesion formation [43] and, Kelly et al. saw that RADIL mediates KRAS-dependent migration [37]. We decided to extend the functional characterisation of this interactor with the different KRAS mutants through functional biological assays. We first tested the possible role of RADIL in KRAS-dependent cell viability by knocking down RADIL with specific siRNA in HKe-3 cells expressing the different KRAS mutants. Although RADIL knockdown by specific siRNA was efficient (Figure S5A), neither expression of the different KRAS proteins nor downregulation of RADIL expression affected cell viability (Figure S5A). Next, we tested the role of RADIL in KRAS-mediated transformation by performing a focus formation assay in NIH3T3 cells. We transfected the cells with KRAS WT and the 4 KRAS mutants and downregulated RADIL protein expression by co-transfection of RADIL siRNA. All KRAS mutants induced the formation of transformed foci, although with clear differences (Figures 4F and S5C). KRASG13D showed the lowest transformation effect similar to KRAS WT, while KRASG12V, KRASG12C, and KRASG12D produced 5–8 times more transformed foci than KRASG13D. This result is in line with previous data showing that KRASG13D has lower transformation efficacy [15,51]. Downregulation of RADIL did not show a statistically significant effect on the number of foci induced by the different mutants (Figures 4F and S5B). Finally, given the known involvement of RADIL in cell motility [43] we studied the role of RADIL in KRAS-dependent migration using a wound healing assay. Downregulation of RADIL alone did not have any effect on the migration of HKe-3 cells (Figures 4G and S5C). Surprisingly, KRASG12D was the only KRAS protein that caused an increase in the migration of these cells. Importantly, this effect is mediated by RADIL as its downregulation reduced KRASG12D-dependent migration. Remarkably, RADIL seems to be a downstream effector of another mutant, KRASG12V, which did not increase migration in these cells. In this case, downregulation of RADIL increased migration of KRASG12V-expressing cells. Thus, RADIL might be playing different roles in the migration of tumour cells that express specific KRAS point mutations. This result strengthens the idea that not all the KRAS mutants have the same functions.

3.5. Sotorasib Shows Off-Target Effects in Cells That Do Not Express KRASG12C

The previous results clearly showed that there are changes in the interactome of KRAS mutants that can be of relevance for the initiation, development, and evolution of KRAS tumours. In light of these results, we tested whether the mechanism of action of the newly approved KRASG12C inhibitors includes the specific regulation of KRASG12C effector pathways mediated by changes in the interactome. For this, we monitored the effects of the KRASG12C mutant inhibitor, Sotorasib, on the regulation of the best characterised RAS effector pathways, RAF and AKT, in cell lines that express different KRAS mutants. Sotorasib inhibits KRASG12C at nM concentrations but in different studies is used at a wide range of concentrations [14,52]. Importantly, different groups are using high concentrations of Sotorasib (5–10 μM) to generate resistant cell lines that can be used to characterise the mechanisms of resistance to this drug [53–55]. For this reason, we decided to use high concentrations of Sotorasib in this part of the study. We used RASless mouse embryonic fibroblasts (MEFs) generated by Drosten et al. [17] that express one of the KRAS mutants and none of the other RAS (HRAS, NRAS, and KRAS) genes. The advantage of these cell lines is that we avoid any unspecific regulation of RAS signalling effector pathways that might be induced by Sotorasib treatment. In addition to MEF-KRASG12C, we assessed the effect of Sotorasib in MEF-KRAS WT, KRASG12V, and KRASG12D cells should not be affected by treatment with this drug even at these high concentrations [55]. Sotorasib treatment of MEF-KRASG12C caused a band shift in this mutant due to the covalent binding of this drug. This band shift did not occur in the other KRAS proteins (Figure 5A and Figure S6A). As expected, Sotorasib only inhibited activating ERK1/2 phosphorylation and ERK induced feedback phosphorylation of RAF1 in MEF-KRASG12C confirming the specificity of this drug for this mutant. Similar results were observed in HEK293 cells where we could see that Sotorasib inhibits KRASG12C-dependent activation of the three core kinases of the ERK pathway (Figure 5B). We also observed an increase in AKT phosphorylation, similar to what has been described before in cancer cells [56]. Surprisingly, Sotorasib upregulated phospho-AKT in all four MEF cell lines, especially in MEF cells expressing KRASG12D (Figure 5A and Figure S6A). The results clearly indicate that Sotorasib must have other targets that regulate AKT activation independent of KRASG12C. Interestingly, these targets seem cell-type-specific, as the induction of pAKT by Sotorasib was observed in MEFs and HKe-3 (Figure S6C), but not in HEK293 cells (Figure 5B and Figure S6B). As pAKT is constitutively upregulated in Sotorasib-resistant lung and pancreatic cancer cell lines [57], the off-target activation of AKT may be relevant for the development of resistance to Sotorasib.

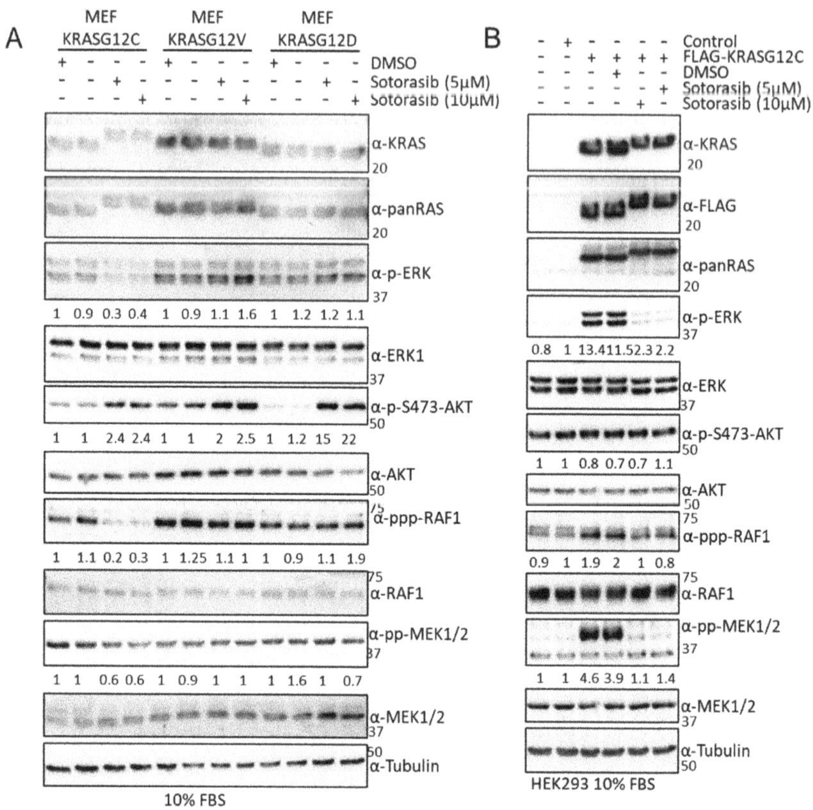

Figure 5. (**A**) MEF-KRASG12C, -KRASG12V, and –KRASG12D were treated with the indicated amount of Sotorasib for 24 h. Cells were lysed, and the indicated proteins were blotted using the specific antibodies. Numbers show fold changes of phosphorylation of the indicated proteins with respect to their total level of expression. Intensity was measured using ImageJ 1.5. (**B**) HEK293 cells were transfected with the indicated constructs; 24 h after transfection the cells were treated with the indicated concentrations of Sotorasib for 24 h. Cell lysates were blotted with the indicated. Numbers show fold changes of phosphorylation of the indicated proteins with respect to their total level of expression. Intensity was measured using ImageJ. Experiments were repeated 3 times. The whole Western blot figure can be found in Supplementary Materials Original Blots and quantification for Figure 5.

3.6. Sotorasib Regulates the KRASG12C Interactome and Has Unspecific Effects on the KRAS WT Interactome

To characterise if Sotorasib's mechanism of action involves changes in the KRASG12C interactome, we used the AP-MS proteomic approach described above. Unfortunately, MEF-KRASG12C cells cannot be used for AP-MS experiments, as the KRASG12C protein is not tagged and there are no suitable antibodies available for immunoprecipitating endogenous RAS and its effector proteins. Therefore, we used HEK293 cells transfected with FLAG-KRASG12C or FLAG-KRAS WT for comparison. Despite the fact that we immunoprecipitated similar amounts of KRAS WT and KRASG12C and Sotorasib treatment does not affect their protein expression, we saw lower level of KRASG12C in AP-MS (Figures 6A,B and S7A and Table S4). This suggested that KRASG12C structure and Sotorasib binding interfered with the trypsin digestion of FLAG-KRASG12C that is part of the sample preparation for MS. It is well-known that some protein conformations can be resistant to trypsin digestion [58], and it seems that Sotorasib can induce such a protease-resistant conformation in KRASG12C. To test this, we performed double digestion with trypsin and LysC, which allowed the full digestion of untreated FLAG-KRASG12C and improved the digestion of Sotorasib-bound KRASG12C (Figure S7A,B). The partial trypsin resistance of KRASG12C, however, did not affect the digestion of interacting proteins co-precipitating with FLAG-KRASG12C (Figure S7C–F). Hence, we used standard trypsin digestion in the AP-MS experiments to study the effect of Sotorasib on its interactome, as the dataset was of better quality.

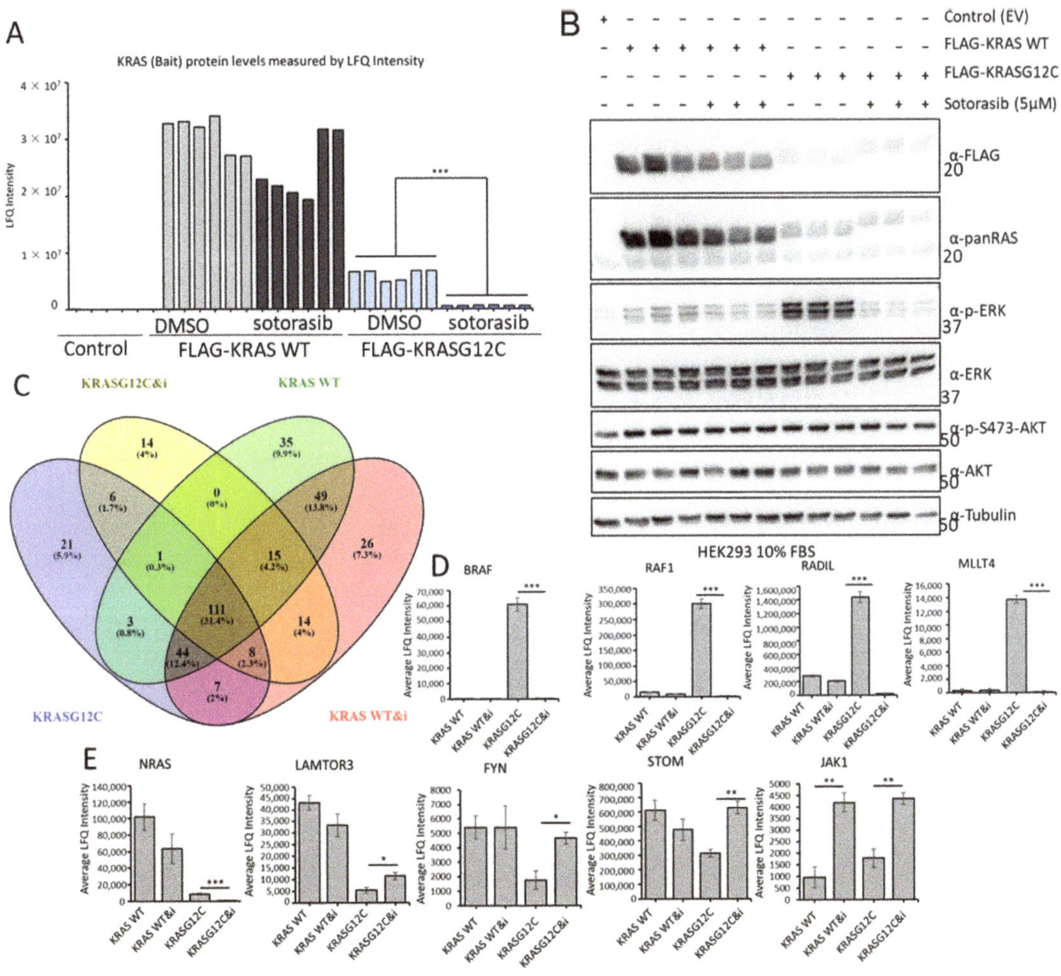

Figure 6. (**A**) HEK293 cells were transfected with the indicated constructs and 24 h later they were treated with Sotorasib (5 μM) or DMSO for 24 h. FLAG IP were tryptic digested and analysed by MS. Graph shows the LFQ intensity of all the samples. *** p value < 0.01. (**B**) A fraction of the cell lysates of HEK293 cells used in A were blotted with the indicated antibodies. The whole Western blot figure can be found in Supplementary materials Original Blots and quantification for (**B**). (**C**) Venn diagram representation of the specific interactors identified in the MS-based proteomic screen in FLAG-KRAS WT (KRAS WT) or FLAG-KRASG12C (KRASG12C) immunoprecipitates treated with DMSO or 5 μM Sotorasib (i) for 24 h. (**D**,**E**) Graphs show average LFQ intensity of the indicated proteins in the different IPs. Error bars show SEM. * p value < 0.05, ** p value < 0.01.

Comparing the interactomes of FLAG-KRAS WT and FLAG-KRASG12C plus/minus Sotorasib treatment showed that 41 proteins were exclusively bound to KRASG12C and 110 interacted only with KRAS WT (Figure 6C). Of the 246 proteins that are specifically interacting with KRASG12C, 43 are exclusively binding when the cells are treated with Sotorasib, and 76 proteins are specifically interacting with KRASG12C in untreated cells (Figures 6C and S7G,I). Remarkably, we also observed that Sotorasib impacts the composition of the KRAS WT interactome; 55 proteins are only present in Sotorasib-treated cells and 39 in untreated cells. Additionally, we could see clear regulation of proteins that are interacting with both baits in treated and untreated conditions (Figures 6D,E and S7H,J) including the best-known interactors of KRAS, RAF1, and BRAF. Thus, our screen shows that Sotorasib changed both the KRASG12C and KRAS WT interactomes. In particular, the drug abolished the strong interactions of RAF1 and BRAF with KRASG12C, which explains the inhibition of ERK activation (Figure 6D). The binding to KRAS WT was much weaker and was not affected by Sotorasib. MLLT4 (Afadin), RADIL, and RAP1GDS showed a similar behaviour (Figures 6D and S7I). Both MLLT4 and RADIL are bona fide RAS interactors that bind to GTP-loaded RAS [41,43]. Therefore, this pattern of binding and inhibition by Sotorasib is expected. Interestingly, Sotorasib also reduced the strong binding of NRAS to KRAS WT and the weaker binding to KRASG12C (Figure 6E). These interactions may rely on activation, i.e., GTP-dependent, but not mediated through the effector domain. These results suggest that Sotorasib interferes with RAS oligomerization and nanoclustering, which are important for RAS function [59]. Binding to LAMTOR3, FYN, and STOM, which was stronger in KRAS WT, was increased by Sotorasib treatment in KRASG12C (Figures 6E and S7J). Interestingly, Sotorasib increased the binding to some proteins. For instance, binding to JAK1 was increased in both KRAS WT and KRASG12C (Figures 6E and S7I,J). Altogether, these results suggest indirect or off-target effects, which can contribute to the Sotorasib mode of action as they affect signalling molecules that play important roles in malignant transformation.

3.7. *Sotorasib Regulates KRASG12C and WT Signalling Networks*

We continued our analysis to obtain a systematic view of the Sotorasib-induced changes on the specific interactomes of KRASG12C and KRAS WT. First, we reconstructed the interaction network of proteins that are binding to each bait using the STRING database (Figure S8A,B). The networks are extremely complex and show a high degree of connection among the nodes, which complicates the identification and representation of functionally relevant findings. For this reason, we represented the sub-network of the proteins that are changing their interaction with RAS upon Sotorasib treatment. In the case of KRASG12C, functional analysis confirms that there is an enrichment of proteins that are part of the AKT and ERK signalling pathways confirming the central role of these pathways in the mechanism of action of Sotorasib (Figure 7A). The changes of interaction of these nodes likely have an effect in some of the proteins that are part of these pathways but do not change their interaction with KRASG12C in our screen including PIK3R1/2 (the regulatory subunits of PI3K) or ILK. Interestingly, this analysis shows an increase in interaction with proteins involved in the functions of the lysosome including STOM, LAMTOR3, and several members of the ATPV6 complex (grey area Figure 7A). An association of KRAS4B with ATPV6 and lysosomal localisation was described before [38]. Thus, our data indicate that Sotorasib induces the localisation of KRAS in the lysosome. Intriguingly, a similar analysis of Sotorasib-regulated KRAS WT network identified a decrease in interaction of this protein with the ATPV6 complex indicating that Sotorasib prevents KRAS WT localisation to the lysosome (Figure 7B). Conversely, proteins involved in RNA recognition motif have an increased binding to KRAS WT treated with Sotorasib. Finally, JAK1 interaction with KRASG12C and KRAS WT is increased in cells treated with Sotorasib confirming that this drug promotes the interaction of this kinase with KRAS proteins. Altogether, these analyses showed that Sotorasib regulates the KRASG12C interactome but also has clear effect in the regulation of the KRAS WT interactome.

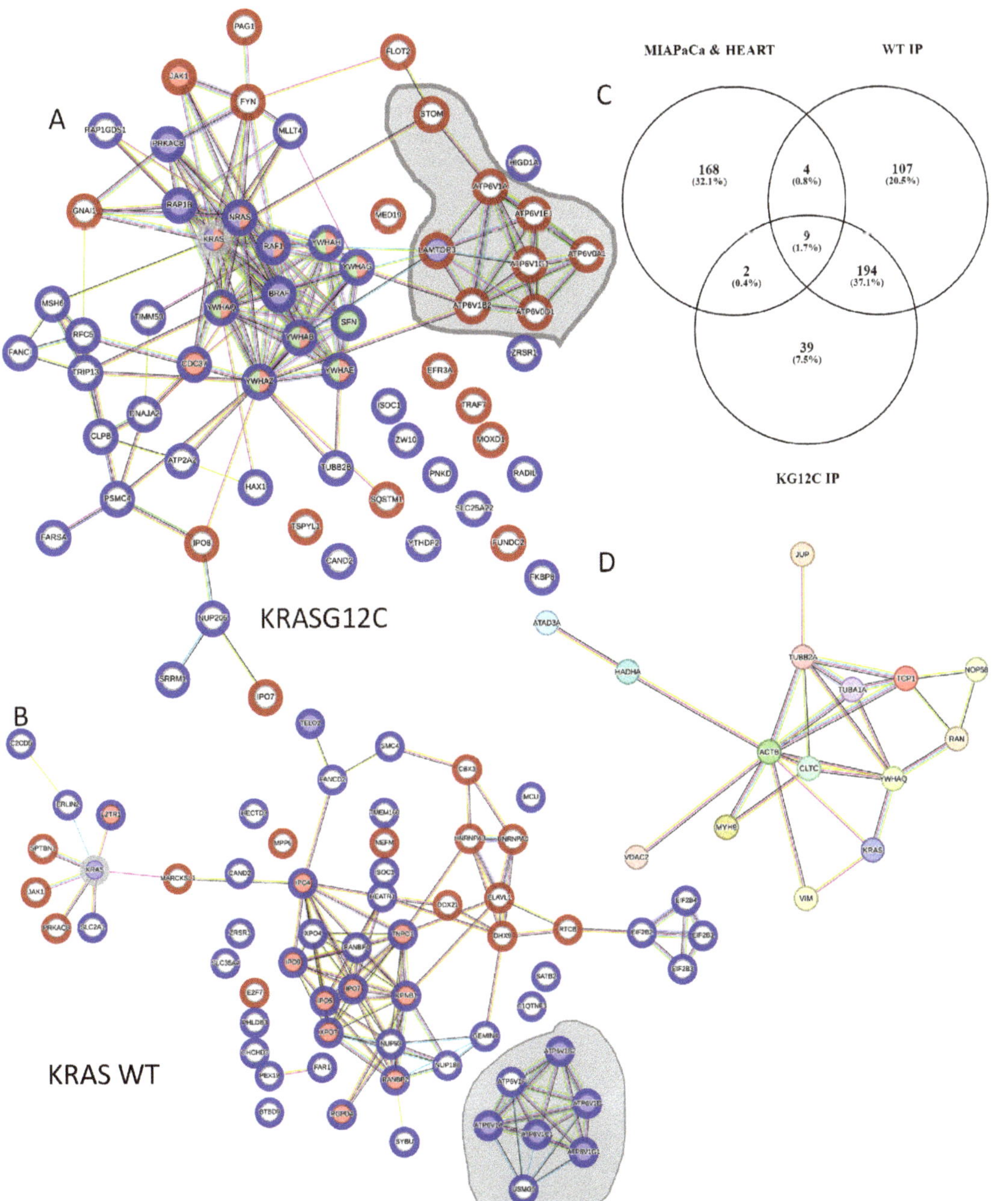

Figure 7. (**A**) Reconstructed network using StringDB of the dynamic interactome of KRASG12C (grey halo) in HEK293 cells identified in the AP-MS screen. Blue halo on proteins shows significantly increased binding to KRASG12C versus KRASG12C treated with Sotorasib $p < 0.05$. Red halos show significantly increased binding to KRASG12C treated with Sotorasib versus untreated KRASG12C $p < 0.05$. Proteins with blue nodes are involved in MAPK, proteins red nodes are PI3K related, and

green nodes 14-3-3 domain superfamily related. Grey area shows proteins involved in lysosomal function and edges show evidence for interaction. (**B**) Reconstructed network using StringDB of the dynamic interactome of KRAS WT (grey halo) identified in the AP-MS screen. Blue halo on protein shows significantly increased binding to KRAS WT versus KRAS WT treated with Sotorasib $p < 0.05$. Red halo show significantly increased binding to KRAS WT treated with Sotorasib versus KRAS WT $p < 0.05$. Proteins with blue nodes are involved in mTOR signalling pathway, proteins red nodes are RAS GTPase binding. Edges show evidence for interaction. (**C**) Venn diagram representation of the proteins identified as specific interactors or KRAS WT and KRASG12C in our study with the proteins identified to be binding Sotorasib in MIAPaCa cells and heart tissue by Wang et al. [14]. (**D**) Network constructed using string DB of the 15 proteins that are in the intersection of the Venn diagram shown in (**C**). Edges show evidence for interaction (high confidence).

Although Sotorasib is approved as a KRASG12C-specific inhibitor, different groups have identified other proteins that bind this drug and we reasoned that some of these targets might also be part of the KRAS interactome [14,52,60,61]. To explore this, we used information from one of the studies that have identified proteins that are targeted by Sotorasib using an MS-based approach [14]. This study identified 183 proteins that bind Sotorasib and using a Venn diagram analysis we saw that 15 Sotorasib target proteins were identified in our screen as part of the interactome of KRAS proteins (Figure 7C). Importantly, these proteins form a highly connected network and include proteins from the cytoskeleton including VIM, ACTB, and TUB1A/B; and the member of the 14-3-3 proteins YWHAQ which are key regulators of KRAS downstream effectors (Figure 7D). These data indicate that at least some of the changes in KRASG12C and KRAS WT interactomes induced by Sotorasib treatment might be caused by Sotorasib directly targeting protein interactors of KRAS.

3.8. *Validation Experiments Show a New Sotorasib-Regulated Crosstalk of RAS Signalling with JAK1*

To experimentally validate our proteomics results, we performed KRAS immunoprecipitation assays and monitored the dynamics of the interaction of KRAS WT and KRASG12C and several of the proteins shown to be changing in the proteomics analysis. We confirmed that RADIL, RAF1, and BRAF interaction with KRASG12C was severely disrupted by Sotorasib in HEK293 cells (Figure 8A). Similar results were observed in HKe-3 cells for the interaction of KRASG12C with RADIL and RIN1 which are also downregulated by Sotorasib (Figure 8B). These validation experiments confirmed that our AP-MS method accurately identified changes induced by Sotorasib in the interactome of these KRAS proteins. They may differentially regulate its effector signalling pathways and therefore need to be understood to fully characterise the full mode of action of Sotorasib.

One of the KRAS effector pathways that our results indicated to be modulated by Sotorasib is the JNK stress pathway, which was shown to be regulated in KRASG12C-expressing HEK293 cells. Indeed, we observed a clear KRASG12C-dependent activation of JNK that was completely abolished by Sotorasib treatment (Figure 8C). JAK1 signalling activation by KRASG12C induced phosphorylation of the STAT3, which was prevented by Sotorasib treatment (Figure 8C). To the best of our knowledge, this is the first demonstration that JAK1-STAT signalling may be a KRAS effector pathway, and that this pathway is modulated by KRASG12C specific inhibitors.

JAK1 has been shown to regulate AKT1 in some cases [62], and we wondered whether JAK1 activation is related to Sotorasib-induced AKT activation. We tested this using Ruxolitinib, a JAK1/2 inhibitor used in the clinic [63], alone, or in combination with Sotorasib. In MEF-KRASG12C cells, there was a much lower level of STAT3 phosphorylation than in MEF-KRAS WT, but in both cell lines, addition of the JAK inhibitor clearly decreased the activation of STAT3 (Figure 8D). No changes in AKT1 phosphorylation were observed upon Sotorasib and Ruxolitinib combination treatment in MEF-KRASG12C cells indicating that JAK1 is not responsible for the Sotorasib-induced AKT activation in this cell line. In the case of MEF-KRAS WT, combination of Sotorasib and Ruxolitinib induced an increase in AKT phosphorylation indicating that the crosstalk of this protein with JAK is different to the KRASG12C-JAK1 crosstalk (Figure 8D). Similarly, we observed an increase in ERK1/2 phosphorylation in MEF-KRAS WT cells treated with Ruxolitinib, while the same treatment did not have any effect in ERK1/2 activation in MEF-KRASG12C. Finally, we observed an increase in caspase three cleavage in cells treated with the combination of KRASG12C and JAK1/2 inhibitors in MEF-KRAS WT but not in cells treated with either one of the treatments. In MEF-KRASG12C caspase cleavage was higher in all treatment conditions, indicating that there is an increase in apoptotic signalling in these cells. Finally, an important observation from these treatments was a conspicuous increase in the expression of KRAS WT and KRASG12C in cells treated with Ruxolitinib. This unexpected observation indicates that JAK signalling has an important role in the regulation of KRAS protein expression. Altogether, these results validate the findings from our proteomic screening and extend the characterisation of Sotorasib's mechanism of action.

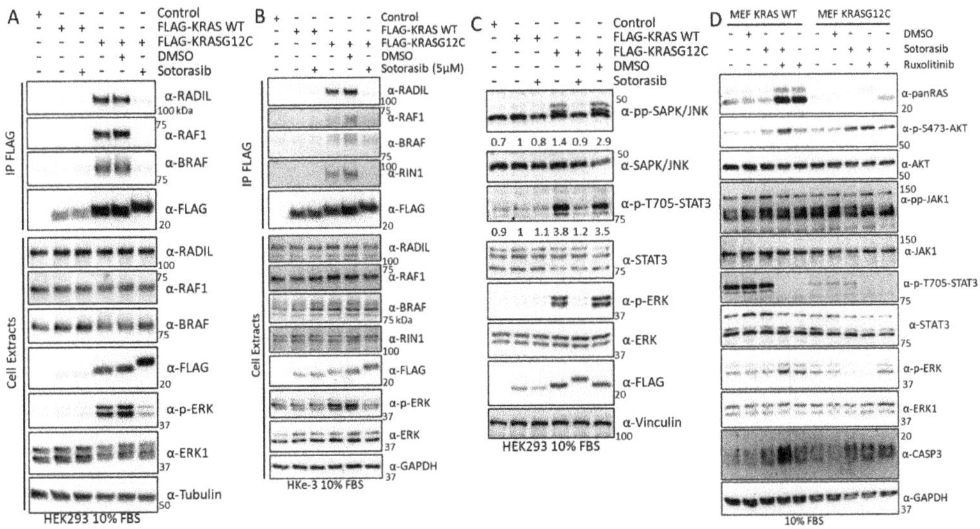

Figure 8. (**A**) HEK293 cells were transfected with the FLAG-KRAS WT or –KRASG12C; 24 h. Cells were treated with DMSO or 5 μM Sotorasib for 24 h. Cells were lysed and immunoprecipitated with FLAG antibody. IPs and cell extract were blotted with the indicated antibodies. (**B**) HKe-3 cells were transfected with the FLAG-KRAS WT or –KRASG12C. The cells were treated with DMSO or 5 μM Sotorasib for 24 h. Cells were lysed and immunoprecipitated with FLAG antibody. Proteins in IPs and cell extract were blotted with the indicated antibodies. (**C**) HEK293 cells treated as in A were blotted for the indicated antibodies. Numbers show fold changes in the level of phosphorylation of the indicated proteins. Intensities were measured using ImageJ. (**D**) MEF-KRAS WT and MEF-KRASG12C were treated with 5 μM Sotorasib and/or 5 μM Ruxolitinib as indicated for 24 h. Cells were lysed and the proteins were blotted with the indicated antibodies. The whole Western blot figure can be found in Supplementary Materials Original Blots and quantification for Figure 8.

4. Discussion

Our study aimed to extend our characterisation of KRAS-interacting proteins and mapping the signalling networks regulated by KRAS WT and oncogenic mutants. The initial proteomics screen clearly showed that KRAS WT and two of the most common KRAS mutants, KRASG12V and KRASG13D, interact with different affinities to specific binding proteins. These results also show that the interactome of these proteins varies in the three cell lines that we used for the study. These findings are in line with several works that have used different MS-based proteomics approaches to identify the interactome of different RAS proteins [10,37–41,46,47,64]. One of the challenges of using AP-MS approaches is that some of the protocols used might not be suitable to identify transient interactions. To overcome this problem, several of the proteomic screens performed so far have used BioID [5,64], which potentially can identify transient and weak interactions of proteins within a ~10 nm radius of the bait [65]. However, our more direct approach clearly identifies interactors that have transient interaction with KRAS including the main effectors RAF1, BRAF, and PI3K. Moreover, we can quantify differences in affinities of a significant number of known and newly identified KRAS interactors including the best characterised GEF of the RAS family SOS1 [2] in HEK293 cells which, to the best of our knowledge, has not been identified using any other MS-based approaches reported to date. In the proteomics dataset SOS1 was shown to have much higher affinity for KRASG13D than KRASG12V and experimentally we show that KRASG12D and KRASG13D have the highest affinity for SOS1 while KRASG12V has very low affinity. This is an unexpected observation as KRASG13D was shown to be less dependent on SOS1 for activation than KRASG12V [9]. Although the functional relevance of this finding needs further studies, it is possible that the mutants with higher affinity can be binding to an allosteric pocket described by Bar Sagi's group which regulates a positive feedback loop that has been proposed to promote tumorigenesis by activating RAS WT isoforms [50]. This might explain the higher level of activation of ERK1/2 when we overexpress KRASG13D.

Our method is also simpler to perform, as it does not rely on the expression of biotin ligase fusion proteins. It must be noted that a disadvantage of both methods is that it is not possible to differentiate between direct and indirect interactors of KRAS which requires further experimental validation or the use of crosslinking protocols suitable for AP-MS proteomics. Nevertheless, our approach is clearly suited to identify dynamic changes in KRAS protein interactomes and shows a great sensitivity to identify true interactors. This is extensively confirmed experimentally with our validation of the interaction of SOS1, RIN1, and RADIL with KRAS WT, KRASG12V, and KRASG13D. Our characterisation of the specific interactomes associated with KRAS mutants has been extended by including in our validation experiments two additional KRAS isoforms, KRASG12C and KRASG12D, which also show differential binding to several of the proteins studied. Hence, the work presented here elucidates a complex signalling network that shows common and specific modules for each KRAS isoform tested and further contributes to identify the growing number of RAS effector pathways. It must be noted that overexpression of RAS proteins and in particular KRAS mutant proteins at similar levels is sometimes challenging as shown in several of our experiments and in the literature [35,66]. Remarkably, this was also shown in the MEFs cells that express different levels of each KRAS mutant. This might be due to activation of pro-apoptotic signals by some mutants [67], rare codon stabilisation of KRAS transcripts [68], or existence of expression sweet spots [69]. To take these variations into account and minimise overinterpretation, we made sure that in all the immunoprecipitation experiments we have similar levels of KRAS proteins and quantified our results. Nevertheless, the findings further support the idea that not all KRAS mutations are equal, and that a given point mutation may be associated with specific pathological effects in tumour cells' signalling networks. Overall, the results from our study must be considered in conjunction with previous reports that focus on mapping RAS protein interactomes and highlight the need to extend these screens to identify cell specific proteomes of all RAS isoforms.

Importantly, our work goes further than previous proteomics screens as, for the first time, we characterised the changes caused by Sotorasib in the KRAS WT and KRASG12C interactomes. This KRASG12C-specific inhibitor is already approved for the treatment of lung cancer but resistance to treatment develops quickly [13]. Additionally, unspecific effects have been reported and it has also been shown to have effects in cell lines regardless of KRAS mutation status [70,71]. Reports also indicate that several patients develop severe side effects which indicate that at therapeutic concentration the drug has a systemic impact affecting cells that do not express KRASG12C [13]. One important consideration is that we used very high concentrations of Sotorasib in our study and some of the changes in

KRASG12C and KRAS WT interactomes that we see in our screen might not occur at nM concentrations of Sotorasib.

Our screen helps delineate the mechanism of action of this drug and can explain some of the side effects of Sotorasib. With respect to the mechanism of action, the dynamic changes shown in AP-MS confirms that it requires the regulation of the interactions of this mutant with effector proteins, something that was shown already for RAF1 and BRAF [37]. As expected, the interaction of these two bona fide effectors is prevented by Sotorasib treatment in our experiments, and this is accompanied by inhibition of ERK1/2 phosphorylation. Interestingly, we saw that Sotorasib also promotes the interaction of KRASG12C with other known KRAS regulators, such as members of the SRC family, i.e., LYN and FYN. With respect to the possible off-target effect of Sotorasib we also characterized an unexpected modulation of the KRAS WT interactome. These experiments indicated that Sotorasib can regulate the interactome of KRAS WT in a way that seems significant but needs further investigations.

Additionally, we consistently observed a cell-type-specific increase in activating AKT phosphorylation in MEFs and HKe-3, but not HEK293. Activation of AKT upon Sotorasib treatment has been associated with resistance to this treatment in patients and is reported in several cell lines [13,57,71,72]. Importantly, our AP-MS approach identified clear changes in the interactomes of both KRAS WT and KRASG12C that potentially can explain the Sotorasib-induced AKT activation. Candidates for mediating this effect detected in our screen include PIK3R1/2, G-coupled receptor proteins, and ILK, which might contribute to the unspecific activation of AKT in these cells. It is worth noting our analysis shows that some of the changes shown in the interactomes of KRASG12C and KRAS WT upon Sotorasib treatment can be explained by the direct effect of Sotorasib on some of the proteins identified which are shown to be targeted by this drug in previous studies (i.e., ACTB, VIM or RAN) [14].

One important new finding from our study is the identification of a Sotorasib-regulated interaction between KRAS and JAK1 that has not been highlighted before. Reassuringly, it must be noted that a recent proteomics screen performed by the Kiel group using a similar protocol also pulled down JAK1 and STAT3 as specific interacting proteins of KRAS WT, KRASG12C, KRASG12V, and KRASG12D [47]. In our AP-MS proteomics experiment, we see that Sotorasib increases JAK1-KRAS interaction and at the molecular level we demonstrate that Sotorasib causes an inhibition of STAT3 phosphorylation in HEK293 cells but not in MEF cells that express only one of the KRAS proteins. This finding is particularly intriguing as it indicates that JAK1 might be specifically interacting with KRAS bound to GDP and this interaction prevents JAK1-dependent activation of STAT signalling. This new connection between KRAS and JAK-STAT signalling might be of clinical relevance as JAK1 inhibition has been shown to reduce murine KRAS mutant adenocarcinoma progression [63]. We show that the clinically approved JAK1/2 inhibitor Ruxolitinib increases caspase three cleavage in MEF-KRASG12C cells but not in MEF-KRAS WT. Remarkably, this experiment led to the unexpected observation that JAK1 inhibition alone causes a significant increase in expression of both KRAS WT and KRASG12C. Taking into account evidence from clinical trials that show an effect of JAK1 inhibitors in EGFR and possible KRAS mutant tumours (NCT02155465, NCT02145637, NCT02917993, and NCT03450330) [63], this serendipitous finding might be of relevance to explain a possible mechanism of action of JAK inhibitors in cancer. One possibility is that JAK1 inhibitors, among other mechanisms, might induce high levels of mutant KRAS expression that leads to the activation of the apoptotic pathways mediated by this RAS isoform [2,73]. It must be noted that this is an acute increase in KRAS expression, which might have very different effects than KRAS overexpression caused by gene amplification that has been associated with resistance to EGFR and KRAS targeted therapies in colorectal cancer cell lines [72]. Finally, our study extends the basic knowledge of the known crosstalk between the JAK and KRAS pathways. The role of JAK-STAT module in the RAS signalling network seems to be at different levels as exemplified by a recent study from Baccarini's group showing a

RAF1 kinase-independent regulation activation of STAT3 signalling in CRC [74]. It would be interesting to test in future studies if Sotorasib can somehow regulate this effect.

5. Conclusions

In summary, this work extends our knowledge about the variety of functional pathways that KRAS proteins are mediating and confirms that the interactome of different KRAS mutants show important differences. Some of these new findings might prove important to explain the differences among KRAS mutant proteins in the promotion of cell transformation, tumour development, drug sensitivity, and maintenance of human tumours [1,2,4,16]. We also show evidence that performing proteomics screens is important to characterise the mechanisms of action of the recent FDA approved RAS targeting drugs and the new ones in drug development. Ultimately, determining if a RAS inhibitor prevents the interaction and/or regulation of specific RAS interactors can predict the mechanism of resistance that can hamper the effect of these novel therapies.

Supplementary Materials: The following supporting information can be downloaded at: https://www.mdpi.com/article/10.3390/cancers15164141/s1, Figure S1: Specific protein interaction network of KRAS isoforms in HEK293 cells; Figure S2: Specific interactome network of KRAS isoform in HKe 3 cells; Figure S3: AP-MS identify KRAS proteins differential interacting proteins; Figure S4: Bioinformatic functional analysis of AP-MS data in HEK293 and HKe-3 cells; Figure S5: Functional assays shows differential regulation of RADIL by KRAS mutants; Figure S6: Sotorasib regulates of KRAS effectors in different cells types; Figure S7: Sotorasib regulates KRAS proteins interactome; Figure S8: Network reconstruction of KRASG12C and KRAS WT specific interactome. Entire Western blot figures for main and supplementary figures are included in supplementary figures PDF as Original Blots and quantification figures with the same numbering as shown in main text and Supplementary figure files. Table S1: HEK293 KRAS interactome; Table S2: HKe-3 KRAS interactome; Table S3: Functional clusters Figure 3B; Table S4: KRASG12C and KRAS WT Sotorasib interactomes.

Author Contributions: Conceptualization, W.K. and D.M.; Data curation, A.N., K.W. and D.M.; Formal analysis, A.N. and D.M.; Funding acquisition, W.K. and D.M.; Investigation, A.N. and D.M.; Methodology, A.N., A.v.K. and K.W.; Project administration, W.K. and D.M.; Resources, D.M.; Supervision, W.K. and D.M.; Validation, A.N. and C.R.; Visualization, A.N. and D.M.; Writing—original draft, A.N. and D.M.; Writing—review and editing, A.N., C.R., W.K., A.v.K., K.W. and D.M. All authors have read and agreed to the published version of the manuscript.

Funding: This project has received funding from the European Union's Horizon 2020 research and innovation programme under grant agreement No 754923 (A.N., W.K. and D.M.). The material presented and views expressed here are the responsibility of the author(s) only. The EU Commission takes no responsibility for any use made of the information set out. D.M. was supported by Science Foundation Ireland (SFI) grant 15-CDA-3495, W.K. by the SFI Precision Oncology Ireland grant 18/SPP/3522, W.K. and K.W. by the SFI Comprehensive Molecular Analytical Platform (CMAP), reference 18/RI/5702.

Institutional Review Board Statement: Not applicable.

Informed Consent Statement: Not applicable.

Data Availability Statement: The mass spectrometry proteomics data have been deposited to the ProteomeXchange Consortium via the PRIDE [75] partner repository with the dataset identifier PXD043536 and PDX043170.

Acknowledgments: Luis Iglesias Martínez for help with analysis of proteomics data. Amaya García Munoz for technical support. Lucía García-Gutiérrez for help in the development of methodology.

Conflicts of Interest: The authors declare no conflict of interest.

References

1. Barbacid, M. Ras genes. *Annu. Rev. Biochem.* **1987**, *56*, 779–827. [CrossRef] [PubMed]
2. Fey, D.; Matallanas, D.; Rauch, J.; Rukhlenko, O.S.; Kholodenko, B.N. The complexities and versatility of the RAS-to-ERK signalling system in normal and cancer cells. *Semin. Cell Dev. Biol.* **2016**, *58*, 96–107. [CrossRef] [PubMed]

3. Buday, L.; Downward, J. Many faces of Ras activation. *Biochim. Biophys. Acta* **2008**, *1786*, 178–187. [CrossRef]
4. Prior, I.A.; Hood, F.E.; Hartley, J.L. The Frequency of Ras Mutations in Cancer. *Cancer Res.* **2020**, *80*, 2969–2974. [CrossRef] [PubMed]
5. Menzies, G.E.; Prior, I.A.; Brancale, A.; Reed, S.H.; Lewis, P.D. Carcinogen-induced DNA structural distortion differences in the RAS gene isoforms; the importance of local sequence. *BMC Chem.* **2021**, *15*, 51. [CrossRef]
6. Smith, M.J.; Neel, B.G.; Ikura, M. NMR-based functional profiling of RASopathies and oncogenic RAS mutations. *Proc. Natl. Acad. Sci. USA* **2013**, *110*, 4574–4579. [CrossRef]
7. Stolze, B.; Reinhart, S.; Bullinger, L.; Frohling, S.; Scholl, C. Comparative analysis of KRAS codon 12, 13, 18, 61, and 117 mutations using human MCF10A isogenic cell lines. *Sci. Rep.* **2015**, *5*, 8535. [CrossRef]
8. Huang, H.; Daniluk, J.; Liu, Y.; Chu, J.; Li, Z.; Ji, B.; Logsdon, C.D. Oncogenic K-Ras requires activation for enhanced activity. *Oncogene* **2014**, *33*, 532–535. [CrossRef]
9. Hunter, J.C.; Manandhar, A.; Carrasco, M.A.; Gurbani, D.; Gondi, S.; Westover, K.D. Biochemical and Structural Analysis of Common Cancer-Associated KRAS Mutations. *Mol. Cancer Res.* **2015**, *13*, 1325–1335. [CrossRef]
10. Kiel, C.; Matallanas, D.; Kolch, W. The Ins and Outs of RAS Effector Complexes. *Biomolecules* **2021**, *11*, 236. [CrossRef] [PubMed]
11. Fernandez-Medarde, A.; De Las Rivas, J.; Santos, E. 40 Years of RAS-A Historic Overview. *Genes* **2021**, *12*, 681. [CrossRef] [PubMed]
12. Huang, L.; Guo, Z.; Wang, F.; Fu, L. KRAS mutation: From undruggable to druggable in cancer. *Signal Transduct. Target. Ther.* **2021**, *6*, 386. [CrossRef]
13. Rosen, J.C.; Sacher, A.; Tsao, M.S. Direct GDP-KRAS(G12C) inhibitors and mechanisms of resistance: The tip of the iceberg. *Ther. Adv. Med. Oncol.* **2023**, *15*, 17588359231160141. [CrossRef]
14. Wang, Y.; Zhong, B.; Xu, C.; Zhan, D.; Zhao, S.; Wu, H.; Liu, M.; Lan, X.; Cai, D.; Ding, Q.; et al. Global profiling of AMG510 modified proteins identified tumor suppressor KEAP1 as an off-target. *iScience* **2023**, *26*, 106080. [CrossRef]
15. Smith, G.; Bounds, R.; Wolf, H.; Steele, R.J.; Carey, F.A.; Wolf, C.R. Activating K-Ras mutations outwith 'hotspot' codons in sporadic colorectal tumours—Implications for personalised cancer medicine. *Br. J. Cancer* **2010**, *102*, 693–703. [CrossRef]
16. De Roock, W.; Jonker, D.J.; Di Nicolantonio, F.; Sartore-Bianchi, A.; Tu, D.; Siena, S.; Lamba, S.; Arena, S.; Frattini, M.; Piessevaux, H.; et al. Association of KRAS p.G13D mutation with outcome in patients with chemotherapy-refractory metastatic colorectal cancer treated with cetuximab. *JAMA* **2010**, *304*, 1812–1820. [CrossRef] [PubMed]
17. Drosten, M.; Dhawahir, A.; Sum, E.Y.; Urosevic, J.; Lechuga, C.G.; Esteban, L.M.; Castellano, E.; Guerra, C.; Santos, E.; Barbacid, M. Genetic analysis of Ras signalling pathways in cell proliferation, migration and survival. *EMBO J.* **2010**, *29*, 1091–1104. [CrossRef] [PubMed]
18. Matallanas, D.; Birtwistle, M.; Romano, D.; Zebisch, A.; Rauch, J.; von Kriegsheim, A.; Kolch, W. Raf family kinases: Old dogs have learned new tricks. *Genes Cancer* **2011**, *2*, 232–260. [CrossRef]
19. Romano, D.; Maccario, H.; Doherty, C.; Quinn, N.P.; Kolch, W.; Matallanas, D. The differential effects of wild-type and mutated K-Ras on MST2 signaling are determined by K-Ras activation kinetics. *Mol. Cell. Biol.* **2013**, *33*, 1859–1868. [CrossRef]
20. Matallanas, D.; Arozarena, I.; Berciano, M.T.; Aaronson, D.S.; Pellicer, A.; Lafarga, M.; Crespo, P. Differences on the inhibitory specificities of H-Ras, K-Ras, and N-Ras (N17) dominant negative mutants are related to their membrane microlocalization. *J. Biol. Chem.* **2003**, *278*, 4572–4581. [CrossRef]
21. Romano, D.; Nguyen, L.K.; Matallanas, D.; Halasz, M.; Doherty, C.; Kholodenko, B.N.; Kolch, W. Protein interaction switches coordinate Raf-1 and MST2/Hippo signalling. *Nat. Cell. Biol.* **2014**, *16*, 673–684. [CrossRef]
22. Turriziani, B.; Garcia-Munoz, A.; Pilkington, R.; Raso, C.; Kolch, W.; von Kriegsheim, A. On-beads digestion in conjunction with data-dependent mass spectrometry: A shortcut to quantitative and dynamic interaction proteomics. *Biology* **2014**, *3*, 320–332. [CrossRef] [PubMed]
23. Rappsilber, J.; Mann, M.; Ishihama, Y. Protocol for micro-purification, enrichment, pre-fractionation and storage of peptides for proteomics using StageTips. *Nat. Protoc.* **2007**, *2*, 1896–1906. [CrossRef]
24. Bache, N.; Geyer, P.E.; Bekker-Jensen, D.B.; Hoerning, O.; Falkenby, L.; Treit, P.V.; Doll, S.; Paron, I.; Muller, J.B.; Meier, F.; et al. A Novel LC System Embeds Analytes in Pre-formed Gradients for Rapid, Ultra-robust Proteomics. *Mol. Cell. Proteom.* **2018**, *17*, 2284–2296. [CrossRef] [PubMed]
25. Howard, J.; Wynne, K.; Moldenhauer, E.; Clarke, P.; Maguire, C.; Bollard, S.; Yin, X.; Brennan, L.; Mooney, L.; Fitzsimons, S.; et al. A comparative analysis of extracellular vesicles (EVs) from human and feline plasma. *Sci. Rep.* **2022**, *12*, 10851. [CrossRef]
26. Meier, F.; Brunner, A.D.; Koch, S.; Koch, H.; Lubeck, M.; Krause, M.; Goedecke, N.; Decker, J.; Kosinski, T.; Park, M.A.; et al. Online Parallel Accumulation-Serial Fragmentation (PASEF) with a Novel Trapped Ion Mobility Mass Spectrometer. *Mol. Cell. Proteom.* **2018**, *17*, 2534–2545. [CrossRef] [PubMed]
27. Cox, J.; Mann, M. MaxQuant enables high peptide identification rates, individualized p.p.b.-range mass accuracies and proteome-wide protein quantification. *Nat. Biotechnol.* **2008**, *26*, 1367–1372. [CrossRef]
28. UniProt, C. UniProt: The Universal Protein Knowledgebase in 2023. *Nucleic Acids Res.* **2023**, *51*, D523–D531. [CrossRef]
29. Cox, J.; Hein, M.Y.; Luber, C.A.; Paron, I.; Nagaraj, N.; Mann, M. Accurate proteome-wide label-free quantification by delayed normalization and maximal peptide ratio extraction, termed MaxLFQ. *Mol. Cell. Proteom.* **2014**, *13*, 2513–2526. [CrossRef]

30. Rodriguez, J.; Pilkington, R.; Garcia Munoz, A.; Nguyen, L.K.; Rauch, N.; Kennedy, S.; Monsefi, N.; Herrero, A.; Taylor, C.T von Kriegsheim, A. Substrate-Trapped Interactors of PHD3 and FIH Cluster in Distinct Signaling Pathways. *Cell. Rep.* **2016** *14*, 2745–2760. [CrossRef]
31. Szklarczyk, D.; Kirsch, R.; Koutrouli, M.; Nastou, K.; Mehryary, F.; Hachilif, R.; Gable, A.L.; Fang, T.; Doncheva, N.T Pyysalo, S.; et al. The STRING database in 2023: Protein-protein association networks and functional enrichment analyses for an sequenced genome of interest. *Nucleic Acids Res.* **2023**, *51*, D638–D646. [CrossRef] [PubMed]
32. Gillespie, M.; Jassal, B.; Stephan, R.; Milacic, M.; Rothfels, K.; Senff-Ribeiro, A.; Griss, J.; Sevilla, C.; Matthews, L.; Gong, C.; et a The reactome pathway knowledgebase 2022. *Nucleic Acids Res.* **2022**, *50*, D687–D692. [CrossRef] [PubMed]
33. Thomas, P.D.; Campbell, M.J.; Kejariwal, A.; Mi, H.; Karlak, B.; Daverman, R.; Diemer, K.; Muruganujan, A.; Narechania, A PANTHER: A library of protein families and subfamilies indexed by function. *Genome Res.* **2003**, *13*, 2129–2141. [CrossRef [PubMed]
34. Aaronson, S.A.; Todaro, G.J.; Freeman, A.E. Human sarcoma cells in culture. Identification by colony-forming ability on monolayers of normal cells. *Exp. Cell. Res.* **1970**, *61*, 1–5. [CrossRef]
35. Herrero, A.; Reis-Cardoso, M.; Jimenez-Gomez, I.; Doherty, C.; Agudo-Ibanez, L.; Pinto, A.; Calvo, F.; Kolch, W.; Crespo, P Matallanas, D. Characterisation of HRas local signal transduction networks using engineered site-specific exchange factors. *Small GTPases* **2017**, *11*, 371–383. [CrossRef] [PubMed]
36. Cappiello, F.; Casciaro, B.; Mangoni, M.L. A Novel In Vitro Wound Healing Assay to Evaluate Cell Migration. *J. Vis. Exp.* **2018** *133*, e56825. [CrossRef]
37. Kelly, M.R.; Kostyrko, K.; Han, K.; Mooney, N.A.; Jeng, E.E.; Spees, K.; Dinh, P.T.; Abbott, K.L.; Gwinn, D.M. Sweet-Cordero, E.A.; et al. Combined Proteomic and Genetic Interaction Mapping Reveals New RAS Effector Pathways and Susceptibilities. *Cancer Discov.* **2020**, *10*, 1950–1967. [CrossRef]
38. Zhang, X.; Cao, J.; Miller, S.P.; Jing, H.; Lin, H. Cyomparative Nucleotide-Dependent Interactome Analysis Reveals Shared and Differential Properties of KRas4a and KRas4b. *ACS Cent. Sci.* **2018**, *4*, 71–80. [CrossRef]
39. Kovalski, J.R.; Bhaduri, A.; Zehnder, A.M.; Neela, P.H.; Che, Y.; Wozniak, G.G.; Khavari, P.A. The Functional Proximal Proteome of Oncogenic Ras Includes mTORC2. *Mol. Cell.* **2019**, *73*, 830–844.e812. [CrossRef]
40. Ritchie, C.; Mack, A.; Harper, L.; Alfadhli, A.; Stork, P.J.S.; Nan, X.; Barklis, E. Analysis of K-Ras Interactions by Biotin Ligase Tagging. *Cancer Genom. Proteom.* **2017**, *14*, 225–239. [CrossRef]
41. Adhikari, H.; Counter, C.M. Interrogating the protein interactomes of RAS isoforms identifies PIP5K1A as a KRAS-specific vulnerability. *Nat. Commun.* **2018**, *9*, 3646. [CrossRef] [PubMed]
42. Huttlin, E.L.; Bruckner, R.J.; Navarrete-Perea, J.; Cannon, J.R.; Baltier, K.; Gebreab, F.; Gygi, M.P.; Thornock, A.; Zarraga, G. Tam, S.; et al. Dual proteome-scale networks reveal cell-specific remodeling of the human interactome. *Cell* **2021**, *184*, 3022–3040.e3028. [CrossRef]
43. Choi, B.H.; Kou, Z.; Colon, T.M.; Chen, C.H.; Chen, Y.; Dai, W. Identification of Radil as a Ras binding partner and putative activator. *J. Biol. Chem.* **2021**, *296*, 100314. [CrossRef] [PubMed]
44. Shirasawa, S.; Furuse, M.; Yokoyama, N.; Sasazuki, T. Altered growth of human colon cancer cell lines disrupted at activated Ki-ras. *Science* **1993**, *260*, 85–88. [CrossRef] [PubMed]
45. Kennedy, S.A.; Jarboui, M.A.; Srihari, S.; Raso, C.; Bryan, K.; Dernayka, L.; Charitou, T.; Bernal-Llinares, M.; Herrera-Montavez, C. Krstic, A.; et al. Extensive rewiring of the EGFR network in colorectal cancer cells expressing transforming levels of KRAS(G13D) *Nat. Commun.* **2020**, *11*, 499. [CrossRef] [PubMed]
46. Beganton, B.; Coyaud, E.; Laurent, E.M.N.; Mange, A.; Jacquemetton, J.; Le Romancer, M.; Raught, B.; Solassol, J. Proximal Protein Interaction Landscape of RAS Paralogs. *Cancers* **2020**, *12*, 3326. [CrossRef]
47. Ternet, C.; Junk, P.; Sevrin, T.; Catozzi, S.; Wahlen, E.; Heldin, J.; Oliviero, G.; Wynne, K.; Kiel, C. Analysis of context-specific KRAS-effector (sub)complexes in Caco-2 cells. *Life Sci. Alliance* **2023**, *6*, e202201670. [CrossRef]
48. Uhlen, M.; Oksvold, P.; Fagerberg, L.; Lundberg, E.; Jonasson, K.; Forsberg, M.; Zwahlen, M.; Kampf, C.; Wester, K. Hober, S.; et al. Towards a knowledge-based Human Protein Atlas. *Nat. Biotechnol.* **2010**, *28*, 1248–1250. [CrossRef]
49. Mi, H.; Lazareva-Ulitsky, B.; Loo, R.; Kejariwal, A.; Vandergriff, J.; Rabkin, S.; Guo, N.; Muruganujan, A.; Doremieux, O. Campbell, M.J.; et al. The PANTHER database of protein families, subfamilies, functions and pathways. *Nucleic Acids Res.* **2005** *33*, D284–D288. [CrossRef]
50. Jeng, H.H.; Taylor, L.J.; Bar-Sagi, D. Sos-mediated cross-activation of wild-type Ras by oncogenic Ras is essential for tumorigenesis *Nat. Commun.* **2012**, *3*, 1168. [CrossRef]
51. Zafra, M.P.; Parsons, M.J.; Kim, J.; Alonso-Curbelo, D.; Goswami, S.; Schatoff, E.M.; Han, T.; Katti, A.; Fernandez, M.T.C. Wilkinson, J.E.; et al. An In Vivo Kras Allelic Series Reveals Distinct Phenotypes of Common Oncogenic Variants. *Cancer Discov* **2020**, *10*, 1654–1671. [CrossRef] [PubMed]
52. Canon, J.; Rex, K.; Saiki, A.Y.; Mohr, C.; Cooke, K.; Bagal, D.; Gaida, K.; Holt, T.; Knutson, C.G.; Koppada, N.; et al. The clinical KRAS(G12C) inhibitor AMG 510 drives anti-tumour immunity. *Nature* **2019**, *575*, 217–223. [CrossRef] [PubMed]
53. Chiou, L.W.; Chan, C.H.; Jhuang, Y.L.; Yang, C.Y.; Jeng, Y.M. DNA replication stress and mitotic catastrophe mediate sotorasib addiction in KRAS(G12C)-mutant cancer. *J. Biomed. Sci.* **2023**, *30*, 50. [CrossRef] [PubMed]

54. Salmon, M.; Alvarez-Diaz, R.; Fustero-Torre, C.; Brehey, O.; Lechuga, C.G.; Sanclemente, M.; Fernandez-Garcia, F.; Lopez-Garcia, A.; Martin-Guijarro, M.C.; Rodriguez-Perales, S.; et al. Kras oncogene ablation prevents resistance in advanced lung adenocarcinomas. *J. Clin. Investig.* **2023**, *133*, e164413. [CrossRef] [PubMed]
55. Koga, T.; Suda, K.; Fujino, T.; Ohara, S.; Hamada, A.; Nishino, M.; Chiba, M.; Shimoji, M.; Takemoto, T.; Arita, T.; et al. KRAS Secondary Mutations That Confer Acquired Resistance to KRAS G12C Inhibitors, Sotorasib and Adagrasib, and Overcoming Strategies: Insights From In Vitro Experiments. *J. Thorac. Oncol.* **2021**, *16*, 1321–1332. [CrossRef]
56. Brown, W.S.; McDonald, P.C.; Nemirovsky, O.; Awrey, S.; Chafe, S.C.; Schaeffer, D.F.; Li, J.; Renouf, D.J.; Stanger, B.Z.; Dedhar, S. Overcoming Adaptive Resistance to KRAS and MEK Inhibitors by Co-targeting mTORC1/2 Complexes in Pancreatic Cancer. *Cell. Rep. Med.* **2020**, *1*, 100131. [CrossRef]
57. Chan, C.H.; Chiou, L.W.; Lee, T.Y.; Liu, Y.R.; Hsieh, T.H.; Yang, C.Y.; Jeng, Y.M. PAK and PI3K pathway activation confers resistance to KRAS(G12C) inhibitor sotorasib. *Br. J. Cancer* **2023**, *128*, 148–159. [CrossRef]
58. McDonald, W.H.; Ohi, R.; Miyamoto, D.T.; Mitchison, T.J.; Yates III, J.R. Comparison of three directly coupled HPLC MS/MS strategies for identification of proteins from complex mixtures: Single-dimension LC-MS/MS, 2-phase MudPIT, and 3-phase MudPIT. *Int. J. Mass Spectrom.* **2002**, *219*, 245–251. [CrossRef]
59. Zhou, Y.; Hancock, J.F. Lipid Profiles of RAS Nanoclusters Regulate RAS Function. *Biomolecules* **2021**, *11*, 1439. [CrossRef]
60. Janes, M.R.; Zhang, J.; Li, L.S.; Hansen, R.; Peters, U.; Guo, X.; Chen, Y.; Babbar, A.; Firdaus, S.J.; Darjania, L.; et al. Targeting KRAS Mutant Cancers with a Covalent G12C-Specific Inhibitor. *Cell* **2018**, *172*, 578–589.e517. [CrossRef]
61. Patricelli, M.P.; Janes, M.R.; Li, L.S.; Hansen, R.; Peters, U.; Kessler, L.V.; Chen, Y.; Kucharski, J.M.; Feng, J.; Ely, T.; et al. Selective Inhibition of Oncogenic KRAS Output with Small Molecules Targeting the Inactive State. *Cancer Discov.* **2016**, *6*, 316–329. [CrossRef] [PubMed]
62. Meyer, S.C.; Levine, R.L. Molecular pathways: Molecular basis for sensitivity and resistance to JAK kinase inhibitors. *Clin. Cancer Res.* **2014**, *20*, 2051–2059. [CrossRef] [PubMed]
63. Mohrherr, J.; Haber, M.; Breitenecker, K.; Aigner, P.; Moritsch, S.; Voronin, V.; Eferl, R.; Moriggl, R.; Stoiber, D.; Gyorffy, B.; et al. JAK-STAT inhibition impairs K-RAS-driven lung adenocarcinoma progression. *Int. J. Cancer* **2019**, *145*, 3376–3388. [CrossRef]
64. Bigenzahn, J.W.; Collu, G.M.; Kartnig, F.; Pieraks, M.; Vladimer, G.I.; Heinz, L.X.; Sedlyarov, V.; Schischlik, F.; Fauster, A.; Rebsamen, M.; et al. LZTR1 is a regulator of RAS ubiquitination and signaling. *Science* **2018**, *362*, 1171–1177. [CrossRef] [PubMed]
65. Roux, K.J.; Kim, D.I.; Raida, M.; Burke, B. A promiscuous biotin ligase fusion protein identifies proximal and interacting proteins in mammalian cells. *J. Cell. Biol.* **2012**, *196*, 801–810. [CrossRef] [PubMed]
66. Matallanas, D.; Sanz-Moreno, V.; Arozarena, I.; Calvo, F.; Agudo-Ibanez, L.; Santos, E.; Berciano, M.T.; Crespo, P. Distinct utilization of effectors and biological outcomes resulting from site-specific Ras activation: Ras functions in lipid rafts and Golgi complex are dispensable for proliferation and transformation. *Mol. Cell. Biol.* **2006**, *26*, 100–116. [CrossRef]
67. Matallanas, D.; Romano, D.; Al-Mulla, F.; O'Neill, E.; Al-Ali, W.; Crespo, P.; Doyle, B.; Nixon, C.; Sansom, O.; Drosten, M.; et al. Mutant K-Ras activation of the proapoptotic MST2 pathway is antagonized by wild-type K-Ras. *Mol. Cell.* **2011**, *44*, 893–906. [CrossRef]
68. Lampson, B.L.; Pershing, N.L.; Prinz, J.A.; Lacsina, J.R.; Marzluff, W.F.; Nicchitta, C.V.; MacAlpine, D.M.; Counter, C.M. Rare codons regulate KRas oncogenesis. *Curr. Biol.* **2013**, *23*, 70–75. [CrossRef] [PubMed]
69. Hood, F.E.; Sahraoui, Y.M.; Jenkins, R.E.; Prior, I.A. Ras protein abundance correlates with Ras isoform mutation patterns in cancer. *Oncogene* **2023**, *42*, 1224–1232. [CrossRef]
70. Barrios-Bernal, P.; Lucio-Lozada, J.; Ramos-Ramirez, M.; Hernandez-Pedro, N.; Arrieta, O. A Novel Combination of Sotorasib and Metformin Enhances Cytotoxicity and Apoptosis in KRAS-Mutated Non-Small Cell Lung Cancer Cell Lines through MAPK and P70S6K Inhibition. *Int. J. Mol. Sci.* **2023**, *24*, 4331. [CrossRef]
71. Ma, Y.; Schulz, B.; Trakooljul, N.; Al Ammar, M.; Sekora, A.; Sender, S.; Hadlich, F.; Zechner, D.; Weiss, F.U.; Lerch, M.M.; et al. Inhibition of KRAS, MEK and PI3K Demonstrate Synergistic Anti-Tumor Effects in Pancreatic Ductal Adenocarcinoma Cell Lines. *Cancers* **2022**, *14*, 4467. [CrossRef] [PubMed]
72. Yaeger, R.; Mezzadra, R.; Sinopoli, J.; Bian, Y.; Marasco, M.; Kaplun, E.; Gao, Y.; Zhao, H.; Paula, A.D.C.; Zhu, Y.; et al. Molecular Characterization of Acquired Resistance to KRASG12C-EGFR Inhibition in Colorectal Cancer. *Cancer Discov.* **2023**, *13*, 41–55. [CrossRef] [PubMed]
73. Cox, A.D.; Der, C.J. The dark side of Ras: Regulation of apoptosis. *Oncogene* **2003**, *22*, 8999–9006. [CrossRef]
74. Dorard, C.; Madry, C.; Buhard, O.; Toifl, S.; Didusch, S.; Ratovomanana, T.; Letourneur, Q.; Dolznig, H.; Garnett, M.J.; Duval, A.; et al. RAF1 contributes to cell proliferation and STAT3 activation in colorectal cancer independently of microsatellite and KRAS status. *Oncogene* **2023**, *42*, 1649–1660. [CrossRef] [PubMed]
75. Perez-Riverol, Y.; Bai, J.; Bandla, C.; Garcia-Seisdedos, D.; Hewapathirana, S.; Kamatchinathan, S.; Kundu, D.J.; Prakash, A.; Frericks-Zipper, A.; Eisenacher, M.; et al. The PRIDE database resources in 2022: A hub for mass spectrometry-based proteomics evidences. *Nucleic Acids Res.* **2022**, *50*, D543–D552. [CrossRef] [PubMed]

cancers

MDPI

Article

Focus on *RAS* Codon 61 Mutations in Metastatic Colorectal Cancer: A Retrospective Analysis

Francesco Schietroma [1,†], Annunziato Anghelone [1,†], Giustina Valente [1], Viria Beccia [1], Giulia Caira [1], Alexia Spring [1], Giovanni Trovato [1], Armando Di Bello [1], Anna Ceccarelli [1], Laura Chiofalo [1], Serena Perazzo [1], Maria Bensi [1], Angelo Minucci [2], Andrea Urbani [3], Luigi Maria Larocca [4], Michele Basso [5], Carmelo Pozzo [5], Lisa Salvatore [1,5], Maria Alessandra Calegari [5,*,‡] and Giampaolo Tortora [1,5,‡]

1 Medical Oncology, Università Cattolica del Sacro Cuore, 00168 Roma, Italy; francesco.schietroma01@icatt.it (F.S.); annunziato.anghelone@ospedalerc.it (A.A.); giustina.valente01@icatt.it (G.V.); viria.beccia01@icatt.it (V.B.); giulia.caira01@icatt.it (G.C.); alexia.spring01@icatt.it (A.S.); giovanni.trovato01@icatt.it (G.T.); armando.dibello01@icatt.it (A.D.B.); anna.ceccarelli01@icatt.it (A.C.); laura.chiofalo01@icatt.it (L.C.); serena.perazzo01@icatt.it (S.P.); maria.bensi01@icatt.it (M.B.); lisa.salvatore@policlinicogemelli.it (L.S.); giampaolo.tortora@policlinicogemelli.it (G.T.)

2 Departmental Unit of Molecular and Genomic Diagnostics, Genomics Core Facility, Gemelli Science and Technology Park (G-STeP), Fondazione Policlinico Universitario Agostino Gemelli IRCCS, 00168 Roma, Italy; angelo.minucci@policlinicogemelli.it

3 Clinical Chemistry, Biochemistry and Molecular Biology Operations, Fondazione Policlinico Universitario Agostino Gemelli IRCCS, 00168 Roma, Italy; andrea.urbani@policlinicogemelli.it

4 Patologia Oncoematologica, Dipartimento di Scienze della Salute della Donna, del Bambino e di Sanità Pubblica, Fondazione Policlinico Universitario Agostino Gemelli IRCCS, 00168 Roma, Italy; luigimaria.larocca@policlinicogemelli.it

5 Medical Oncology, Comprehensive Cancer Center, Fondazione Policlinico Universitario Agostino Gemelli IRCCS, 00168 Roma, Italy; michele.basso@policlinicogemelli.it (M.B.); carmelo.pozzo@policlinicogemelli.it (C.P.)

* Correspondence: mariaalessandra.calegari@policlinicogemelli.it; Tel.: +39-0630156318

† These authors contributed equally to this work and are co-first authors.

‡ These authors contributed equally to this work and share the last co-authorship.

Simple Summary: Codon 61 *RAS* mutations are rare in metastatic colorectal cancer. Despite being associated with primary and acquired resistance to anti-EGFR agents, little is known about their phenotype and prognostic impact. We retrospectively investigated the clinicopathological features and prognoses of 50 patients with tumors harboring codon 61 *RAS* mutations compared to 648 codon 61 *RAS* wild-type tumors. We identified a significant correlation between codon 61 *RAS* mutations and metastatic involvement of the peritoneum and ovary and a negative prognostic impact. This is the first evidence of an impact of *RAS* mutational status on the metastatization pattern. These results are of great interest given the high frequency of codon 61 *RAS* mutations as mechanisms of secondary resistance to anti-EGFRs and the advent of RAS inhibitors. This is the widest codon 61 *RAS*-mutated cohort reported so far; nevertheless, these findings must be validated in larger studies.

Abstract: *RAS* mutations involving codon 61 are rare in metastatic colorectal cancer (mCRC), accounting for only 1–4%, but they have recently been identified with high frequency in the circulating tumor DNA (ctDNA) of patients with secondary resistance to anti-EGFRs. This retrospective monocentric study aimed to investigate the clinical phenotype and prognostic performance of codon 61 *RAS*-mutated mCRC. Fifty patients with codon 61 *RAS*-mutated mCRC treated at our institution between January 2013 and December 2021 were enrolled. Additional datasets of codon 61 *RAS* wild-type mCRCs (648 patients) were used as comparators. The endpoint for prognostic assessment was overall survival (OS). Metastatic involvement of the peritoneum or ovary was significantly more frequent in codon 61 *RAS*-mutated mCRC compared to codon 61 *RAS* wild-type (54 vs. 28.5%), non-codon 61 *RAS*-mutated (35.6%), *BRAF* V600E-mutated (25%), and *RAS/BRAF* wild-type (20.5%) cohorts. At a median follow up of 96.2 months, the median OS for codon 61 *RAS*-mutated patients was significantly shorter compared to *RAS/BRAF* wild-type (26.9 vs. 36.0 months, HR 0.56) patients, while no significant difference was observed compared to non-codon 61 *RAS*-mutated and *BRAF* V600E-mutated patients. We showed a negative prognostic impact and a statistically significant correlation between codon 61 *RAS* mutations and metastatic involvement of the peritoneum and ovary.

Keywords: colorectal adenocarcinoma; RAS; metastatic colorectal cancer; RAS signaling; Codon 61; MAPK pathway; RAS effectors; KRAS inhibitors; resistance

1. Introduction

Nowadays, while the molecular classification of colorectal cancer (CRC) is becoming more and more complex [1], rat sarcoma virus (*RAS*) mutational status remains a key determinant in every turning point in patients' therapeutic algorithm [2]. Together with *NRAS* and *HRAS*, *KRAS* is a gene belonging to the *RAS* family, which encodes guanosine-5′-triphosphate (GTP)-binding proteins, important effectors of ligand-bound epidermal growth factor receptor (EGFR) signaling through the mitogen-activated protein kinase (MAPK) axis [3]. *KRAS* mutations affect approximately 30–40% of metastatic CRC (mCRC), with mutations involving codons 12 and 13 being the most represented, occurring in about 85–90% of cases [3,4]. Several studies have demonstrated their role as predictive biomarkers of resistance to anti-EGFR agents [5–7]. Hence, all patients diagnosed with mCRC require *RAS* profiling before the administration of anti-EGFRs agents (cetuximab or panitumumab) [8–14].

Codon 61 mutations are less prevalent, affecting 1–4% of patients with mCRC. Similarly to other *RAS* mutations, these alterations are responsible for a constitutive activation of the RAS/RAF/MAPKs pathway, resulting in oncogenic activity and cell proliferation [15]. Furthermore, in *KRAS* codon 12 and 13, wild-type mCRCs codon 61 mutations have been linked to resistance to anti-EGFR therapies [16,17]. Recently, codon 61 variants have been identified with high frequency in the circulating tumor DNA (ctDNA) of patients with mCRC with secondary resistance to anti-EGFR agents [18–21], with a prevalence of 50% in the CHRONOS trial [18]. Other rare *KRAS* mutations involve exon 4, codon 117, and codon 146. Similarly to more frequent *RAS* mutations, mutations involving codon 117 and 146 have been associated with resistance to anti-EGFRs therapies [22,23]. Moreover, a large analysis showed a higher incidence of codon 117 and 146 in older patients [24].

Despite its growing clinical relevance, little is known about the clinicopathological and molecular features and prognosis of mCRCs harboring *RAS* codon 61 mutations and their differences with more common codon 12 and 13 mutations, as well as and their impact on prognosis. In 2014, a cohort study by Imamura et al. [25] reported the clinicopathological and molecular features of 19 *KRAS* codon 61-mutated mCRC to be similar to *KRAS* codon 12- and 13-mutated mCRCs. Another study found a weak tendency for peritoneum localization in a population of 14 patients with codon 61 *RAS*-mutated CRC [26]. In our study, we aimed to further investigate the clinical characteristics and prognosis of patients with mCRC harboring *RAS* codon 61 mutations treated at our institution compared to those harboring other non-codon 61 *RAS*-mutated and wild-type tumors.

2. Materials and Methods

This is an observational, retrospective, monocentric study. The study was approved by the local Ethics Committee of Fondazione Policlinico Universitario Agostino Gemelli IRCCS, Rome, Italy (protocol number 0054049/2019 18 December 2019). The objective of the study was to investigate and describe clinical phenotype and prognostic performance of mCRCs harboring *RAS* codon 61 mutations.

We examined the medical records of patients diagnosed with mCRC who were treated at our center from January 2013 through December 2021. Eligible subjects were those patients whose tumors carried mutations involving codon 61 of *RAS* gene and were evaluable for survival after at least one line of therapy. We collected data regarding bbaseline demographic and clinical characteristics, first-line treatment, and survival from medical records, while histological reports were used to gather pathological and molecular data. The following baseline demographic and clinical characteristics were collected: sex, age, Eastern Cooperative Oncology Group performance status (ECOG PS) at diagnosis, primary tumor location, onset of metastatic disease, number of metastatic sites, site of metastases, presence of peritoneal and/or adnexal metastases, mucinous histology, grade of differentiation, *RAS/BRAF* mutational status, microsatellite instability/mismatch repair (MSI/MMR) status, treatments received (surgery, neoadjuvant or adjuvant chemotherapy, first-line chemotherapy), investigator-assessed best response according to Response Evaluation Criteria in Solid Tumors (RECIST) 1.1 criteria, and survival. *RAS* and *BRAF* mutational status was assessed by means of next-generation sequencing (NGS) or pyrosequencing on formalin-fixed, paraffin-embedded (FFPE) archival tumor tissue samples from primary tumor or metastases. Expression of MMR proteins (MLH1, MSH2, MSH6, and PMS2) was performed via immunohistochemistry. MSI status was assessed via NGS.

Additional datasets of patients affected by mCRC without codon 61 *RAS* mutations (codon 61 *RAS* wild-type) treated at out center during the same time frame were used as comparators. Among this group of patients, we identified three different molecular subgroups, which included, respectively, patients with an *RAS*-mutated disease not involving codon 61 (non-codon 61 *RAS*-mutated mCRCs), patients with mCRC harboring a *BRAF* V600E mutation (*BRAF* V600E-mutated mCRCs), and patients with an *RAS* and *BRAF* wild-type disease (*RAS/BRAF* wild-type mCRCs).

For categorial data, counts and percentages were reported using a descriptive method; for continuous variables, median and range were provided. Fisher's exact test or the chi-square test, when applicable, were used to compare group differences for categorical variables. Overall survival (OS), defined as the time occurring between the diagnosis of metastatic disease to the date of death from any cause, was the endpoint for prognostic analysis. All patients were followed up until death or the time of database lock (January 2023). Patients not experiencing events were censored at the date of last follow up. Survivals were estimated with the Kaplan–Meier method and compared using the log-rank test. Statistical significance was set at $p = 0.05$. Statistical analyses were performed using MedCalc version 14.8.1.

3. Results

Between January 2013 and January 2023, a total of 50 patients with a diagnosis of mCRC harboring an *RAS* codon 61 mutation were included in our analysis. Of those, 28 mutations (56%) affected *KRAS* and 22 (44%) *NRAS*. Patients and disease characteristics are summarized in Table 1.

Median age at diagnosis was 65 years (range 34–86 years). Nineteen patients were males (38%) and thirty-one were females (62%). Patients were mainly in good clinical conditions at the time of diagnosis (88% with an ECOG PS 0 or 1). Thirty-six patients (72%) had a left-sided primary tumor, and thirty-three patients (66%) had a synchronous metastatic disease. The most frequent site of metastases was liver (24 patients, 48%), followed by peritoneum or ovary (16 patients, 32%), lymph nodes (15 patients, 30%), and lungs (11 patients, 22%). Moreover, 27 patients (54%) developed metastases involving the peritoneum or ovary during their clinical history. The majority of patients received resection of primary tumor (40 patients, 80%). Twenty nine patients (58%) underwent a first-line therapy which included bevacizumab. As chemotherapy regimen, twenty nine patients (58%) received mFOLFOX6 (with or without bevacizumab), while FOLFIRI (with or without bevacizumab) was administered in nine patients (18%). Only three patients were treated with FOLFOXIRI plus bevacizumab (6%), whereas nine patients (18%) received other regimens (such as a fluoropyrimidine, alone or in combination with bevacizumab). Twenty patients (40%) received only one line of therapy, while ten patients (20%) received two lines, thirteen patients (26%) received three lines, five patients (10%) received four lines and, only two patients (4%) received five lines of therapy.

Table 1. Codon 61 *RAS*-mutated patients characteristics.

Characteristics		N = 50 (%)	*KRAS* (n = 28)	*NRAS* (n = 22)
Age (at metastatic diagnosis), median (range)		65 yrs (34–86 yrs)	65 yrs (41–86 yrs)	63 yrs (34–84 yrs)
ECOG PS	**0**	25 (50%)	15 (53%)	10 (45%)
	1	19 (38%)	8 (28%)	11 (50%)
	2	6 (12%)	5 (19%)	1 (5%)
Sex	**Male**	19 (38%)	10 (36%)	9 (41%)
	Female	31 (62%)	18 (64%)	13 (59%)
Previous surgery	**Y**	40 (80%)	22 (79%)	18 (82%)
	N	10 (20%)	6 (21%)	4 (18%)
Metastatic at diagnosis	**Y**	33 (66%)	19 (68%)	14 (64%)
	N	17 (34%)	9 (32%)	8 (36%)
Primary tumor location	**Right**	14 (28%)	8 (29%)	6 (27%)
	Left	36 (72%)	20 (71%)	16 (73%)
Sites of metastatic disease at diagnosis	**Liver**	24 (48%)	12 (43%)	12 (54%)
	Lung	11 (22%)	7 (25%)	4 (18%)
	Nodes	15 (30%)	8 (28%)	7 (32%)
	Peritoneum/Ovary	16 (32%)	7 (25%)	9 (41%)
	Other	5 (10%)	3 (10%)	2 (9%)
Peritoneal and/or ovarian metastasis	**Y**	27 (54%)	13 (46%)	14 (64%)
	N	23 (46%)	15 (54%)	8 (36%)
First line chemotherapy regimen	**FOLFOXIRI +/− bevacizumab**	3 (6%)	0	3 (14%)
	FOLFOX +/− bevacizumab	29 (58%)	20 (71%)	9 (41%)
	FOLFIRI +/− bevacizumab	9 (18%)	3 (11%)	6 (27%)
	Other	9 (18%)	5 (18%)	4 (18%)
Total number of treatment lines	**1**	20 (40%)	15 (53%)	5 (23%)
	2	10 (20%)	5 (18%)	5 (23%)
	3	13 (26%)	6 (21%)	7 (32%)
	4	5 (10%)	1 (4%)	4 (18%)
	5	2 (4%)	1 (4%)	1 (4%)
***RAS* mutation**	*KRAS*	28 (56%)		
	Q61X	15 (30%)		
	Q61H	6 (12%)		
	Q61L	3 (6%)		
	Q61R	2 (4%)		
	G61X	2 (4%)		
	NRAS	22 (44%)		
	Q61R	8 (16%)		
	Q61K	8 (16%)		
	Q61L	5 (10%)		
	G61H	1 (2%)		

ECOG PS: Eastern Cooperative Oncology Group performance status; N: no; Y: yes; yrs: years.

The comparator dataset included 648 consecutive patients with codon 61 *RAS* wild-type mCRC treated at our institution during the same time frame. This group included

326 patients (50.3%) with an *RAS*-mutated disease not involving codon 61 (non-codon 61 *RAS*-mutated mCRCs), 254 patients (39.2%) with an *RAS* and *BRAF* wild-type disease (*RAS/BRAF* wild-type mCRCs), and 68 patients (10.5%) with a *BRAF* V600E-mutated disease (*BRAF* V600E-mutated mCRCs). The probability of experiencing peritoneal or ovarian metastases was statistically significantly higher in patients with codon 61 *RAS*-mutated mCRC than in patients with codon 61 *RAS* wild-type mCRC (54% vs. 28.5%, $p = 0.000163$) (Figure 1). More specifically, the rate of peritoneal or ovarian metastases was higher in the codon 61 *RAS*-mutated cohort also when compared to the non-codon 61 *RAS*-mutated cohort (54% vs. 35.6%, $p = 0.012495$), *BRAF* V600E-mutated cohort (54% vs. 25%, $p = 0.001286$), and *RAS/BRAF* all wild-type cohort (54% vs. 20.5%, $p < 0.00001$) (Figure 1).

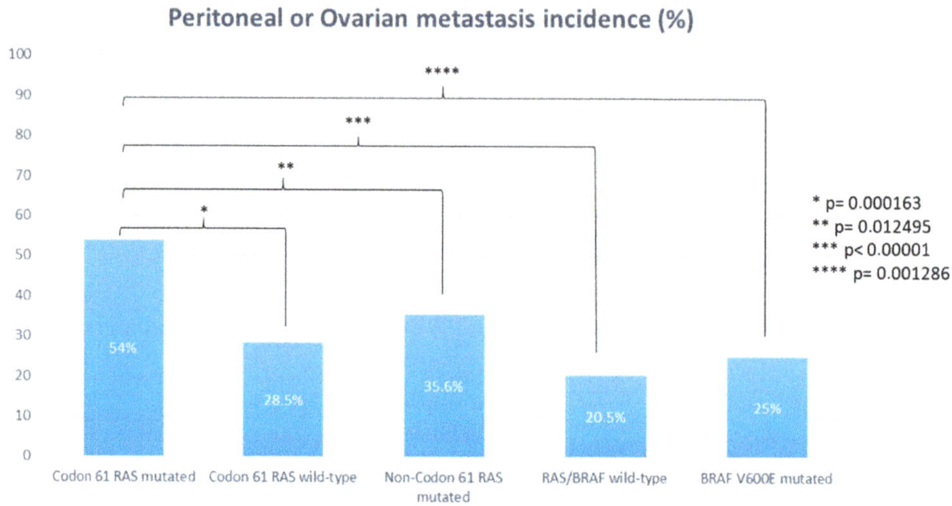

Figure 1. Probability of experiencing peritoneal or ovarian metastases according to *RAS* and *BRAF* mutational status.

At a median follow up of 96.2 months (95% confidence interval (CI), 92.4–109.0 months), 40 death events were reported in the codon 61 *RAS*-mutated cohort and 556 in the comparator dataset. Median OS (mOS) was 26.9 months (95%CI 21.6–31.4 months) for the codon 61 *RAS*-mutated cohort and 31.5 months (95%CI 30.0–33.8 months) for the codon 61 *RAS* wild-type dataset (hazard ratio (HR) 0.69, 95%CI 0.47–1.00; $p = 0.0221$) (Figure 2).

Moreover, dissecting the comparator dataset in accordance with *RAS* and *BRAF* mutational status, mOS was confirmed to be significantly shorter for the codon 61 *RAS*-mutated cohort compared to the *RAS* and *BRAF* wild-type cohort (mOS 36.0 months, 95%CI 32.1–41.7 months; HR 0.56, 95%CI 0.37–0.85; $p = 0.0006$) (Figure 3A). On the contrary, no statistically significant difference was observed compared to the non-codon 61 *RAS*-mutated cohort (mOS 30.2 months, 95%CI 27.5–33.1 months; HR 0.76, 95%CI 0.52–1.09; $p = 0.0993$) (Figure 3B) and the *BRAF* V600E-mutated cohort (mOS 22.6 months, 95%CI 17.8–31.1 months; HR 0.97, 95%CI 0.64–1.48; $p = 0.9124$) (Figure 3C).

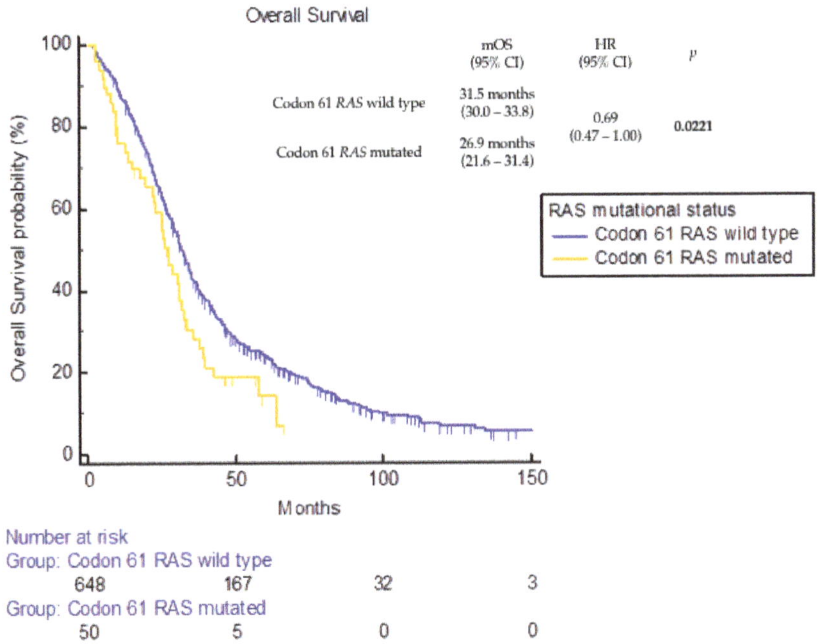

Figure 2. Overall survival according to codon 61 *RAS* mutational status.

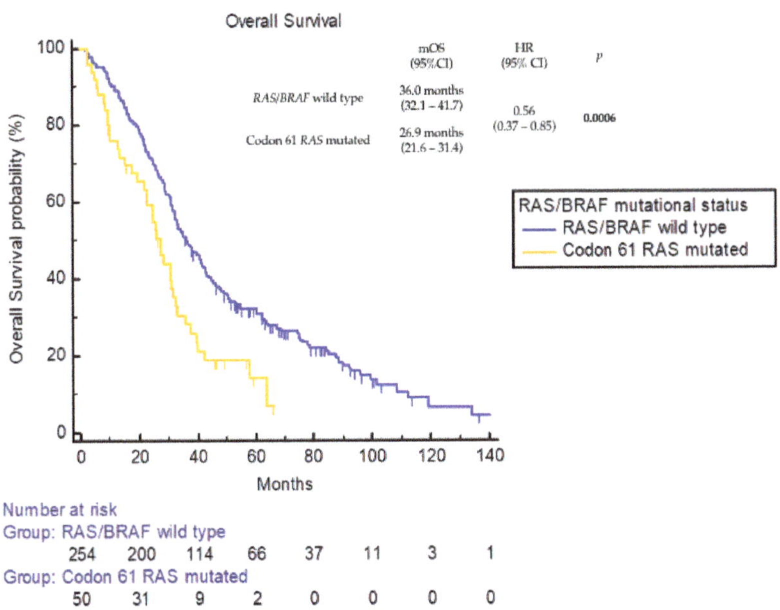

A

Figure 3. *Cont.*

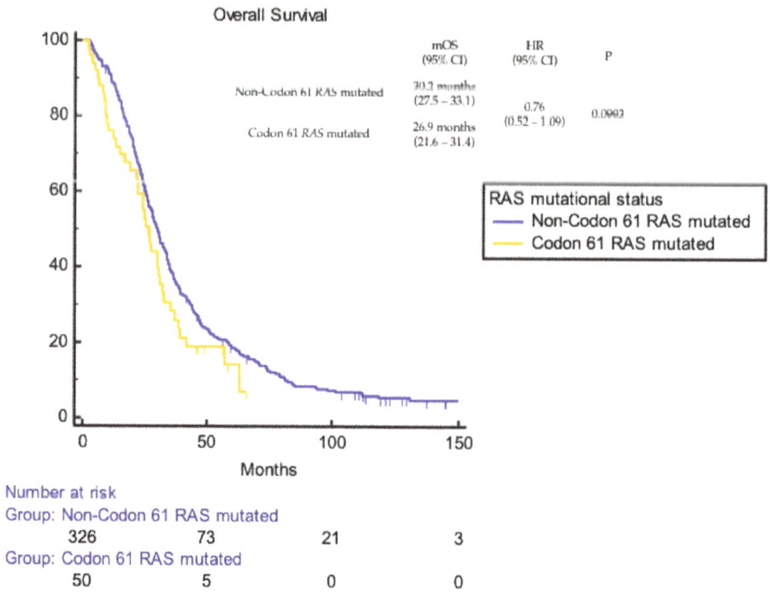

B

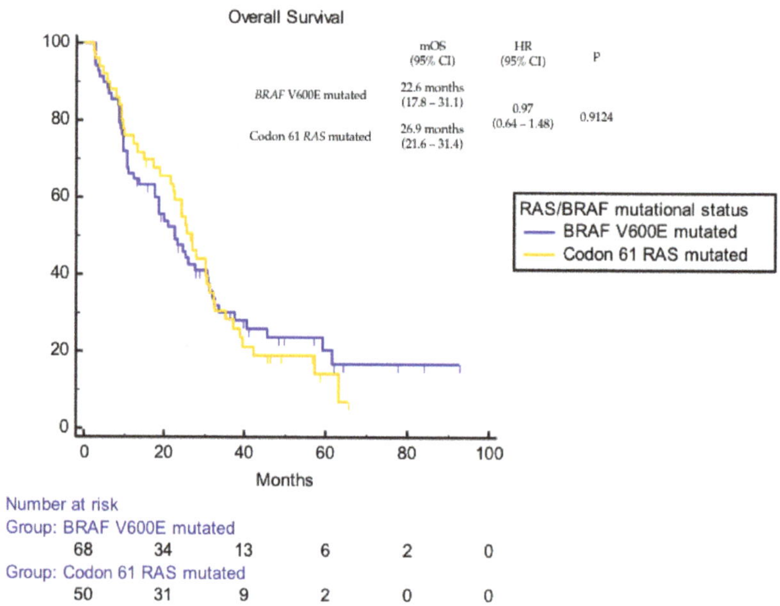

C

Figure 3. Overall survival according to codon 61 *RAS* mutational status. (**A**) Codon 61 *RAS*-mutated vs. *RAS/BRAF* wild-type patients. (**B**) Codon 61 *RAS*-mutated vs. non-codon 61 *RAS*-mutated patients. (**C**) Codon 61 *RAS*-mutated vs. *BRAF* V600E-mutated patients.

4. Discussion

In our study, we demonstrated that codon 61 *RAS*-mutated mCRCs display a tropism for metastatic spread to the peritoneum and ovary and have a negative prognostic impact.

We found out that patients with mCRC harboring codon 61 *RAS* mutation are more likely to experience peritoneal or ovarian metastases during their clinical history. Indeed, the incidence of peritoneal or ovarian involvement was significantly higher in the codon 61 *RAS*-mutated cohort than in the comparator dataset including mCRC without codon 61 *RAS* mutations (54 vs. 28.5%, $p = 0.000163$). The higher tropism for the peritoneum and ovary of codon 61 *RAS*-mutated mCRCs retained statistical significance when compared to all molecular subgroups of the control dataset ($p = 0.012495$, $p = 0.001286$, and $p < 0.00001$, compared to non-codon 61 *RAS*-mutated, *BRAF* V600E-mutated, and *RAS/BRAF* all wild-type cohort, respectively). This feature might be related to the worse prognostic impact. To our knowledge, this is the first evidence of an impact of *RAS* mutational status on metastatization pattern in colorectal tumors. Although involving a small population, this evidence might lead to a more accurate surveillance for peritoneal spread, such as diagnostic laparoscopy before primary tumor resection or routine peritoneal washing sampling. Moreover, this evidence might have pivotal implication in the era of neoadjuvant treatment of colon cancer that we are currently approaching [27]. Confirmation of a peritoneal or ovarian tropism could support therapeutic approaches such as prophylactic hyperthermic intraperitoneal chemotherapy (HIPEC) in combination for stage II–III primary tumor resection or in combination with cytoreductive surgery (CRS) for a stage IV disease in this category of patients. Thus, whether this evidence were validated, codon 61 *RAS* status should be taken into account in a routine clinical approach and might be used as a stratification factor when planning surgical trials (either prophylactic or therapeutic). The COLOPEC trial failed to show the efficacy of adjuvant HIPEC with oxaliplatin, delivered at the time of primary tumor resection or within 5–8 weeks, for T4 or perforated stage II–III colon cancer [28]. Compared to the control arm, there was no difference in the peritoneal-free survival rate at 18-months [28]. Accordingly, the PROPHYLOCHIP trial did not show benefit in terms of disease free-survival for second surgical look combined with HIPEC compared to surveillance in patients at a high risk of developing peritoneal metastases [29]. Concerning stage IV disease, the PRODIGE 7 trial failed to show an additional benefit, in terms of OS and disease-free survival, of combining oxaliplatin-based HIPEC with CRS [30]. Based on this evidence, HIPEC is not currently recommended, neither in adjuvant settings nor in combination with CRS for stage IV disease [2,31]. We postulate that codon 61 *RAS* mutations might be used as stratification factors or even inclusion criteria to optimize the selection of patients that can benefit from adjuvant or therapeutic HIPEC in future studies.

Furthermore, recently published analyses of colorectal peritoneal metastases microenvironment demonstrated a predominance of the consensus molecular subtype (CMS) 4, which is associated with the infiltration of regulatory T cells and macrophages that inhibit immune response [32]. This could unveil a role for immunotherapy regimens in this setting in order to overcome these inhibitory mechanisms and to control peritoneal disease. Patterns of tumor-infiltrating lymphocyte expression in peritoneal nodes seem also to be associated with a better surgical outcome and improved OS, particularly in the case of low-volume disease, providing a possible patient selection for peritoneal cytoreductive surgery and HIPEC, as well as novel pathways for effective immunotherapy [33].

Our data showed a negative prognostic impact of codon 61 *RAS* mutations compared to *RAS/BRAF* wild-type disease, while no difference in terms of OS was observed compared to other non-codon 61 *RAS*-mutated tumors and *BRAF* V600E-mutated tumors. After a median follow up of 96.2 months (95%CI 92.4–109.0), median OS was significantly shorter in tumors harboring *RAS* codon 61 mutations compared to those harboring wild-type codon 61 (26.9 vs. 31.5 months, $p = 0.0221$). The negative prognostic impact of codon 61 *RAS* mutations was retained compared to *RAS/BRAF* wild-type tumors (26.9 vs. 36.0 months, $p = 0.0006$). This negative prognostic role in colon cancer differs from what is observed in other diseases such as pancreatic adenocarcinoma, where *RAS* codon 61 mutations showed a significantly improved survival [34].

We showed that mCRCs harboring *RAS* codon 61 mutations have distinct clinical and biological behaviors. This is of great interest given the high frequency of codon 61 *RAS* mutations as mechanism of secondary resistance to anti-EGFR agents and the advent of RAS inhibitors [35]. The acquired *RAS* codon 61 mutations could play a role in developing resistance to EGFR inhibitors, being enriched in the setting of secondary resistance in mCRCs treated with anti-EGFR agents [36]. Notably, the incidence of acquired *RAS* codon 61 mutations differs according to the treatment line and to the administration in combination with doublet cytotoxic chemotherapy. Indeed, the analysis of paired plasma samples from patients with *RAS/BRAF* wild-type mCRC treated with anti-EGFR agents showed a low incidence of acquired *KRAS* codon 61 mutations in patients treated in the first line in combination with chemotherapy. On the contrary, patients treated with single-agent anti-EGFR in the third line were more likely to develop acquired mutations. Of those, 63% were *KRAS* codon 61 mutations [37]. In the CRICKET trial [38], *RAS* mutations were identified in 48% of liquid biopsy samples collected at the baseline of the anti-EGFR rechallenge; of those, 17% involved codon 61. Furthermore, codon 61 variants have been recently identified with high frequency in the ctDNA of patients with mCRC with secondary resistance to anti-EGFR agents [18–21], with a prevalence of 50% in the CHRONOS trial [18].

Despite many years of effort, only lately have anti-RAS therapies reached clinical application. This is probably linked to the great complexity of RAS, not only in CRC but also in other tumors. The *RAS* gene isoforms display notable variations in the frequency of mutations at each of the three hotspots (G12, G13, and Q61), which have distinct structural and biochemical defects [39]. Recently, novel KRAS G12C inhibitors, alone or in combination with EGFR inhibitors, showed promising results [40–44]. Finally, the phase III CodeBreaK 300 trial showed that dual KRAS G12C and EGFR blockade with sotorasib and panitumumab in refractory *RAS* G12C-mutated mCRCs is associated with longer progression-free survival and a higher response rate than the standard treatment [45].

Our study has several limitations. First of all, the small sample size and the mono-institutional design do not allow us to extend our conclusions to the general population. Nevertheless, we should point out that this is the widest codon 61 *RAS*-mutated cohort reported so far. Moreover, selection biases are inevitable, given the retrospective nature of our analysis. Wider, multicentric, and prospective trials are warranted to confirm our results and investigate the possible prophylactic and therapeutic implications.

5. Conclusions

In our study, we identified a statistically significant correlation between codon 61 *RAS* mutations and metastatic involvement of the peritoneum and ovary. This is the first evidence of an impact of *RAS* mutational status on the metastatization pattern in colorectal tumors. This evidence could lead to new prophylactic applications in preventing peritoneal spreading in this specific group of patients.

Differently to what we have thought for years, "not all RAS mutants are created equal", as Hobbs et al. stated [36], and our aim in the future is to better characterize each of them, leading to new therapeutic strategies.

Author Contributions: Conceptualization: M.A.C., M.B. (Michele Basso), F.S., A.A. and G.T. (Giampaolo Tortora); Methodology: M.A.C., M.B. (Michele Basso), F.S. and A.A.; Software: M.A.C., M.B. (Michele Basso), F.S. and A.A.; Validation: M.A.C., M.B. (Michele Basso), F.S. and A.A.; Formal Analysis: M.A.C., M.B. (Michele Basso), F.S. and A.A.; Investigation: M.A.C., M.B. (Michele Basso), F.S., A.A., A.M., A.U. and L.M.L.; Resources: M.A.C., M.B. (Michele Basso), F.S., A.A., A.M., A.U. and L.M.L.; Data Curation: all authors; Writing—Original Draft Preparation: M.A.C., F.S. and A.A.; Writing—Review and Editing: all authors; Visualization: M.A.C., F.S. and A.A.; Supervision: M.A.C., M.B. (Michele Basso), F.S., A.A., C.P., L.S. and G.T. (Giampaolo Tortora); Project Administration: M.A.C., M.B. (Michele Basso), F.S., A.A., C.P., L.S. and G.T. (Giampaolo Tortora); Funding Acquisition: L.S. and G.T. (Giampaolo Tortora). All authors have read and agreed to the published version of the manuscript.

Funding: This research was supported by the Associazione Italiana per la Ricerca sul Cancro (AIRC) under My first AIRC (grant number: 27367) to L.S.

Institutional Review Board Statement: The study was conducted in accordance with the Declaration of Helsinki and approved by the local Ethics Committee of Fondazione Policlinico Universitario Agostino Gemelli IRCCS, Rome, Italy (protocol code: 0054049/2019; date of approval 18 December 2019).

Informed Consent Statement: Patient consent was waived due to the retrospective nature of the analysis. This is in accordance with Italian legislation ("Linee guida per i trattamenti di dati personali nell'ambito delle sperimentazioni cliniche di medicinali" 24 July 2008 GUN 190 14 August 2008) and the Ethics Committee of Fondazione Policlinico Universitario "Agostino Gemelli" IRCCS.

Data Availability Statement: Data are contained within the article.

Acknowledgments: The authors thank all the patients and their relatives.

Conflicts of Interest: The authors declare no conflicts of interest.

References

1. Guinney, J.; Dienstmann, R.; Wang, X.; de Reyniès, A.; Schlicker, A.; Soneson, C.; Marisa, L.; Roepman, P.; Nyamundanda, G.; Angelino, P.; et al. The consensus molecular subtypes of colorectal cancer. *Nat. Med.* **2015**, *21*, 1350–1356. [CrossRef] [PubMed]
2. Cervantes, A.; Adam, R.; Roselló, S.; Arnold, D.; Normanno, N.; Taïeb, J.; Seligmann, J.; De Baere, T.; Osterlund, P.; Yoshino, T.; et al. Metastatic colorectal cancer: ESMO Clinical Practice Guideline for diagnosis, treatment and follow-up. *Ann. Oncol.* **2023**, *34*, 10–32. [CrossRef] [PubMed]
3. De Roock, W.; De Vriendt, V.; Normanno, N.; Ciardiello, F.; Tejpar, S. KRAS, BRAF, PIK3CA, and PTEN mutations: Implications for targeted therapies in metastatic colorectal cancer. *Lancet Oncol.* **2011**, *12*, 594–603. [CrossRef]
4. Rosty, C.; Young, J.P.; Walsh, M.D.; Clendenning, M.; Walters, R.J.; Pearson, S.; Pavluk, E.; Nagler, B.; Pakenas, D.; Jass, J.R.; et al. Colorectal carcinomas with KRAS mutation are associated with distinctive morphological and molecular features. *Mod. Pathol.* **2013**, *26*, 825–834. [CrossRef] [PubMed]
5. Karapetis, C.S.; Khambata-Ford, S.; Jonker, D.J.; O'Callaghan, C.J.; Tu, D.; Tebbutt, N.C.; Simes, R.J.; Chalchal, H.; Shapiro, J.D.; Robitaille, S.; et al. *K-ras*Mutations and Benefit from Cetuximab in Advanced Colorectal Cancer. *N. Engl. J. Med.* **2008**, *359*, 1757–1765. [CrossRef] [PubMed]
6. Amado, R.G.; Wolf, M.; Peeters, M.; Van Cutsem, E.; Siena, S.; Freeman, D.J.; Juan, T.; Sikorski, R.; Suggs, S.; Radinsky, R.; et al. Wild-type *KRAS* is required for panitumumab efficacy in patients with metastatic colorectal cancer. *J. Clin. Oncol.* **2008**, *26*, 1626–1634. [CrossRef] [PubMed]
7. De Roock, W.; Claes, B.; Bernasconi, D.; De Schutter, J.; Biesmans, B.; Fountzilas, G.; Kalogeras, K.T.; Kotoula, V.; Papamichael, D.; Laurent-Puig, P.; et al. Effects of KRAS, BRAF, NRAS, and PIK3CA mutations on the efficacy of cetuximab plus chemotherapy in chemotherapy-refractory metastatic colorectal cancer: A retrospective consortium analysis. *Lancet Oncol.* **2010**, *11*, 753–762. [CrossRef] [PubMed]
8. Bardelli, A.; Siena, S. Molecular mechanisms of resistance to cetuximab and panitumumab in colorectal cancer. *J. Clin. Oncol.* **2010**, *28*, 1254–1261. [CrossRef]
9. Tran, N.H.; Cavalcante, L.L.; Lubner, S.J.; Mulkerin, D.L.; LoConte, N.K.; Clipson, L.; Matkowskyj, K.A.; Deming, D.A. Precision medicine in colorectal cancer: The molecular profile alters treatment strategies. *Ther. Adv. Med. Oncol.* **2015**, *7*, 252–262. [CrossRef]
10. Van Cutsem, E.; Lenz, H.-J.; Köhne, C.-H.; Heinemann, V.; Tejpar, S.; Melezínek, I.; Beier, F.; Stroh, C.; Rougier, P.; van Krieken, J.H.; et al. Fluorouracil, leucovorin, and irinotecan plus cetuximab treatment and *RAS* mutations in colorectal cancer. *J. Clin. Oncol.* **2015**, *33*, 692–700. [CrossRef]
11. Allegra, C.J.; Jessup, J.M.; Somerfield, M.R.; Hamilton, S.R.; Hammond, E.H.; Hayes, D.F.; McAllister, P.K.; Morton, R.F.; Schilsky, R.L. American society of clinical oncology provisional clinical opinion: Testing for *KRAS* gene mutations in patients with metastatic colorectal carcinoma to predict response to anti–epidermal growth factor receptor monoclonal antibody therapy. *J. Clin. Oncol.* **2009**, *27*, 2091–2096. [CrossRef]
12. Tejpar, S.; Celik, I.; Schlichting, M.; Sartorius, U.; Bokemeyer, C.; Van Cutsem, E. Association of *KRAS* G13D tumor mutations with outcome in patients with metastatic colorectal cancer treated with first-line chemotherapy with or without cetuximab. *J. Clin. Oncol.* **2012**, *30*, 3570–3577. [CrossRef]
13. Bokemeyer, C.; Köhne, C.-H.; Ciardiello, F.; Lenz, H.-J.; Heinemann, V.; Klinkhardt, U.; Beier, F.; Duecker, K.; van Krieken, J.; Tejpar, S. FOLFOX4 plus cetuximab treatment and RAS mutations in colorectal cancer. *Eur. J. Cancer* **2015**, *51*, 1243–1252. [CrossRef]
14. Douillard, J.Y.; Siena, S.; Cassidy, J.; Tabernero, J.; Burkes, R.; Barugel, M.; Humblet, Y.; Bodoky, G.; Cunningham, D.; Jassem, J.; et al. Final results from PRIME: Randomized phase III study of panitumumab with FOLFOX4 for first-line treatment of metastatic colorectal cancer. *Ann. Oncol.* **2014**, *25*, 1346–1355. [CrossRef]
15. Buhrman, G.; Wink, G.; Mattos, C. Transformation Efficiency of RasQ61 Mutants Linked to Structural Features of the Switch Regions in the Presence of Raf. *Structure* **2007**, *15*, 1618–1629. [CrossRef] [PubMed]
16. Loupakis, F.; Ruzzo, A.; Cremolini, C.; Vincenzi, B.; Salvatore, L.; Santini, D.; Masi, G.; Stasi, I.; Canestrari, E.; Rulli, E.; et al. KRAS codon 61, 146 and BRAF mutations predict resistance to cetuximab plus irinotecan in KRAS codon 12 and 13 wild-type metastatic colorectal cancer. *Br. J. Cancer* **2009**, *101*, 715–721. [CrossRef]

17. Douillard, J.-Y.; Oliner, K.S.; Siena, S.; Tabernero, J.; Burkes, R.; Barugel, M.; Humblet, Y.; Bodoky, G.; Cunningham, D.; Jassem, J.; et al. Panitumumab–FOLFOX4 Treatment and RAS Mutations in Colorectal Cancer. *N. Engl. J. Med.* **2013**, *36*5, 1023–1034. [CrossRef] [PubMed]
18. Sartore-Bianchi, A.; Pietrantonio, F.; Lonardi, S.; Mussolin, B.; Rua, F.; Crisafulli, G.; Bartolini, A.; Fenocchio, E.; Amatu, A.; Manca, P.; et al. Circulating tumor DNA to guide rechallenge with panitumumab in metastatic colorectal cancer: The phase 2 CHRONOS trial. *Nat. Med.* **2022**, *28*, 1612–1618. [CrossRef] [PubMed]
19. Misale, S.; Yaeger, R.; Hobor, S.; Scala, E.; Janakiraman, M.; Liska, D.; Valtorta, E.; Schiavo, R.; Buscarino, M.; Siravegna, G.; et al. Emergence of KRAS mutations and acquired resistance to anti-EGFR therapy in colorectal cancer. *Nature* **2012**, *486*, 532–536. [CrossRef] [PubMed]
20. Siena, S.; Sartore-Bianchi, A.; Garcia-Carbonero, R.; Karthaus, M.; Smith, D.; Tabernero, J.; Van Cutsem, E.; Guan, X.; Boedigheimer, M.; Ang, A.; et al. Dynamic molecular analysis and clinical correlates of tumor evolution within a phase II trial of panitumumab-based therapy in metastatic colorectal cancer. *Ann. Oncol.* **2018**, *29*, 119–126. [CrossRef] [PubMed]
21. Ali, M.; Kaltenbrun, E.; Anderson, G.R.; Stephens, S.J.; Arena, S.; Bardelli, A.; Counter, C.M.; Wood, K.C. Codon bias imposes a targetable limitation on KRAS-driven therapeutic resistance. *Nat. Commun.* **2017**, *8*, 15617. [CrossRef]
22. Lavacchi, D.; Fancelli, S.; Roviello, G.; Castiglione, F.; Caliman, E.; Rossi, G.; Venturini, J.; Pellegrini, E.; Brugia, M.; Vannini, A.; et al. Mutations matter: An observational study of the prognostic and predictive value of KRAS mutations in metastatic colorectal cancer. *Front. Oncol.* **2022**, *12*, 1055019. [CrossRef]
23. Zeng, J.; Fan, W.; Li, J.; Wu, G.; Wu, H. KRAS/NRAS Mutations Associated with Distant Metastasis and BRAF/PIK3CA Mutations Associated with Poor Tumor Differentiation in Colorectal Cancer. *Int. J. Gen. Med.* **2023**, *16*, 4109–4120. [CrossRef]
24. Serebriiskii, I.G.; Connelly, C.; Frampton, G.; Newberg, J.; Cooke, M.; Miller, V.; Ali, S.; Ross, J.S.; Handorf, E.; Arora, S.; et al. Comprehensive characterization of RAS mutations in colon and rectal cancers in old and young patients. *Nat. Commun.* **2019**, *10*, 1–12. [CrossRef] [PubMed]
25. Imamura, Y.; Lochhead, P.; Yamauchi, M.; Kuchiba, A.; Qian, Z.R.; Liao, X.; Nishihara, R.; Jung, S.; Wu, K.; Nosho, K.; et al. Analyses of clinicopathological, molecular, and prognostic associations of KRAS codon 61 and codon 146 mutations in colorectal cancer: Cohort study and literature review. *Mol. Cancer* **2014**, *13*, 135. [CrossRef]
26. Morris, V.K.; Lucas, F.A.S.; Overman, M.J.; Eng, C.; Morelli, M.P.; Jiang, Z.-Q.; Luthra, R.; Meric-Bernstam, F.; Maru, D.; Scheet, P.; et al. Clinicopathologic characteristics and gene expression analyses of non-KRAS 12/13, RAS-mutated metastatic colorectal cancer. *Ann. Oncol.* **2014**, *25*, 2008–2014. [CrossRef] [PubMed]
27. Morton, D.; Seymour, M.; Magill, L.; Handley, K.; Glasbey, J.; Glimelius, B.; Palmer, A.; Seligmann, J.; Laurberg, S.; Murakami, K.; et al. Preoperative Chemotherapy for Operable Colon Cancer: Mature Results of an International Randomized Controlled Trial. *J. Clin. Oncol.* **2023**, *41*, 1541–1552. [CrossRef] [PubMed]
28. Klaver, C.E.L.; Wisselink, D.D.; Punt, C.J.A.; Snaebjornsson, P.; Crezee, J.; Aalbers, A.G.J.; Brandt, A.; Bremers, A.J.A.; Burger, J.W.A.; Fabry, H.F.J.; et al. Adjuvant hyperthermic intraperitoneal chemotherapy in patients with locally advanced colon cancer (COLOPEC): A multicentre, open-label, randomised trial. *Lancet Gastroenterol. Hepatol.* **2019**, *4*, 761–770. [CrossRef]
29. Goéré, D.; Glehen, O.; Quenet, F.; Guilloit, J.-M.; Bereder, J.-M.; Lorimier, G.; Thibaudeau, E.; Ghouti, L.; Pinto, A.; Tuech, J.-J.; et al. Second-look surgery plus hyperthermic intraperitoneal chemotherapy versus surveillance in patients at high risk of developing colorectal peritoneal metastases (PROPHYLOCHIP–PRODIGE 15): A randomised, phase 3 study. *Lancet Oncol.* **2020**, *21*, 1147–1154. [CrossRef]
30. Quénet, F.; Elias, D.; Roca, L.; Goéré, D.; Ghouti, L.; Pocard, M.; Facy, O.; Arvieux, C.; Lorimier, G.; Pezet, D.; et al. Cytoreductive surgery plus hyperthermic intraperitoneal chemotherapy versus cytoreductive surgery alone for colorectal peritoneal metastases (PRODIGE 7): A multicentre, randomised, open-label, phase 3 trial. *Lancet Oncol.* **2021**, *22*, 256–266. [CrossRef]
31. Argilés, G.; Tabernero, J.; Labianca, R.; Hochhauser, D.; Salazar, R.; Iveson, T.; Laurent-Puig, P.; Quirke, P.; Yoshino, T.; Taieb, J.; et al. Localised colon cancer: ESMO Clinical Practice Guidelines for diagnosis, treatment and follow-up. *Ann. Oncol.* **2020**, *31*, 1291–1305. [CrossRef]
32. Lenos, K.J.; Bach, S.; Moreno, L.F.; Hoorn, S.T.; Sluiter, N.R.; Bootsma, S.; Braga, F.A.V.; Nijman, L.E.; Bosch, T.v.D.; Miedema, D.M.; et al. Molecular characterization of colorectal cancer related peritoneal metastatic disease. *Nat. Commun.* **2022**, *13*, 4443. [CrossRef]
33. Garland-Kledzik, M.; Uppal, A.; Naeini, Y.B.; Stern, S.; Erali, R.; Scholer, A.J.; Khader, A.M.; Santamaria-Barria, J.A.; Cummins, Perry, K.; Zhou, Y.; et al. Prognostic Impact and Utility of Immunoprofiling in the Selection of Patients with Colorectal Peritoneal Carcinomatosis for Cytoreductive Surgery (CRS) and Heated Intraperitoneal Chemotherapy (HIPEC). *J. Gastrointest. Surg.* **2021**, *25*, 233–240. [CrossRef] [PubMed]
34. Witkiewicz, A.K.; McMillan, E.A.; Balaji, U.; Baek, G.; Lin, W.-C.; Mansour, J.; Mollaee, M.; Wagner, K.-U.; Koduru, P.; Yopp, A.; et al. Whole-exome sequencing of pancreatic cancer defines genetic diversity and therapeutic targets. *Nat. Commun.* **2015**, *6*, 6744. [CrossRef]
35. Nusrat, M.; Yaeger, R. KRAS inhibition in metastatic colorectal cancer: An update. *Curr. Opin. Pharmacol.* **2023**, *68*, 102343. [CrossRef] [PubMed]
36. Vangala, D.; Ladigan, S.; Liffers, S.T.; Noseir, S.; Maghnouj, A.; Götze, T.-M.; Verdoodt, B.; Klein-Scory, S.; Godfrey, L.; Zowada, M.K.; et al. Secondary resistance to anti-EGFR therapy by transcriptional reprogramming in patient-derived colorectal cancer models. *Genome Med.* **2021**, *13*, 116. [CrossRef]

37. Parseghian, C.M.; Sun, R.; Woods, M.; Napolitano, S.; Lee, H.M.; Alshenaifi, J.; Willis, J.; Nunez, S.; Raghav, K.P.; Morris, V.K.; et al. Resistance Mechanisms to Anti–Epidermal Growth Factor Receptor Therapy in *RAS/RAF* Wild-Type Colorectal Cancer Vary by Regimen and Line of Therapy. *J. Clin. Oncol.* **2023**, *41*, 460–471. [CrossRef]
38. Cremolini, C.; Rossini, D.; Dell'aquila, E.; Lonardi, S.; Conca, E.; Del Re, M.; Busico, A.; Pietrantonio, F.; Danesi, R.; Aprile, G.; et al. Rechallenge for Patients with *RAS* and *BRAF* Wild-Type Metastatic Colorectal Cancer with Acquired Resistance to First-line Cetuximab and Irinotecan: A Phase 2 Single-Arm Clinical Trial. *JAMA Oncol.* **2019**, *5*, 343–350. [CrossRef]
39. Hobbs, G.A.; Der, C.J.; Rossman, K.L. RAS isoforms and mutations in cancer at a glance. *J. Cell Sci.* **2016**, *129*, 1287–1292. [CrossRef] [PubMed]
40. Sacher, A.; LoRusso, P.; Patel, M.R.; Miller, W.H.; Garralda, E.; Forster, M.D.; Santoro, A.; Falcon, A.; Kim, T.W.; Paz-Ares, L.; et al. Single-Agent Divarasib (GDC-6036) in Solid Tumors with a *KRAS* G12C Mutation. *N. Engl. J. Med.* **2023**, *389*, 710–721. [CrossRef]
41. Yaeger, R.; Weiss, J.; Pelster, M.S.; Spira, A.I.; Barve, M.; Ou, S.-H.I.; Leal, T.A.; Bekaii-Saab, T.S.; Paweletz, C.P.; Heavey, G.A.; et al. Adagrasib with or without Cetuximab in Colorectal Cancer with Mutated *KRAS* G12C. *N. Engl. J. Med.* **2023**, *388*, 44–54. [CrossRef] [PubMed]
42. Fakih, M.G.; Kopetz, S.; Kuboki, Y.; Kim, T.W.; Munster, P.N.; Krauss, J.C.; Falchook, G.S.; Han, S.-W.; Heinemann, V.; Muro, K.; et al. Sotorasib for previously treated colorectal cancers with KRASG12C mutation (CodeBreaK100): A prespecified analysis of a single-arm, phase 2 trial. *Lancet Oncol.* **2022**, *23*, 115–124. [CrossRef]
43. Kuboki, Y.; Yaeger, R.; Fakih, M.; Strickler, J.; Masuishi, T.; Kim, E.; Bestvina, C.; Langer, C.; Krauss, J.; Puri, S.; et al. 315O Sotorasib in combination with panitumumab in refractory KRAS G12C-mutated colorectal cancer: Safety and efficacy for phase Ib full expansion cohort. *Ann. Oncol.* **2022**, *33*, S680–S681. [CrossRef]
44. Hong, D.S.; Fakih, M.G.; Strickler, J.H.; Desai, J.; Durm, G.A.; Shapiro, G.I.; Falchook, G.S.; Price, T.J.; Sacher, A.; Denlinger, C.S.; et al. $KRAS^{G12C}$ Inhibition with Sotorasib in Advanced Solid Tumors. *N. Engl. J. Med.* **2020**, *383*, 1207–1217. [CrossRef] [PubMed]
45. Fakih, M.G.; Salvatore, L.; Esaki, T.; Modest, D.P.; Lopez-Bravo, D.P.; Taieb, J.; Karamouzis, M.V.; Ruiz-Garcia, E.; Kim, T.-W.; Kuboki, Y.; et al. Sotorasib plus Panitumumab in Refractory Colorectal Cancer with Mutated *KRAS* G12C. *N. Engl. J. Med.* **2023**, *389*, 2125–2139. [CrossRef] [PubMed]

cancers

MDPI

Article

Preclinical Therapeutic Efficacy of RAF/MEK/ERK and IGF1R/AKT/mTOR Inhibition in Neuroblastoma

Stacey Stauffer [1], Jacob S. Roth [2], Edjay R. Hernandez [3], Joshua T. Kowalczyk [3], Nancy E. Sealover [4], Katie E. Hebron [1], Amy James [5], Kristine A. Isanogle [5], Lisa A. Riffle [6], Lilia Ileva [6], Xiaoling Luo [7], Jin-Qiu Chen [7], Noemi Kedei [7], Robert L. Kortum [4], Haiyan Lei [3], Jack F. Shern [3], Joseph D. Kalen [6], Elijah F. Edmondson [8], Matthew D. Hall [2], Simone Difilippantonio [5], Carol J. Thiele [3] and Marielle E. Yohe [1,3,*]

1 Laboratory of Cell and Developmental Signaling, Center for Cancer Research, National Cancer Institute, NIH, 8560 Progress Drive, Frederick, MD 21701, USA
2 Early Translation Branch, Division of Preclinical Innovation, National Center for Advancing Translational Sciences, 9800 Medical Center Drive, Rockville, MD 20850, USA; jacob.roth@einsteinmed.edu (J.S.R.)
3 Pediatric Oncology Branch, Center for Cancer Research, National Cancer Institute, NIH, 9000 Rockville Pike, Bethesda, MD 20892, USA
4 Department of Pharmacology and Molecular Therapeutics, Uniformed Services University of the Health Services, Bethesda, MD 20814, USA; robert.kortum@usuhs.edu (R.L.K.)
5 Animal Research Technical Support, Laboratory Animal Sciences Program, Leidos Biomedical Research, Inc., Frederick National Laboratory for Cancer Research, Frederick, MD 21702, USA
6 Small Animal Imaging Program, Laboratory Animal Sciences Program, Leidos Biomedical Research, Inc., Frederick National Laboratory for Cancer Research, Frederick, MD 21702, USA
7 Collaborative Protein Technology Resource, National Cancer Institute, NIH, Bethesda, MD 20892, USA
8 Molecular Histopathology Laboratory, Laboratory Animal Sciences Program, Leidos Biomedical Research, Inc., Frederick National Laboratory for Cancer Research, Frederick, MD 21702, USA
* Correspondence: yoheme@mail.nih.gov; Tel.: +1-240-760-7436

Simple Summary: The prognosis for patients with relapsed neuroblastoma is poor, and novel treatment options for these patients are needed. Some relapsed neuroblastoma tumors harbor activating mutations in the RAS/MAPK pathway. In prior studies, single agent MEK or IGF1R inhibitors induced transient responses as single agents in neuroblastoma models. In this study, we tested the efficacy of a combination of the MEK inhibitor trametinib and the IGF1R inhibitor ganitumab in RAS-mutated neuroblastoma models. While the trametinib/ganitumab combination decreased cell viability and tumor growth, the combination did not prevent metastasis of RAS-mutated neuroblastoma. Therefore, further studies on the effect of trametinib and ganitumab on neuroblastoma metastasis are necessary before initiating clinical trials of this combination of targeted agents in patients with relapsed neuroblastoma.

Abstract: Activating mutations in the RAS/MAPK pathway are observed in relapsed neuroblastoma. Preclinical studies indicate that these tumors have an increased sensitivity to inhibitors of the RAS/MAPK pathway, such as MEK inhibitors. MEK inhibitors do not induce durable responses as single agents, indicating a need to identify synergistic combinations of targeted agents to provide therapeutic benefit. We previously showed preclinical therapeutic synergy between a MEK inhibitor, trametinib, and a monoclonal antibody specific for IGF1R, ganitumab in RAS-mutated rhabdomyosarcoma. Neuroblastoma cells, like rhabdomyosarcoma cells, are sensitive to the inhibition of the RAS/MAPK and IGF1R/AKT/mTOR pathways. We hypothesized that the combination of trametinib and ganitumab would be effective in RAS-mutated neuroblastoma. In this study, trametinib and ganitumab synergistically suppressed neuroblastoma cell proliferation and induced apoptosis in cell culture. We also observed a delay in tumor initiation and prolongation of survival in heterotopic and orthotopic xenograft models treated with trametinib and ganitumab. However, the growth of both primary and metastatic tumors was observed in animals receiving the combination of trametinib and ganitumab. Therefore, more preclinical work is necessary before testing this combination in patients with relapsed or refractory RAS-mutated neuroblastoma.

Keywords: RAS; MEK; neuroblastoma; IGF1R

1. Introduction

Neuroblastoma is an embryonal tumor of the peripheral sympathetic nervous system. Neuroblastoma accounts for 6% of all cancers in children and 15% of the deaths in children due to cancer. Patients with high-risk neuroblastoma, including patients with metastatic disease at diagnosis, have an overall survival of less than 50%, despite aggressive multi-modality treatment, including chemotherapy, surgery, radiation, differentiation therapy, immunotherapy, and stem cell transplant. Up to 60% of patients with high-risk neuroblastoma relapse following up-front therapy, and long-term survivors have significant morbidities, indicating a need for additional therapeutic options, including molecularly targeted therapies [1].

The advent of next-generation sequencing (NGS) has facilitated the discovery of the genomic drivers of cancer development and, thus, relevant targets for drug discovery for many human malignancies. Despite significant effort, however, very few recurrent targetable mutations have been identified in high-risk neuroblastoma. Focal amplification of the oncogene *MYCN* and hemizygous deletion of chromosomes 1p and 11q are associated with high-risk disease [2]. Activating mutations in the receptor tyrosine kinase *ALK* and loss of function mutations or deletions of genes important for chromatin remodeling, including *ATRX, ARID1A*, and *ARID1B*, are observed in high-risk neuroblastoma [2]. While somatic variations in the genes of the RAS/MAP kinase pathway are rarely observed in high-risk neuroblastomas at diagnosis [2–4], variants that lead to increased RAS/MAP kinase activity are enriched in relapsed neuroblastomas [5]. These variants include gain-of-function mutations in the RAS isoforms *HRAS, KRAS*, and *NRAS*; gain-of-function mutations in the protein phosphatase *PTPN11*; and loss-of-function mutations and deletions in the RAS GTPase activating protein *NF1*. RAS/MAP kinase pathway alterations are associated with poor overall survival in neuroblastoma [4]. The presence of a RAS mutation in neuroblastoma cell lines introduces a functional dependency on RAS [6]. The RAS/MAP kinase pathway, then, is a potential therapeutic target for treating relapsed neuroblastoma.

The RAS/MAP kinase pathway consists of a cascade of serine/threonine kinases. In this cascade, GTP-bound and active RAS recruits RAF kinase to the plasma membrane, where it is activated. Active RAF phosphorylates and activates MEK1 and 2, which in turn phosphorylate and activate the MAP kinases, ERK1 and ERK2. ERK1 and ERK2 phosphorylate many additional substrates, which impact diverse cellular processes, such as proliferation, differentiation, migration, and survival. Initial efforts to target the RAS/MAP kinase pathway in neuroblastoma focused on the use of MEK inhibitors. As is the case in many other RAS-driven cancer models, MEK inhibitors as single agents have short-lived efficacy in preclinical models of neuroblastoma [5,7–10]. To improve the efficacy of MEK inhibitors in RAS-driven neuroblastoma, investigators have combined these agents with other molecularly targeted agents. The combination of a MEK inhibitor with a CDK4/6 inhibitor was synergistic in preclinical models of neuroblastoma [11], resulting in the initiation of a phase I clinical trial (NCT02780128). Combinations of MEK inhibitors with inhibitors of the PI3 kinase/AKT/mTOR pathway are also effective in preclinical models of neuroblastoma [12,13]. These combinations have been effective in models of other RAS-driven cancers, but their clinical success has been limited by toxicity [14]. Other promising combinations in RAS-driven neuroblastoma include a MEK inhibitor with retinoic acid [7,15,16], a CENPE inhibitor [17], a Hippo pathway modulator [18], a SHP2 inhibitor [19], and a CDK8 inhibitor [20]. Preclinical evaluations of combinations of a MEK inhibitor with a BET bromodomain inhibitor [21] or an ALK inhibitor [9] have been unsuccessful in neuroblastoma.

In addition to the alterations in the RAS/MAPK pathway described above, most neuroblastoma tumors express IGF1R. The IGF1R pathway is aberrantly activated by autocrine and paracrine signaling loops in neuroblastoma tumors through the release of

IGF1 and IGF2, ligands for IGF1R, from the neuroblastoma and stromal cells within the tumor [22–25]. The increased activity of the IGF1R/PI3 kinase/AKT/mTOR pathway induces proliferation, prevents apoptosis, increases motility, induces differentiation, and drives MYCN expression in neuroblastoma cells [26–29]. In addition, signaling through the IGF1R/PI3 kinase/AKT/mTOR pathway facilitates neuroblastoma metastasis to bone [30] and induces resistance to cytotoxic chemotherapy, retinoic acid, and ALK inhibitors [31–33]. Decreasing the expression of IGF1R [34] or AKT2 [12] in neuroblastoma cells decreases proliferation, migration, and invasion, validating the IGF1R/PI3 kinase/AKT pathway as a potential therapeutic target in neuroblastoma. Importantly, small-molecule and monoclonal antibody inhibitors of IGF1R potently decrease neuroblastoma cell viability and delay tumor growth in xenograft models of neuroblastoma [35–44]. Monoclonal antibodies specific for the IGF1R ligands [45], as well as inhibitors of PI3 kinase and mTOR, which are activated by signaling through IGF1R, are also effective in neuroblastoma models [46–51].

In our previous studies, we evaluated the efficacy of the combination of the MEK inhibitor, trametinib, with a monoclonal antibody specific for IGF1R, ganitumab, in RAS-mutated rhabdomyosarcoma [52]. Combinations of a MEK inhibitor and an IGF1R antibody have been effective in models of other RAS-mutated malignancies, such as lung cancer and leukemia [53,54]. In contrast to combinations of MEK inhibitors and PI3 kinase/mTOR/AKT inhibitors, the combination of a MEK inhibitor and an IGF1R antibody was well tolerated in adults [55]. Because both MEK and IGF1R inhibitors are effective in RAS-driven neuroblastoma, we sought to evaluate the combination of trametinib and ganitumab in this malignancy.

2. Materials and Methods

2.1. Cell Lines and Reagents

The neuroblastoma cell lines CHP-212 and LA-N-6 cell lines were obtained from the COG/ALSF Childhood Cancer Repository. SK-N-AS, NB-Eb-C1, NBL-S, SK-N-FI, and SK-N-BE(2)-C cells were obtained from the NCI Pediatric Oncology Branch. We ensured that our cells were free of mycoplasma contamination by testing with the MycoAlert kit (Lonza, Basel, Switzerland). In addition, we confirmed the identity of our cell lines by STR fingerprinting (Genetica/LapCorp, Burlington, NC, USA) (Supplementary Table S1). The neuroblastoma cell lines were grown in Roswell Park Memorial Institute 1640 medium (RPMI) with added fetal bovine serum (10% *v*/*v*), penicillin (100 IU/mL), streptomycin (100 mg/mL), and glutamine (2 mM). We obtained trametinib from the NIH Developmental Therapeutics Program (DTP) and ganitumab from the Cancer Therapy Evaluation Program (CTEP).

2.2. Whole-Exome Sequencing

Genomic DNA was extracted from neuroblastoma cell lines using Allprep DNA/RNA mini kits (Qiagen, Hilden, Germany). Sequencing libraries were prepared using the Agilent SureSelectXT human all exon V5 target enrichment kit (Agilent, Santa Clara, CA, USA). Paired-end 150 bp read sequencing was performed on a HiSeq3000 system using Illumina TruSeq V3 chemistry (Illumina, San Diego, CA, USA) at the CCR Sequencing Facility. Paired-end reads were mapped to the human genome (Hg38) using MWA-MEM 0.7.12 with default parameters. Data analysis was accomplished as previously described [52].

2.3. Cell Viability Assay

Cells were plated at a density of 10,000 cells/well in white, clear bottom 96-well plates. The next day, the cells were treated with trametinib at the indicated concentrations, with or without 1 μM ganitumab. The plates were incubated at 37 °C, with 5% CO2, for 72 h. To determine cell viability, 25 μL of CellTiter-Glo reagent (Promega, Madison, WI, USA) was added to each well. Luminescence was read on a SpectraMax iD3 plate reader (Molecular Devices, San Jose, CA, USA). The background signal from blank reactions (CellTiter-Glo reagent with cell culture media and no cells) was subtracted from the raw signal. The resulting values were normalized to DMSO control.

2.4. Matrix Combination Assay

The matrix combination assay was performed as previously described [52]. In brief, assay-ready plates were prepared by acoustic-droplet spotting of 25 nL DMSO-solvated trametinib and 1 μL RPMI-solvated ganitumab to each well. SK-N-AS cells were harvested and dispensed into the prepared plates to yield 500 cells/well. The plates were incubated at 37 °C, with 5% CO2, for 72 h, and then 2.5 μL CellTiter-Glo reagent (Promega) was dispensed into each well. We read the luminescence on a ViewLux instrument (Perkin-Elmer, Waltham, MA, USA). We normalized the resulting values to the DMSO control and performed the synergy calculations as previously reported [52].

2.5. Apoptosis Assays

For caspase 3/7 Glo assays, SK-N-AS or CHP-212 cells were plated at a density of 30,000 cells/well in white, clear-bottom 96-well plates. The next day, the cells were treated with vehicle (DMSO), trametinib, ganitumab, or trametinib/ganitumab. The plates were incubated at 37 °C, 5% CO2, for 18 h, and then 25 μL of Caspase 3/7-Glo reagent (Promega) was added to each well. Luminescence was read on a SpectraMax iD3 plate reader (Molecular Devices). The background signal from blank reactions (Caspase 3/7-Glo reagent with cell culture media and no cells) was subtracted from the raw signal. The resulting values were normalized to DMSO control.

For the Annexin V assays, SK-N-AS or CHP-212 cells were plated at a density of 500,000 cells/well in 6-well plates. The following day, the cells were treated with vehicle (DMSO), trametinib, ganitumab, or trametinib/ganitumab. The plates were incubated at 37 °C, 5% CO2, for 48 h. The resulting suspension cells were harvested, and the resulting adherent cells were detached with Accutase (Thermo, Waltham, MA, USA). The suspension and adherent cells were combined and incubated with APC-labeled human recombinant Annexin V and 7-Aminoactinomycin D (7-AAD) according to the manufacturers' instructions (BD Biosciences, Franklin Lakes, NJ, USA). Samples were read on an Accuri C6 Plus flow cytometer (BD Biosciences).

2.6. Capillary Immunoassays

Fresh frozen tumor samples were prepared in TPER (Thermo) using a TissueRuptor. Lysates were analyzed as described [56]. Primary antibodies included pERK (Cell Signaling Technologies # 9101, Danvers, MA, USA), total ERK (Cell Signaling Technologies # 9102), and IGF1R (Cell Signaling Technologies #3027)

2.7. Immunoblot Assays

Cells were treated as indicated, and lysates were analyzed essentially, as described in [56]. Primary antibodies include those detailed in Section 2.6, as well as NF1 (Bethyl # A300-140A), pS6 (Cell Signaling Technologies #2211), total S6 (Cell Signaling Technologies #2217), α-tubulin (Cell Signaling Technologies #2217), and vinculin (Sigma #h-vin1, Burlington, MA, USA). Secondary antibodies included anti-rabbit HRP (Cell Signaling Technologies #7074) and anti-mouse HRP (Cell Signaling Technologies #7076).

2.8. Subcutaneous Xenograft Experiments

For both the subcutaneous and orthotopic xenograft studies, 6-week-old female SCID beige mice were purchased from Charles River laboratories. For the subcutaneous xenograft experiments, 2 million neuroblastoma cells (SK-N-AS or NB-Eb-C1) in a 1:1 mixture of Matrigel (BioTechne, Minneapolis, MN, USA) and HBSS (Sigma) were injected subcutaneously in the left flank of the mouse. Two weeks following tumor cell injection, the mice were randomized into treatment groups (n = 10 mice per group). No formal power calculation was performed prior to the start of these studies.

Trametinib suspensions were prepared in a vehicle, as previously described [52]. Trametinib was dosed in the indicated groups by oral gavage (OG) at the dosage reported previously [57]. Ganitumab was administered to indicated groups by intraperitoneal injection (IP), as reported previously [58].

In these experiments, the tumor dimensions were measured twice a week with digital calipers to obtain two diameters of the tumor sphere, from which the tumor volume was determined using the equation $(D \times d^2)/6 \times 3.14$ (where D = the maximum diameter, and d = the minimum diameter). Two mice per group were euthanized 4 h after the fifth dose of vehicle or trametinib for assessment of pharmacodynamic markers of response, and the remaining mice were observed for tumor response. These animals were euthanized when they reached the tumor endpoint.

2.9. Orthotopic Xenograft Experiments

For the orthotopic xenograft experiments, intra-adrenal tumor cell injections were achieved via survival surgery. Mice ($n = 20$) were anesthetized with isoflurane and then placed laterally with the left flank facing upward. A lateral incision 0.5 inches below the spine on the animal's right flank was made, followed by a small peritoneal incision to expose the kidney and visualize the adrenal gland. A Hamilton syringe was used to inject 3×10^5 SK-N-AS cells in 7 uL of PBS. The kidney was then returned to the abdomen, the peritoneum was closed with absorbable suture, and the skin was closed with surgical staples. The staples were removed after 10–14 days. After the staples were removed, the mice were able to undergo ultrasound and MRI for tumor monitoring. Three weeks following tumor cell injection, the mice were randomized based on tumor volume and body weight into treatment groups ($n = 10$ mice per group). In these experiments, mice were euthanized when they gained more than 10% of their initial body weight due to tumor burden, or when the tumor volume exceeded 2000 mm^3, as determined by ultrasound.

2.10. Mouse Imaging Preparation

Standard imaging and animal-handling protocols for both ultrasound and MRI required maintaining the rodent's internal temperature and monitoring anesthesia administration. Animals were anesthetized at 3% isoflurane with a carrier gas of oxygen at a flow of 1 L/min in the induction chamber, and during imaging, the isoflurane was administered via a nose cone at 1.5–2.0%. Pulmonary function was monitored by a multichannel physiology system (MP150 Biopac System, Inc., Goleta, CA, USA), and the percent isoflurane administered via the nose cone was modified to maintain a pulmonary rate of 36–45 breaths per min (bpm) to reduce MRI motion artifacts. The animal body temperature was maintained by a thermostat-controlled heated table in the range of 34–37 °C during preparation, imaging, and post-imaging recovery.

2.11. Ultrasound

Ultrasound imaging was performed weekly after tumor cell injection, as previously described [59]. In brief, the area to be imaged was shaved with hair clippers, and additional hair was removed with depilatory cream (SurgiCream, American International Industries, Los Angeles, CA, USA), followed by 1% acetic acid. Heated gel (Aqua-Gel, Parker Laboratory, Inc., Fairfield, NJ, USA) was applied to match the tissue acoustic characteristics between the animal and the transducer, and the acoustic focus was placed at the center of the adrenal gland. B-mode images (Vevo2100, VisualSonics, Toronto, ON, Canada) were acquired using the 40 MHz transducer (MS-550S, VisualSonics, Toronto, ON, Canada), with an axial and lateral image spatial resolution of 40 and 90 µm, respectively. Three-dimensional volumes were calculated using vendor-supplied software (Vevo Lab v. 1.7.1, VisualSonics, Toronto, ON, Canada), utilizing the maximum axis linear X, Y, and Z measurements and the volume calculation $(0.523 \times (X \times Y \times Z))$.

2.12. Magnetic Resonance Imaging

Magnetic Resonance Imaging (MRI) was performed weekly on a 3.0T clinical scanner (Philips Intera Achieva, Best, The Netherlands), as previously described [60], but with the following modifications. Custom-built volume receive array coils were utilized for simultaneously imaging of three mice to achieve high throughput. After a navigational survey scan with slices in sagittal, coronal, and axial view, a multi-slice T2 weighted turbo spin echo sequence (T2w-TSE) was applied. An 18 mm thick slab in coronal view was arranged to cover the whole mouse body. Image acquisition parameters: field of view, 160×78 mm^2; flip angle, 90°; in-plane resolution, 0.18×0.18 mm^2; slice thickness, 0.5 mm; repetition time (TR), 6000 ms; and echo time, (TE) 45 ms. A fat suppression imaging technique, Spectral Presaturation with Inversion Recovery (SPIR), was used to improve tissue contrast and facilitate the detection of metastases.

2.13. Electrochemiluminescence Assays

Neuroblastoma cells were lysed in Tris buffer with added protease and phosphatase inhibitors (Meso Scale Discovery, Rockville, MD, USA). The resulting lysates were used in the insulin signaling panel (total protein kit, Meso Scale Discovery, K15152C-1). The assay was read on a Meso Sector S600 (Meso Scale Discovery). Lysate from MCF-7 cells without growth factor stimulation (Meso Scale Discovery, Insulin Signaling Panel Whole Cell Lysate Set, C1151-1) was used as a positive control.

2.14. Histological Analysis

Tumor tissue was fixed in 10% neutral buffered formalin (NBF) and paraffin embedded. Paraffin-embedded sections of 5 µm thickness were prepared and stained with hematoxylin and eosin (H&E). The resulting slides were digitized with an Aperio ScanScope XT (Leica, Wetzlar, Germany) at 200× in a single z-plane. The whole-slide images were evaluated and annotated by a board-certified veterinary pathologist (EFE).

3. Results

3.1. Drugs That Specifically Inhibit MEK1/2 and IGF1R Synergistically Inhibit Proliferation of RAS-Mutated Neuroblastoma Cells

The combination of the MEK inhibitor trametinib and the IGF1R monoclonal antibody ganitumab is effective in murine models of RAS-mutated rhabdomyosarcoma [52]. We hypothesized that this combination would be similarly effective in neuroblastoma with hyperactivity of the RAS/MAPK pathway, either through mutation of one of the RAS isoforms or through inactivation of the RAS GTPase activating protein, NF1. To test this hypothesis, we assembled a panel of RAS- and NF1-altered neuroblastoma cell lines (Figure 1A). A majority of these cell lines were derived from neuroblastoma tumors at relapse (SK-N-AS, LAN-6, NB-Eb-C1, SK-N-FI, and SK-N-BE(2)-C), with one cell line derived from a patient with newly diagnosed neuroblastoma (NBL-S) and one cell line for which the clinical information is unknown (CHP-212) [61]. We confirmed the RAS mutational status of the RAS-altered cell lines by whole-exome sequencing (Supplementary Table S2) and loss of NF1 expression at the protein level in NF1-altered cell lines via immunoblot (Figure 1B). We also confirmed that the cells expressed IGF1R at the protein level, using an electrochemiluminescence assay (Figure 1C) and immunoblot (Supplementary Figure S1). These results established that IGF1R is highly expressed in SK-N-AS. In contrast, the expression of IGF1R in SK-N-FI cells is low. The remaining RAS/MAPK pathway-altered neuroblastoma lines, namely NB-Eb-C1, LAN-6, CHP-212, NBL-S, and SK-N-BE(2)-C, expressed IGF1R to a similar extent as the positive control cell line, MCF7. In contrast, all the RAS/MAPK pathway-altered neuroblastoma lines expressed the insulin receptor, which is highly homologous to IGF1R, to a similar extent as MCF7 cells, except CHP-212.

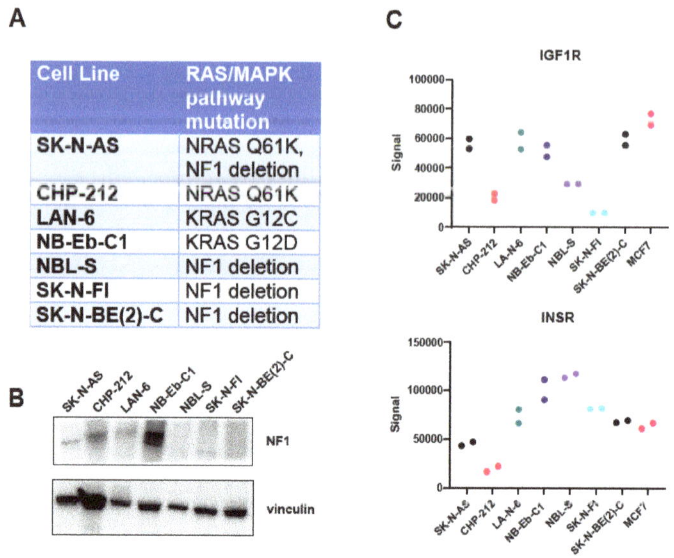

Cell Line	RAS/MAPK pathway mutation
SK-N-AS	NRAS Q61K, NF1 deletion
CHP-212	NRAS Q61K
LAN-6	KRAS G12C
NB-Eb-C1	KRAS G12D
NBL-S	NF1 deletion
SK-N-FI	NF1 deletion
SK-N-BE(2)-C	NF1 deletion

Figure 1. Neuroblastoma cell lines with alterations in the RAS/MAPK pathway express IGF1R. (**A**) RAS/MAPK pathway mutation status of the neuroblastoma cell line panel used in this study. (**B**) NF1 expression in the neuroblastoma cell line panel was determined by immunoblot. Vinculin immunoblot served as a loading control. (**C**) Expression of IGF1R (**top**) or the insulin receptor (InsR, **bottom**) in the panel of neuroblastoma cell lines was determined by an ECL-based sandwich immunoassay (MSD). MCF7 lysate was included as a positive control. Technical duplicates are displayed. Each cell line is represented by different color circles. The original immunoblot images can be found in the Supplementary File S1.

Next, we confirmed that the MEK inhibitor trametinib was potent and efficacious in this panel of RAS/MAPK-altered neuroblastoma cell lines. Trametinib was potent in all the RAS-altered neuroblastoma cell lines, with IC50 values for each cell line in the nanomolar range (Figure 2A, left). However, the efficacy of trametinib was poor in NB-Eb-C1, which showed a maximal effect of only 30% decreased viability. In contrast, trametinib was potent in only two of the three NF1-altered neuroblastoma cell lines (Figure 2B, right). SK-N-BE(2)-C was resistant to trametinib, while SK-N-FI and NBL-S had IC50 values in the nanomolar range.

Despite these differences in the impact of trametinib on the viability in the RAS/MAPK pathway-altered neuroblastoma cell line panel, trametinib was able to decrease ERK phosphorylation in each of the cell lines studied (Figure 2B). Because they are viable despite decreased ERK phosphorylation in the presence of trametinib, SK-N-BE(2)-C cells may not be functionally dependent on the RAS/MAPK pathway for cell viability.

In contrast to previous reports using the IGF1R monoclonal antibody EM164 [37], the anti-proliferative effects of ganitumab as a single agent on neuroblastoma cells were modest (Supplemental Figure S2). Ganitumab was most efficacious in CHP-212 cells, where it showed a maximal effect of 40% decreased viability. Ganitumab was most potent in SK-N-AS and NB-Eb-C1 cells, with IC50 values of 400 nM (Supplementary Figure S2). These three RAS/MAPK-altered neuroblastoma cell lines were selected for further study based on their sensitivity to trametinib and ganitumab as single agents. To determine if the combination of MEK inhibition and IGF1R inhibition would synergistically decrease viability in RAS-mutated neuroblastoma cells, we performed a matrix combination assay in which the viability of SK-N-AS cells was determined in the presence of 10 different concentrations of either trametinib or ganitumab (Figure 3A). Synergy was observed between trametinib and ganitumab in SK-N-AS, according to the Bliss independence model. To assess if there was synergy in additional neuroblastoma cell lines, we tested the ability of a fixed concentration of ganitumab to affect the viability decrease achieved by trametinib alone. In NB-Eb-C1 cells, ganitumab did not alter the efficacy or potency of trametinib (Figure 3B, top); however, in CHP-212 cells, synergy with ganitumab was observed by a shift in the IC50 to the left in the presence of 1000 nM ganitumab (Figure 3B, bottom). We also assessed the impact of the combination of trametinib and ganitumab on apoptosis in RAS-altered neuroblastoma cells using both a luminescence-based assessment of caspase 3/7 activity (Figure 3C) and a flow cytometric assessment of annexin positivity (Figure 3D). Neither trametinib alone, ganitumab alone, nor the combination of trametinib and ganitumab increased caspase 3/7 activity or annexin positivity in SK-N-AS cells. However, trametinib in the presence and absence of ganitumab increased both caspase 3/7 activity and annexin positivity consistent with induction of apoptosis in CHP-212 cells (Figure 3C,D).

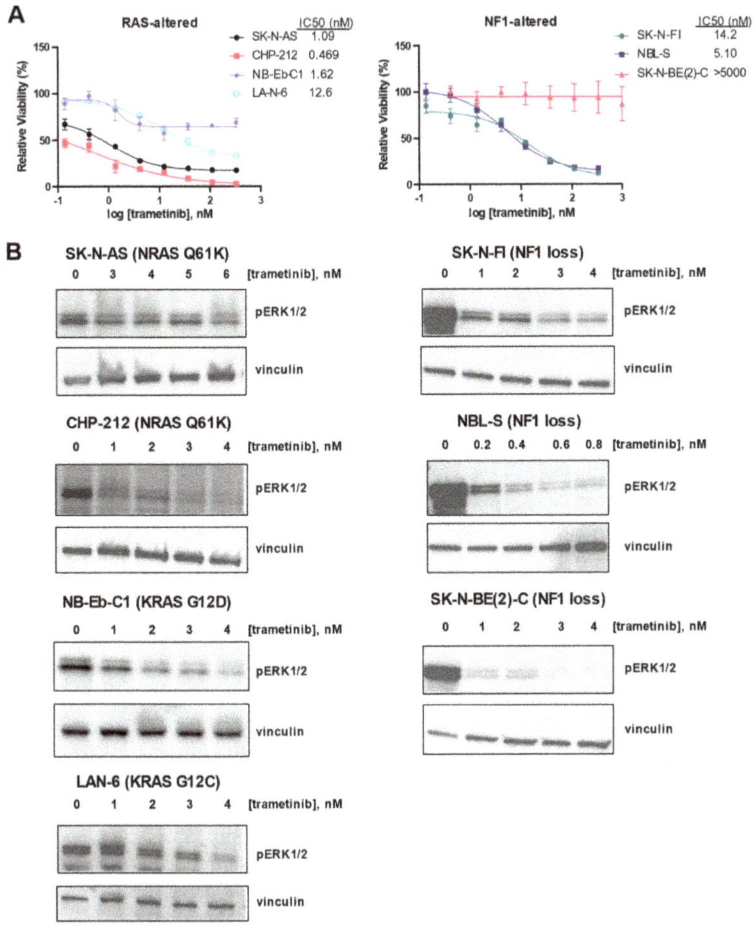

Figure 2. RAS mutation correlates with increased sensitivity to trametinib in neuroblastoma cells. (**A**) Efficacy and potency of trametinib were determined in RAS-altered (**left**) and NF1-altered (**right**) neuroblastoma cells using CellTiter-Glo signal 72 h after treatment as a marker of viability. Means of technical triplicates are displayed. Error bars indicate the standard deviation. (**B**) Erk1/2 phosphorylation after 24 h of treatment with the indicated concentrations of trametinib was determined by immunoblot in 4 RAS-altered neuroblastoma cell lines (**left**) and 3 NF1-altered neuroblastoma cell lines (**right**). The original immunoblot images can be found in the Supplementary File S1.

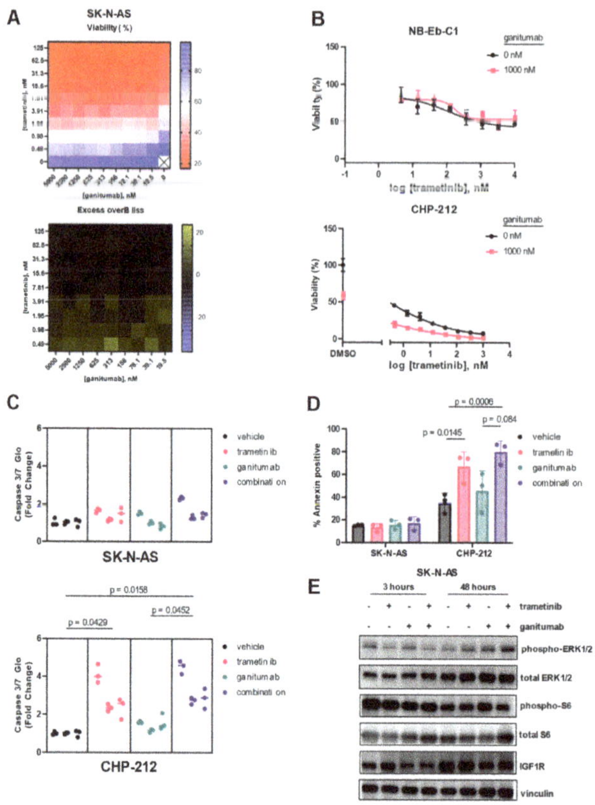

Figure 3. Trametinib and ganitumab synergistically inhibit RAS-mutant neuroblastoma viability. (**A**) Matrix (10 × 10) plot for the combination of trametinib (0 to 125 nM) and ganitumab (0 to 5000 nM) in both viability (CellTiter-Glo, **top**) and Excess-over-Bliss (**bottom**) format for SK-N-AS cells. (**B**) Efficacy of trametinib and the trametinib/ganitumab combination was determined in NB-Eb-C1 (**top**) and CHP-212 (**bottom**) neuroblastoma cells using CellTiter-Glo signal 72 h after treatment as a marker of viability. Cells were treated with varying concentrations of trametinib in the presence or absence of ganitumab (1 µM). Means of technical triplicates are displayed. (**C**) Caspase 3/7 activity of SK-N-AS (**top**) or CHP-212 (**bottom**) cells 18 h after treatment with vehicle, trametinib, ganitumab, or the combination of trametinib and ganitumab. SK-N-AS cells were treated with 10 nM trametinib; CHP-212 cells were treated with 4 nM trametinib. Both cell lines were treated with 1 µM ganitumab. Nested plots of three biological replicates (experimental units) comprising 3 technical replicates are displayed. Individual data points are shown; lines denote the median for that experimental unit. P-values were determined by nested 1-way ANOVA performed on transformed data. (**D**) SK-N-AS (**left**) or CHP-212 (**right**) cells were treated with vehicle, trametinib, ganitumab, or trametinib/ganitumab for 48 h. SK-N-AS cells were treated with 10 nM trametinib; CHP-212 cells were treated with 4 nM trametinib. Both cell lines were treated with 1 µM ganitumab. The treated cells were stained with Annexin V-APC and 7-AAD and analyzed by flow cytometry. The percentage of Annexin-positive cells is defined as the percentage of cells that were Annexin V positive and 7-AAD positive or negative. Individual biological replicates are shown. P-values were determined by 3-way ANOVA. (**E**) SK-N-AS cells were treated with vehicle, 10 nM trametinib, 1000 nM ganitumab, or the combination for 3 h (**left**) or 48 h (**right**), after which cells were harvested and the resulting cells were analyzed for phosphorylated and total forms of ERK, as well as IGF1R. Vinculin blot is included as a loading control. Representative blots are shown. The original immunoblot images can be found in the Supplementary File S1.

We used immunoblot experiments to confirm the on-target activity of trametinib and ganitumab at the doses used in our SK-N-AS apoptosis experiments (Figure 3E). After 3 h of treatment, 10 nM trametinib decreased ERK phosphorylation in SK-N-AS cells in the presence or absence of ganitumab; however, rebound ERK phosphorylation was observed in SK-N-AS cells treated with 10 nM trametinib for 48 h. Ganitumab at a dose of 1000 nM, both alone and in combination with trametinib, decreased total IGF1R expression in SK-N-AS after both 3 h and 48 h of treatment. Notably, the combination of trametinib and ganitumab decreased the phosphorylation of S6 ribosomal protein, a point of convergence of the MAPK and mTOR pathways [62], after 48 h of treatment. These results indicate that trametinib and ganitumab had on target effects in SK-N-AS.

3.2. Combined Trametinib and Ganitumab Treatment Is Efficacious in Murine Xenograft Models of RAS-Mutated Neuroblastoma

We next wanted to evaluate if ganitumab would provide therapeutic enhancement to trametinib monotherapy in xenograft models of RAS-mutated neuroblastoma. In SK-N-AS, consistent with prior reports [5], we observed a tumor growth delay for mice treated with either trametinib or ganitumab as single agents as compared to the mice that received the vehicle. We also observed a tumor growth delay for mice treated with the trametinib/ganitumab combination as compared to either single agent (Figure 4A, left). The tumor volumes on day 15 in mice treated with trametinib and ganitumab were smaller than those from mice treated with the vehicle or trametinib alone (Figure 4B, left). The tumor growth delay was associated with a survival advantage for the mice treated with the combination compared to trametinib alone but not ganitumab alone. Importantly, the trametinib/ganitumab combination was associated with the prolongation of overall survival compared to the vehicle (Figure 4C, left). Similarly, in NB-Eb-C1 xenografts, we observed a tumor growth delay (Figure 4A, right), decreased tumor volume on day 36 (Figure 4B, right), and prolongation of overall survival (Figure 4C, right) for the mice treated with the combination of trametinib and ganitumab compared to the vehicle. The phosphorylation of ERK (Figure 4D, left) and expression of IGF1R (Figure 4D, right) was decreased in tumor lysates from SK-N-AS and NB-Eb-C1 xenografts treated with trametinib and ganitumab, indicating that the in vitro mechanisms of action of each of these drugs was preserved in vivo. In summary, these results suggest that the addition of an IGF1R inhibitor to a MEK inhibitor provided modest therapeutic enhancement in heterotopic RAS-mutated neuroblastoma models.

Adrenal SK-N-AS xenografts are more vascular, more locally invasive, and more metastatic as compared to subcutaneous SK-N-AS xenografts [63]. We hypothesized that the combination of trametinib and ganitumab would be more effective in adrenal than subcutaneous SK-N-AS xenografts because the orthotopic site (adrenal) is a better model of the human tumor microenvironment. To test this hypothesis, we injected unmodified SK-N-AS cells into the right adrenal gland of SCID beige mice and monitored tumor growth via once-weekly ultrasound and MRI imaging. When the adrenal tumors had engrafted as determined by ultrasound, we randomized the mice to receive either the vehicle or the trametinib/ganitumab combination. Similar to the subcutaneous xenografts, trametinib/ganitumab caused a tumor growth delay in adrenal SK-N-AS xenografts (Figure 5A). Upon necropsy for humane endpoints, the xenograft weight was decreased in the trametinib/ganitumab-treated group as compared to the vehicle-treated group (Figure 5B). However, the combination of trametinib and ganitumab was not associated with the prolongation of overall survival compared to the vehicle alone in adrenal xenografts ($p = 0.0946$, log-rank test) (Figure 5C). Surprisingly, we also observed that four of the nine mice (44%) in the trametinib/ganitumab group developed new nodules in the liver or lungs during treatment (Figure 5D). One of these nodules was confirmed by histopathology to be neuroblastoma (Figure 5E), consistent with metastasis. In contrast, only one of the nine mice (11%) in the vehicle group developed new nodules. The presence of metastatic disease could account for the lack of survival advantage conferred by trametinib/ganitumab for mice bearing adrenal tumors.

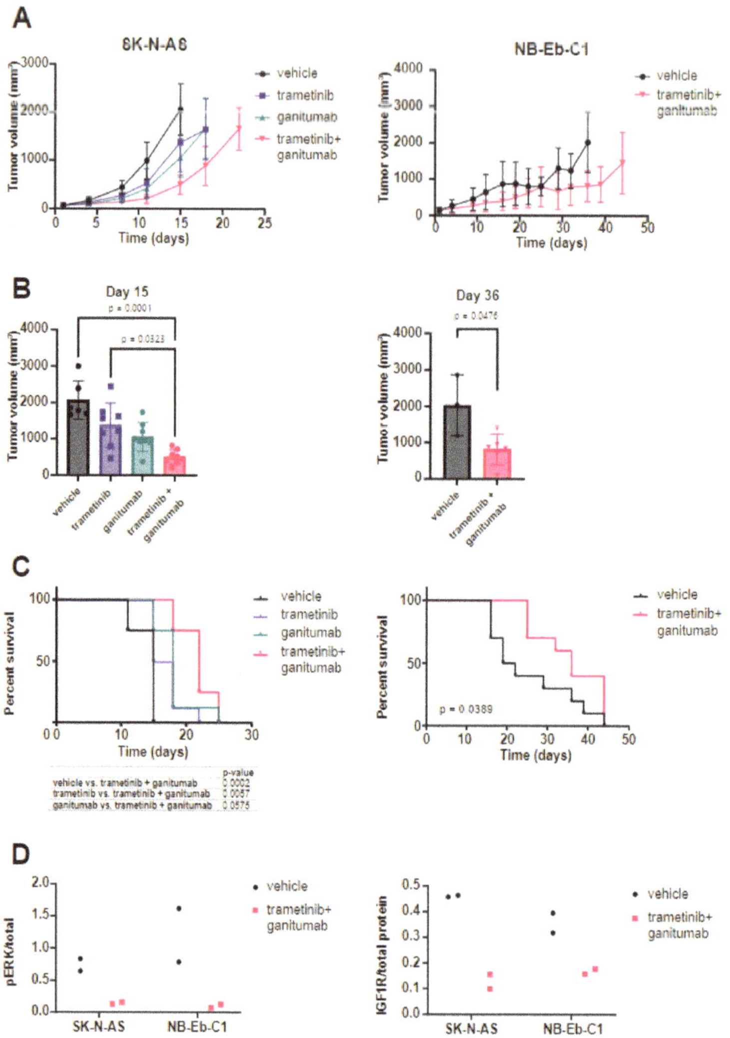

Figure 4. Trametinib and ganitumab delay tumor progression in heterotopic (subcutaneous) neuroblastoma cell line xenograft models. (**A**) SCID beige mice bearing 100–200 mm^3 SK-N-AS (**left**) or NB-Eb-C1 (**right**) tumors were randomized to receive the indicated treatments (8 mice per group). Mean tumor volume ± one standard deviation is plotted. Means are displayed only if there are three or more tumor volume measurements at that time point. (**B**) Tumor volume at the study endpoint for mice bearing SK-N-AS (**left**) or NB-Eb-C1 (**right**) xenografts. P-values were determined by the Kruskal–Wallis test with Dunn's correction for multiple comparisons for SK-N-AS and by the Mann–Whitney test for NB-Eb-C1. (**C**) Overall survival of mice bearing SK-N-AS (**left**) or NB-Eb-C1 (**right**) tumors treated as indicated. *p*-values determined by log-rank test. (**D**) Pharmacodynamic assessment was performed on tumors harvested from mice during treatment with vehicle or trametinib/ganitumab (4 h after the dose of vehicle or trametinib on the 5th day of treatment). The ratio of phosphorylated to total ERK and total IGF1R to total protein was determined by capillary immunoassay.

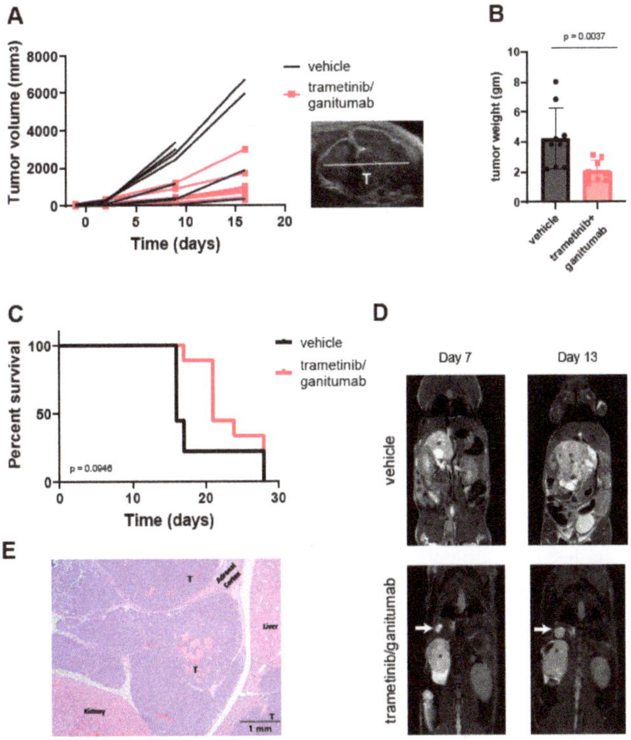

Figure 5. Trametinib and ganitumab delay tumor progression but do not prevent metastasis in an orthotopic (adrenal) SK-N-AS xenograft model. (**A**) Line graphs of tumor volume as a function of treatment duration for mice bearing adrenal SK-N-AS xenografts receiving vehicle or trametinib/ganitumab. Each line represents data from a single mouse (n = 9 per group). Tumor volume was estimated by ultrasound. A representative ultrasound image is shown (inset). "T" indicates the tumor. (**B**) Weight of xenografted tumors at study endpoint, as determined at necropsy. Bars indicate the mean; error bars indicate the standard deviation. p-value determined by Mann–Whitney test. (**C**) Overall survival of mice bearing SK-N-AS adrenal xenografts treated with either vehicle or trametinib/ganitumab. p-value determined by log-rank test. (**D**) Serial MRIs of SCID beige mice with SK-N-AS adrenal xenografts (indicated by an asterisk) receiving vehicle (**top**) or trametinib/ganitumab (**bottom**). White arrows indicate metastatic tumors. (**E**) Hematoxylin and eosin staining of the xenograft from the mouse treated with trametinib/ganitumab shown in panel (**D**). "T" denotes small, round blue tumor cells, consistent with neuroblastoma effacing the adrenal medulla and infiltrating into the adjacent peritoneal space. The upper-right liver lobe contains multifocal nodules of neoplastic cells within the hepatic parenchyma, one of which is also labeled with a "T".

4. Discussion

In this study, we showed that the MEK inhibitor trametinib and the IGF1R inhibitor ganitumab synergistically inhibited RAS-mutated neuroblastoma cell proliferation and induced apoptosis in cell culture models. The response to ganitumab was not associated with the expression level of IGF1R in these cells. The trametinib/ganitumab combination delayed tumor growth in both of the heterotopic cell line xenograft models tested and delayed tumor growth in an orthotopic cell line xenograft model of neuroblastoma. Interestingly, the trametinib/ganitumab combination did not prevent neuroblastoma metastasis in the orthotopic xenograft model.

Several types of tumor cell resistance to molecularly targeted agents have been identified: intrinsic, adaptive, and acquired [64]. Intrinsic resistance refers to resistance that is present within tumor cells before drug exposure. Adaptive resistance refers to the changes occurring within tumor cells after short-term exposure to a targeted agent that can compensate for its action. In contrast, acquired resistance occurs in tumors that initially respond to the targeted agent and then regrow [64]. Intrinsic, adaptive, and acquired resistance to MEK inhibition in RAS- or NF1-altered cells are all caused by cellular changes that either result in hyperactivation of the MAPK pathway or activation of alternate signaling pathways, such as the IGF1R/PI3K/AKT pathway [65]. Both RAS mutations and NF1 deletion confer sensitivity to MEK inhibition [66]. However, here we confirmed that SK-N-BE(2)-C cells display intrinsic resistance to MEK inhibition despite demonstrating loss of NF1 expression, a result that has been shown previously [5]. Intrinsic resistance to MAPK pathway inhibitor in cell lines with loss of NF1 expression has also been seen in melanoma [67] and glioblastoma [68], suggesting that there are additional biomarkers of MEK inhibitor sensitivity that require elucidation.

One approach toward the prevention of adaptive resistance is combining MEK inhibitors with molecularly targeted agents that either augment the inhibition of the RAS/MAPK pathway achieved by the MEK inhibitor (vertical pathway inhibition) or block signaling through the compensatory signaling pathways (horizontal pathway inhibition) [66]. In this study, we attempted to accomplish both vertical and horizontal pathway inhibition by combining a MEK inhibitor with an inhibitor of a receptor tyrosine kinase (RTK), IGF1R, which signals through both the RAS/MAPK pathway and a potential compensatory signaling pathway, the PI3K/AKT pathway. IGF1R is a functional dependency in neuroblastoma cells [69]. However, in our neuroblastoma xenografts, the trametinib/ganitumab combination delayed but did not prevent tumor growth, indicating the development of adaptive resistance. This resistance could be due to the activation of an additional RTK. Several RTKs, including ALK, KIT, MET, NTRK2, and RET, play a major role in neuroblastoma pathogenesis [70–72], and their activation could be responsible for adaptive resistance to trametinib and ganitumab in neuroblastoma. Notably, strategies to inhibit RAS activation downstream of RTKs, such as with an SHP2 inhibitor, as has been previously reported [19], would not completely prevent the activation of compensatory signaling pathways such as the PI3K/AKT [73] or JAK/STAT [74] pathways, which can be activated in a RAS-independent manner.

We also evaluated the impact of trametinib/ganitumab on tumor invasion in an orthotopic SK-N-AS xenograft. We did not assess for tumor invasion and metastasis in the heterotopic xenograft models. Previous studies have shown that orthotopic SK-N-AS xenograft tumors directly extend into the ipsilateral lobes of the liver [63]. We observed this direct extension, as well as the evidence of new liver and/or lung nodules, consistent with the development of distant metastases. These nodules developed during treatment with trametinib and ganitumab even though the primary tumors were smaller in mice receiving treatment. In fact, a higher percentage of mice receiving trametinib and ganitumab experienced nodule formation as compared to mice receiving vehicle. Our study was not designed to determine the individual effects of MEK inhibition and IGF1R inhibition on neuroblastoma tumor dissemination. We have not assessed the impact of trametinib and ganitumab on the dissemination of additional orthotopic xenograft models and cannot rule out the possibility that our observations are specific to SK-N-AS. Previous studies have shown that the genetic knockdown of IGF1R expression using shRNA decreased the transwell invasion of neuroblastoma cells such as IMR32 and SH-SY5Y [34]; however, the effect of IGF1R inhibition on RAS- or NF1-altered neuroblastoma cell invasion is unknown. Interestingly, in some melanoma cell lines, treatment with the MEK inhibitor selumetinib (AZD6244) increased invasion in a spheroid assay [75], but the effects of MEK inhibition on neuroblastoma cell invasion has yet to be determined. Future studies will define the effects of MEK or IGF1R inhibition on the invasive and metastatic capacity of RAS- and NF1-mutated neuroblastoma models.

5. Conclusions

In this study, we assembled a panel of RAS- and NF1-altered neuroblastoma cell lines with variable IGF1R expression to evaluate the efficacy of the combination of a MEK inhibitor (trametinib) and an IGF1R inhibitor (ganitumab) in neuroblastoma. We found that some neuroblastoma cell lines were resistant to single-agent MEK inhibition despite the presence of a RAS or NF1 alteration. In addition, we showed that trametinib and ganitumab inhibited proliferation in RAS-mutated neuroblastoma cells, even cells with a low expression of IGF1R. We further showed that the trametinib/ganitumab combination inhibited the growth of heterotopic and orthotopic neuroblastoma xenografts. However, the trametinib/ganitumab combination appeared to increase metastasis of a RAS-mutated neuroblastoma orthotopic xenograft, which is concerning and merits further investigation. This study highlights the importance of using orthotopic xenograft models to investigate the effects of targeted agents on metastasis preclinically.

Supplementary Materials: The following are available online at https://www.mdpi.com/article/10.3390/cancers16132320/s1, Figure S1: Expression of IGF1R in the panel of RAS/MAPK-altered neuroblastoma cell lines as determined by immunoblot; Figure S2: Efficacy and potency of ganitumab was determined in RAS/MAPK-altered neuroblastoma cells using CellTiter-Glo signal 72 h after treatment as a marker of viability. Table S1: STR genotyping of neuroblastoma cell lines used in this study; Table S2: Gene sequence variants of neuroblastoma cell lines used in this study; File S1: Original immunoblot figures.

Author Contributions: Conceptualization, M.E.Y. and C.J.T.; methodology, M.D.H. and J.D.K.; software, H.L.; validation, S.S., K.E.H. and M.E.Y.; formal analysis, M.E.Y., M.D.H., J.S.R., H.L., J.-Q.C., E.F.E. and N.K.; investigation, S.S., J.S.R., E.R.H., J.T.K., A.J., K.A.I., L.A.R., L.I. and X.L.; resources, M.E.Y.; data curation, J.D.K., H.L., J.F.S., K.A.I. and S.D.; writing—original draft preparation, M.E.Y.; writing—review and editing, all; visualization, M.E.Y., S.S., J.S.R., J.D.K., L.A.R., E.F.E. and N.E.S.; supervision, M.E.Y., J.F.S., J.D.K., M.D.H., S.D., C.J.T., N.K. and J.-Q.C.; project administration, M.E.Y., J.D.K., M.D.H., S.D., C.J.T., N.K. and J.F.S.; funding acquisition, M.E.Y. All authors have read and agreed to the published version of the manuscript.

Funding: This project was funded in part with federal funds from the National Cancer Institute, National Institutes of Health, under Contract No. HHSN261201500003I. The content of this publication does not necessarily reflect the views or policies of the Department of Health and Human Services, nor does the mention of trade names, commercial products, or organizations imply endorsement by the U.S. Government. This project was supported by intramural funding given to MEY (ZIA BC011892).

Institutional Review Board Statement: Xenograft studies were approved by the Animal Care and Use Committee (ACUC) of the Frederick National Laboratory for Cancer Research (FNLCR). FNLCR is accredited by AAALAC International and follows the Public Health Service Policy for the Care and Use of Laboratory Animals. Animal care was provided in accordance with the procedures outlined in the Guide for the Care and Use of Laboratory Animals (ASP # 21-449 to M.E.Y.).

Informed Consent Statement: Not applicable.

Data Availability Statement: The data generated in this study are available within the manuscript and its Supplementary Materials. Sequencing data are available through dbGaP, accession number PRJNA1083467 (https://www.ncbi.nlm.nih.gov/sra/PRJNA1083467, accessed on 25 April 2024).

Acknowledgments: The authors are grateful to Xiaolin Wan, Norris Lam, Zhihui Lui, Rosa Nguyen, Christine Heske, Subhra Dash, and Lucas Stauffer for helpful discussions and critical reading of the manuscript.

Conflicts of Interest: The authors declare no conflict of interest.

References

1. Maris, J.M. Recent advances in neuroblastoma. *N. Engl. J. Med.* **2010**, *362*, 2202–2211. [CrossRef] [PubMed]
2. Pugh, T.J.; Morozova, O.; Attiyeh, E.F.; Asgharzadeh, S.; Wei, J.S.; Auclair, D.; Carter, S.L.; Cibulskis, K.; Hanna, M.; Kiezun, A.; et al. The genetic landscape of high-risk neuroblastoma. *Nat. Genet.* **2013**, *45*, 279–284. [CrossRef] [PubMed]

3. Ackermann, S.; Cartolano, M.; Hero, B.; Welte, A.; Kahlert, Y.; Roderwieser, A.; Bartenhagen, C.; Walter, E.; Gecht, J.; Kerschke, L.; et al. A mechanistic classification of clinical phenotypes in neuroblastoma. *Science* **2018**, *362*, 1165–1170. [CrossRef]
4. Brady, S.W.; Liu, Y.; Ma, X.; Gout, A.M.; Hagiwara, K.; Zhou, X.; Wang, J.; Macias, M.; Chen, X.; Easton, J.; et al. Pan-neuroblastoma analysis reveals age- and signature-associated driver alterations. *Nat. Commun.* **2020**, *11*, 5183. [CrossRef] [PubMed]
5. Eleveld, T.F.; Oldridge, D.A.; Bernard, V.; Koster, J.; Colmet Daage, L.; Diskin, S.J.; Schild, L.; Bentahar, N.B.; Bellini, A.; Chicard, M.; et al. Relapsed neuroblastomas show frequent RAS-MAPK pathway mutations. *Nat. Genet.* **2015**, *47*, 864–871. [CrossRef] [PubMed]
6. Vaseva, A.V.; Yohe, M.E. Targeting RAS in pediatric cancer: Is it becoming a reality? *Curr. Opin. Pediatr.* **2020**, *32*, 48–56. [CrossRef] [PubMed]
7. Singh, A.; Ruan, Y.; Tippett, T.; Narendran, A. Targeted inhibition of MEK1 by cobimetinib leads to differentiation and apoptosis in neuroblastoma cells. *J. Exp. Clin. Cancer Res.* **2015**, *34*, 104. [CrossRef]
8. Tanaka, T.; Higashi, M.; Kimura, K.; Wakao, J.; Fumino, S.; Iehara, T.; Hosoi, H.; Sakai, T.; Tajiri, T. MEK inhibitors as a novel therapy for neuroblastoma: Their in vitro effects and predicting their efficacy. *J. Pediatr. Surg.* **2016**, *51*, 2074–2079. [CrossRef] [PubMed]
9. Umapathy, G.; Guan, J.; Gustafsson, D.E.; Javanmardi, N.; Cervantes-Madrid, D.; Djos, A.; Martinsson, T.; Palmer, R.H.; Hallberg, B. MEK inhibitor trametinib does not prevent the growth of anaplastic lymphoma kinase (ALK)-addicted neuroblastomas. *Sc[illegible] Signal* **2017**, *10*, eaam7550. [CrossRef]
10. Woodfield, S.E.; Zhang, L.; Scorsone, K.A.; Liu, Y.; Zage, P.E. Binimetinib inhibits MEK and is effective against neuroblastoma tumor cells with low NF1 expression. *BMC Cancer* **2016**, *16*, 172. [CrossRef]
11. Hart, L.S.; Rader, J.; Raman, P.; Batra, V.; Russell, M.R.; Tsang, M.; Gagliardi, M.; Chen, L.; Martinez, D.; Li, Y.; et al. Preclinical Therapeutic Synergy of MEK1/2 and CDK4/6 Inhibition in Neuroblastoma. *Clin. Cancer Res.* **2017**, *23*, 1785–1796. [CrossRef] [PubMed]
12. Kim, K.W.; Kim, J.Y.; Qiao, J.; Clark, R.A.; Powers, C.M.; Correa, H.; Chung, D.H. Dual-Targeting AKT2 and ERK in cancer stem-like cells in neuroblastoma. *Oncotarget* **2019**, *10*, 5645–5659. [CrossRef] [PubMed]
13. Xu, D.Q.; Toyoda, H.; Qi, L.; Morimoto, M.; Hanaki, R.; Iwamoto, S.; Komada, Y.; Hirayama, M. Induction of MEK/ERK activity by AZD8055 confers acquired resistance in neuroblastoma. *Biochem. Biophys. Res. Commun.* **2018**, *499*, 425–432. [CrossRef] [PubMed]
14. Shimizu, T.; Tolcher, A.W.; Papadopoulos, K.P.; Beeram, M.; Rasco, D.W.; Smith, L.S.; Gunn, S.; Smetzer, L.; Mays, T.A.; Kaiser, B.; et al. The clinical effect of the dual-targeting strategy involving PI3K/AKT/mTOR and RAS/MEK/ERK pathways in patients with advanced cancer. *Clin. Cancer Res.* **2012**, *18*, 2316–2325. [CrossRef]
15. Duffy, D.J.; Krstic, A.; Halasz, M.; Schwarzl, T.; Fey, D.; Iljin, K.; Mehta, J.P.; Killick, K.; Whilde, J.; Turriziani, B.; et al. Integrative omics reveals MYCN as a global suppressor of cellular signalling and enables network-based therapeutic target discovery in neuroblastoma. *Oncotarget* **2015**, *6*, 43182–43201. [CrossRef] [PubMed]
16. Holzel, M.; Huang, S.; Koster, J.; Ora, I.; Lakeman, A.; Caron, H.; Nijkamp, W.; Xie, J.; Callens, T.; Asgharzadeh, S.; et al. NF1 is a tumor suppressor in neuroblastoma that determines retinoic acid response and disease outcome. *Cell* **2010**, *142*, 218–229. [CrossRef] [PubMed]
17. Mayes, P.A.; Degenhardt, Y.Y.; Wood, A.; Toporovskya, Y.; Diskin, S.J.; Haglund, E.; Moy, C.; Wooster, R.; Maris, J.M. Mitogen activated protein kinase (MEK/ERK) inhibition sensitizes cancer cells to centromere-associated protein E inhibition. *Int. J. Cancer* **2013**, *132*, E149–E157. [CrossRef] [PubMed]
18. Coggins, G.E.; Farrel, A.; Rathi, K.S.; Hayes, C.M.; Scolaro, L.; Rokita, J.L.; Maris, J.M. YAP1 Mediates Resistance to MEK1/2 Inhibition in Neuroblastomas with Hyperactivated RAS Signaling. *Cancer Res.* **2019**, *79*, 6204–6214. [CrossRef] [PubMed]
19. Valencia-Sama, I.; Ladumor, Y.; Kee, L.; Adderley, T.; Christopher, G.; Robinson, C.M.; Kano, Y.; Ohh, M.; Irwin, M.S. NRAS Status Determines Sensitivity to SHP2 Inhibitor Combination Therapies Targeting the RAS-MAPK Pathway in Neuroblastoma. *Cancer Res.* **2020**, *80*, 3413–3423. [CrossRef]
20. Malone, C.F.; Kim, M.; Alexe, G.; Engel, K.; Forman, A.B.; Robichaud, A.; Conway, A.S.; Goodale, A.; Meyer, A.; Khalid, D.; et a[illegible] Transcriptional Antagonism by CDK8 Inhibition Improves Therapeutic Efficacy of MEK Inhibitors. *Cancer Res.* **2023**, *83*, 285–300. [CrossRef]
21. Healy, J.R.; Hart, L.S.; Shazad, A.L.; Gagliardi, M.E.; Tsang, M.; Elias, J.; Ruden, J.; Farrel, A.; Rokita, J.L.; Li, Y.; et al. Limited antitumor activity of combined BET and MEK inhibition in neuroblastoma. *Pediatr. Blood Cancer* **2020**, *67*, e28267. [CrossRef]
22. El-Badry, O.M.; Helman, L.J.; Chatten, J.; Steinberg, S.M.; Evans, A.E.; Israel, M.A. Insulin-like growth factor II-mediated proliferation of human neuroblastoma. *J. Clin. Investig.* **1991**, *87*, 648–657. [CrossRef]
23. El-Badry, O.M.; Romanus, J.A.; Helman, L.J.; Cooper, M.J.; Rechler, M.M.; Israel, M.A. Autonomous growth of a human neuroblastoma cell line is mediated by insulin-like growth factor II. *J. Clin. Investig.* **1989**, *84*, 829–839. [CrossRef]
24. Leventhal, P.S.; Randolph, A.E.; Vesbit, T.E.; Schenone, A.; Windebank, A.; Feldman, E.L. Insulin-like growth factor-II as a paracrine growth factor in human neuroblastoma cells. *Exp. Cell Res.* **1995**, *221*, 179–186. [CrossRef]
25. Singleton, J.R.; Randolph, A.E.; Feldman, E.L. Insulin-like growth factor I receptor prevents apoptosis and enhances neuroblastoma tumorigenesis. *Cancer Res.* **1996**, *56*, 4522–4529.
26. Kim, B.; van Golen, C.M.; Feldman, E.L. Insulin-like growth factor-I signaling in human neuroblastoma cells. *Oncogene* **2004**, *2[illegible]*, 130–141. [CrossRef]

27. Meyer, G.E.; Shelden, E.; Kim, B.; Feldman, E.L. Insulin-like growth factor I stimulates motility in human neuroblastoma cells. *Oncogene* **2001**, *20*, 7542–7550. [CrossRef]
28. Misawa, A.; Hosoi, H.; Arimoto, A.; Shikata, T.; Akioka, S.; Matsumura, T.; Houghton, P.J.; Sawada, T. N-Myc induction stimulated by insulin-like growth factor I through mitogen-activated protein kinase signaling pathway in human neuroblastoma cells. *Cancer Res.* **2000**, *60*, 64–69.
29. van Golen, C.M.; Castle, V.P.; Feldman, E.L. IGF-I receptor activation and BCL-2 overexpression prevent early apoptotic events in human neuroblastoma. *Cell Death Differ.* **2000**, *7*, 654–665. [CrossRef]
30. van Golen, C.M.; Schwab, T.S.; Kim, B.; Soules, M.E.; Su Oh, S.; Fung, K.; van Golen, K.L.; Feldman, E.L. Insulin-like growth factor-I receptor expression regulates neuroblastoma metastasis to bone. *Cancer Res.* **2006**, *66*, 6570–6578. [CrossRef]
31. MacFarland, S.P.; Naraparaju, K.; Iyer, R.; Guan, P.; Kolla, V.; Hu, Y.; Tan, K.; Brodeur, G.M. Mechanisms of Entrectinib Resistance in a Neuroblastoma Xenograft Model. *Mol. Cancer Ther.* **2020**, *19*, 920–926. [CrossRef] [PubMed]
32. Matsumoto, K.; Gaetano, C.; Daughaday, W.H.; Thiele, C.J. Retinoic acid regulates insulin-like growth factor II expression in a neuroblastoma cell line. *Endocrinology* **1992**, *130*, 3669–3676. [CrossRef]
33. Matsumoto, K.; Lucarelli, E.; Minniti, C.; Gaetano, C.; Thiele, C.J. Signals transduced via insulin-like growth factor I receptor (IGF(R)) mediate resistance to retinoic acid-induced cell growth arrest in a human neuroblastoma cell line. *Cell Death Differ.* **1994**, *1*, 49–58. [PubMed]
34. Wang, X.H.; Wu, H.Y.; Gao, J.; Wang, X.H.; Gao, T.H.; Zhang, S.F. IGF1R facilitates epithelial-mesenchymal transition and cancer stem cell properties in neuroblastoma via the STAT3/AKT axis. *Cancer Manag. Res.* **2019**, *11*, 5459–5472. [CrossRef]
35. Coulter, D.W.; Wilkie, M.B.; Moats-Staats, B.M. Inhibition of IGF-I receptor signaling in combination with rapamycin or temsirolimus increases MYC-N phosphorylation. *Anticancer. Res.* **2009**, *29*, 1943–1949.
36. DeNardo, B.D.; Holloway, M.P.; Ji, Q.; Nguyen, K.T.; Cheng, Y.; Valentine, M.B.; Salomon, A.; Altura, R.A. Quantitative phosphoproteomic analysis identifies activation of the RET and IGF-1R/IR signaling pathways in neuroblastoma. *PLoS ONE* **2013**, *8*, e82513. [CrossRef]
37. Geoerger, B.; Brasme, J.F.; Daudigeos-Dubus, E.; Opolon, P.; Venot, C.; Debussche, L.; Vrignaud, P.; Vassal, G. Anti-insulin-like growth factor 1 receptor antibody EM164 (murine AVE1642) exhibits anti-tumour activity alone and in combination with temozolomide against neuroblastoma. *Eur. J. Cancer* **2010**, *46*, 3251–3262. [CrossRef]
38. Guerreiro, A.S.; Boller, D.; Shalaby, T.; Grotzer, M.A.; Arcaro, A. Protein kinase B modulates the sensitivity of human neuroblastoma cells to insulin-like growth factor receptor inhibition. *Int. J. Cancer* **2006**, *119*, 2527–2538. [CrossRef]
39. Houghton, P.J.; Morton, C.L.; Gorlick, R.; Kolb, E.A.; Keir, S.T.; Reynolds, C.P.; Kang, M.H.; Maris, J.M.; Wu, J.; Smith, M.A. Initial testing of a monoclonal antibody (IMC-A12) against IGF-1R by the Pediatric Preclinical Testing Program. *Pediatr. Blood Cancer* **2010**, *54*, 921–926. [CrossRef] [PubMed]
40. Kolb, E.A.; Gorlick, R.; Lock, R.; Carol, H.; Morton, C.L.; Keir, S.T.; Reynolds, C.P.; Kang, M.H.; Maris, J.M.; Billups, C.; et al. Initial testing (stage 1) of the IGF-1 receptor inhibitor BMS-754807 by the pediatric preclinical testing program. *Pediatr. Blood Cancer* **2011**, *56*, 595–603. [CrossRef]
41. Pappano, W.N.; Jung, P.M.; Meulbroek, J.A.; Wang, Y.C.; Hubbard, R.D.; Zhang, Q.; Grudzien, M.M.; Soni, N.B.; Johnson, E.F.; Sheppard, G.S.; et al. Reversal of oncogene transformation and suppression of tumor growth by the novel IGF1R kinase inhibitor A-928605. *BMC Cancer* **2009**, *9*, 314. [CrossRef]
42. Singh, A.; Meier-Stephenson, V.; Jayanthan, A.; Narendran, A. In Vitro Sensitivity Profiling of Neuroblastoma Cells Against A Comprehensive Small Molecule Kinase Inhibitor Library to Identify Agents for Future Therapeutic Studies. *Curr. Cancer Drug Targets* **2017**, *17*, 569–584. [CrossRef]
43. Tanno, B.; Mancini, C.; Vitali, R.; Mancuso, M.; McDowell, H.P.; Dominici, C.; Raschella, G. Down-regulation of insulin-like growth factor I receptor activity by NVP-AEW541 has an antitumor effect on neuroblastoma cells in vitro and in vivo. *Clin. Cancer Res.* **2006**, *12*, 6772–6780. [CrossRef]
44. Wojtalla, A.; Salm, F.; Christiansen, D.G.; Cremona, T.; Cwiek, P.; Shalaby, T.; Gross, N.; Grotzer, M.A.; Arcaro, A. Novel agents targeting the IGF-1R/PI3K pathway impair cell proliferation and survival in subsets of medulloblastoma and neuroblastoma. *PLoS ONE* **2012**, *7*, e47109. [CrossRef]
45. Zhao, Q.; Tran, H.; Dimitrov, D.S.; Cheung, N.K. A dual-specific anti-IGF-1/IGF-2 human monoclonal antibody alone and in combination with temsirolimus for therapy of neuroblastoma. *Int. J. Cancer* **2015**, *137*, 2243–2252. [CrossRef] [PubMed]
46. Boller, D.; Schramm, A.; Doepfner, K.T.; Shalaby, T.; von Bueren, A.O.; Eggert, A.; Grotzer, M.A.; Arcaro, A. Targeting the phosphoinositide 3-kinase isoform p110delta impairs growth and survival in neuroblastoma cells. *Clin. Cancer Res.* **2008**, *14*, 1172–1181. [CrossRef]
47. Chanthery, Y.H.; Gustafson, W.C.; Itsara, M.; Persson, A.; Hackett, C.S.; Grimmer, M.; Charron, E.; Yakovenko, S.; Kim, G.; Matthay, K.K.; et al. Paracrine signaling through MYCN enhances tumor-vascular interactions in neuroblastoma. *Sci. Transl. Med.* **2012**, *4*, 115ra113. [CrossRef]
48. Chesler, L.; Schlieve, C.; Goldenberg, D.D.; Kenney, A.; Kim, G.; McMillan, A.; Matthay, K.K.; Rowitch, D.; Weiss, W.A. Inhibition of phosphatidylinositol 3-kinase destabilizes Mycn protein and blocks malignant progression in neuroblastoma. *Cancer Res.* **2006**, *66*, 8139–8146. [CrossRef]

49. Johnsen, J.I.; Segerstrom, L.; Orrego, A.; Elfman, L.; Henriksson, M.; Kagedal, B.; Eksborg, S.; Sveinbjornsson, B.; Kogner, P. Inhibitors of mammalian target of rapamycin downregulate MYCN protein expression and inhibit neuroblastoma growth in vitro and in vivo. *Oncogene* **2008**, *27*, 2910–2922. [CrossRef] [PubMed]
50. Mohlin, S.; Hamidian, A.; von Stedingk, K.; Bridges, E.; Wigerup, C.; Bexell, D.; Pahlman, S. PI3K-mTORC2 but not PI3K-mTORC1 regulates transcription of HIF2A/EPAS1 and vascularization in neuroblastoma. *Cancer Res.* **2015**, *75*, 4617–4628. [CrossRef] [PubMed]
51. Vaughan, L.; Clarke, P.A.; Barker, K.; Chanthery, Y.; Gustafson, C.W.; Tucker, E.; Renshaw, J.; Raynaud, F.; Li, X.; Burke, R.; et al. Inhibition of mTOR-kinase destabilizes MYCN and is a potential therapy for MYCN-dependent tumors. *Oncotarget* **2016**, *7*, 57525–57544. [CrossRef] [PubMed]
52. Hebron, K.E.; Wan, X.; Roth, J.S.; Liewehr, D.J.; Sealover, N.E.; Frye, W.J.E.; Kim, A.; Stauffer, S.; Perkins, O.L.; Sun, W.; et al. The Combination of Trametinib and Ganitumab is Effective in RAS-Mutated PAX-Fusion Negative Rhabdomyosarcoma Models. *Clin. Cancer Res.* **2023**, *29*, 472–487. [CrossRef]
53. Molina-Arcas, M.; Hancock, D.C.; Sheridan, C.; Kumar, M.S.; Downward, J. Coordinate direct input of both KRAS and IGF1 receptor to activation of PI3 kinase in KRAS-mutant lung cancer. *Cancer Discov.* **2013**, *3*, 548–563. [CrossRef] [PubMed]
54. Weisberg, E.; Nonami, A.; Chen, Z.; Nelson, E.; Chen, Y.; Liu, F.; Cho, H.; Zhang, J.; Sattler, M.; Mitsiades, C.; et al. Upregulation of IGF1R by mutant RAS in leukemia and potentiation of RAS signaling inhibitors by small-molecule inhibition of IGF1R. *Clin. Cancer Res.* **2014**, *20*, 5483–5495. [CrossRef]
55. Wilky, B.A.; Rudek, M.A.; Ahmed, S.; Laheru, D.A.; Cosgrove, D.; Donehower, R.C.; Nelkin, B.; Ball, D.; Doyle, L.A.; Chen, H.; et al. A phase I trial of vertical inhibition of IGF signalling using cixutumumab, an anti-IGF-1R antibody, and selumetinib, an MEK 1/2 inhibitor, in advanced solid tumours. *Br. J. Cancer* **2015**, *112*, 24–31. [CrossRef]
56. Yohe, M.E.; Gryder, B.E.; Shern, J.F.; Song, Y.K.; Chou, H.C.; Sindiri, S.; Mendoza, A.; Patidar, R.; Zhang, X.; Guha, R.; et al. MEK inhibition induces MYOG and remodels super-enhancers in RAS-driven rhabdomyosarcoma. *Sci. Transl. Med.* **2018**, *10*, eaan4470. [CrossRef]
57. Gilmartin, A.G.; Bleam, M.R.; Groy, A.; Moss, K.G.; Minthorn, E.A.; Kulkarni, S.G.; Rominger, C.M.; Erskine, S.; Fisher, K.E.; Yang, J.; et al. GSK1120212 (JTP-74057) is an inhibitor of MEK activity and activation with favorable pharmacokinetic properties for sustained in vivo pathway inhibition. *Clin. Cancer Res.* **2011**, *17*, 989–1000. [CrossRef]
58. Beltran, P.J.; Calzone, F.J.; Mitchell, P.; Chung, Y.A.; Cajulis, E.; Moody, G.; Belmontes, B.; Li, C.M.; Vonderfecht, S.; Velculescu, V.E.; et al. Ganitumab (AMG 479) inhibits IGF-II-dependent ovarian cancer growth and potentiates platinum-based chemotherapy. *Clin. Cancer Res.* **2014**, *20*, 2947–2958. [CrossRef]
59. Tatum, J.L.; Kalen, J.D.; Jacobs, P.M.; Ileva, L.V.; Riffle, L.A.; Hollingshead, M.G.; Doroshow, J.H. A spontaneously metastatic model of bladder cancer: Imaging characterization. *J. Transl. Med.* **2019**, *17*, 425. [CrossRef] [PubMed]
60. Kalen, J.D.; Ileva, L.V.; Riffle, L.A.; Keita, S.; Tatum, J.L.; Jacobs, P.M.; Sanders, C.; James, A.; Difilippantonio, S.; Thang, L.; et al. Serial Non-Contrast Non-Gated T2w MRI Datasets of Patient Derived Xenograft Cancer Models for Development of Tissue Characterization Algorithms (PDMR-Texture Analysis) (Version 1) [Data Set]. The Cancer Imaging Archive. 2023. Available online: https://www.cancerimagingarchive.net/collection/pdmr-texture-analysis/ (accessed on 25 April 2024).
61. Thiele, C.J. Neuroblastoma. In *Human Cell Culture: Cancer Cell Lines Part 1*; Masters, J.R.W., Palsson, B., Eds.; Springer: Dordrecht, The Netherlands, 1999; pp. 21–53.
62. Gao, M.-Z.; Wang, H.-B.; Chen, X.-L.; Cao, W.-T.; Fu, L.; Li, Y.; Quan, H.-T.; Xie, C.-Y.; Lou, L.-G. Aberrant modulation of ribosomal protein S6 phosphorylation confers acquired resistance to MAPK pathway inhibitors in BRAF-mutant melanoma. *Acta Pharmacol. Sin.* **2019**, *40*, 268–278. [CrossRef]
63. Khanna, C.; Jaboin, J.J.; Drakos, E.; Tsokos, M.; Thiele, C.J. Biologically relevant orthotopic neuroblastoma xenograft models: Primary adrenal tumor growth and spontaneous distant metastasis. *In Vivo* **2002**, *16*, 77–85. [PubMed]
64. Kugel III, C.H.; Aplin, A.E. Adaptive resistance to RAF inhibitors in melanoma. *Pigment Cell Melanoma Res.* **2014**, *27*, 1032–1038. [CrossRef]
65. Kun, E.; Tsang, Y.T.M.; Ng, C.W.; Gershenson, D.M.; Wong, K.K. MEK inhibitor resistance mechanisms and recent developments in combination trials. *Cancer Treat. Rev.* **2021**, *92*, 102137. [CrossRef] [PubMed]
66. Moore, A.R.; Rosenberg, S.C.; McCormick, F.; Malek, S. RAS-targeted therapies: Is the undruggable drugged? *Nat. Rev. Drug Discov.* **2020**, *19*, 533–552. [CrossRef]
67. Whittaker, S.R.; Theurillat, J.P.; Van Allen, E.; Wagle, N.; Hsiao, J.; Cowley, G.S.; Schadendorf, D.; Root, D.E.; Garraway, L.A. A genome-scale RNA interference screen implicates NF1 loss in resistance to RAF inhibition. *Cancer Discov.* **2013**, *3*, 350–362. [CrossRef]
68. See, W.L.; Tan, I.L.; Mukherjee, J.; Nicolaides, T.; Pieper, R.O. Sensitivity of glioblastomas to clinically available MEK inhibitors is defined by neurofibromin 1 deficiency. *Cancer Res.* **2012**, *72*, 3350–3359. [CrossRef]
69. Dharia, N.V.; Kugener, G.; Guenther, L.M.; Malone, C.F.; Durbin, A.D.; Hong, A.L.; Howard, T.P.; Bandopadhayay, P.; Wechsler, C.S.; Fung, I.; et al. A first-generation pediatric cancer dependency map. *Nat. Genet.* **2021**, *53*, 529–538. [CrossRef]
70. Hua, Z.; Gu, X.; Dong, Y.; Tan, F.; Liu, Z.; Thiele, C.J.; Li, Z. PI3K and MAPK pathways mediate the BDNF/TrkB-increased metastasis in neuroblastoma. *Tumour Biol.* **2016**, *37*, 16227–16236. [CrossRef]
71. Li, Z.; Oh, D.Y.; Nakamura, K.; Thiele, C.J. Perifosine-induced inhibition of Akt attenuates brain-derived neurotrophic factor/TrkB-induced chemoresistance in neuroblastoma in vivo. *Cancer* **2011**, *117*, 5412–5422. [CrossRef]

2. Rozen, E.J.; Shohet, J.M. Systematic review of the receptor tyrosine kinase superfamily in neuroblastoma pathophysiology. *Cancer Metastasis Rev.* **2022**, *41*, 33–52. [CrossRef]
3. Castellano, E.; Downward, J. RAS Interaction with PI3K: More Than Just Another Effector Pathway. *Genes. Cancer* **2011**, *2*, 261–274. [CrossRef]
4. Rah, B.; Rather, R.A.; Bhat, G.R.; Baba, A.B.; Mushtaq, I.; Farooq, M.; Yousuf, T.; Dar, S.B.; Parveen, S.; Hassan, R.; et al. JAK/STAT Signaling: Molecular Targets, Therapeutic Opportunities, and Limitations of Targeted Inhibitions in Solid Malignancies. *Front. Pharmacol.* **2022**, *13*, 821344. [CrossRef]
5. Ferguson, J.; Arozarena, I.; Ehrhardt, M.; Wellbrock, C. Combination of MEK and SRC inhibition suppresses melanoma cell growth and invasion. *Oncogene* **2013**, *32*, 86–96. [CrossRef]

cancers

MDPI

Article

Fibroblasts Promote Resistance to KRAS Silencing in Colorectal Cancer Cells

Susana Mendonça Oliveira [1,2,3,4], Patrícia Dias Carvalho [1,2,5], André Serra-Roma [1,2,†], Patrícia Oliveira [1,2], Andreia Ribeiro [1,2], Joana Carvalho [1,2], Flávia Martins [1,2,3], Ana Luísa Machado [1,3,4], Maria José Oliveira [1,3,5,6] and Sérgia Velho [1,2,*]

1 i3S—Instituto de Investigação e Inovação em Saúde, Universidade do Porto, Rua Alfredo Allen 208, 4200-135 Porto, Portugal; smendonca@ipatimup.pt (S.M.O.); poliveira@ipatimup.pt (P.O.); jcarvalho@ipatimup.pt (J.C.); flaviam@ipatimup.pt (F.M.); almachado@i3s.up.pt (A.L.M.); mariajo@i3s.up.pt (M.J.O.)

2 IPATIMUP—Instituto de Patologia e Imunologia Molecular, Universidade do Porto, Rua Júlio Amaral de Carvalho 45, 4200-135 Porto, Portugal

3 FMUP—Faculdade de Medicina da Universidade do Porto, Alameda Prof. Hernâni Monteiro, 4200-319 Porto, Portugal

4 ESS | P.PORTO—Escola Superior de Saúde, Instituto Politécnico do Porto, Rua Dr. António Bernardino de Almeida 400, 4200-072 Porto, Portugal

5 ICBAS—Instituto de Ciências Biomédicas Abel Salazar, Universidade do Porto, Rua Jorge de Viterbo Ferreira 228, 4050-313 Porto, Portugal

6 INEB—Instituto Nacional de Engenharia Biomédica, Universidade do Porto, Rua do Campo Alegre 823, 4150-177 Porto, Portugal

* Correspondence: svelho@i3s.up.pt; Tel.: +351-226-074-900

† Current address: Biomedical Research, Novartis Pharma AG, Fabrikstrasse 2, Novartis Campus, CH-4056 Basel, Switzerland.

Simple Summary: Novel therapies targeting KRAS offer treatment options for previously untreatable patients. However, in colorectal cancer (CRC), resistance to KRAS-targeted therapy develops rapidly, making it imperative to understand its underlying mechanisms. Cancer-associated fibroblasts (CAFs) contribute to therapy resistance by generating and maintaining the cancer stem cell niche. This study investigates whether CAF-secreted factors induce resistance to KRAS inhibition by enhancing cancer stemness. Our findings demonstrate that while KRAS silencing reduced the expression of stem cell markers and stemness, CAF-secreted factors counteracted those effects by activating pro-tumorigenic pathways, such as epithelial-to-mesenchymal transition, and increasing cell proliferation. Overall, we provide novel mechanistic insights into how CAF-secreted factors oppose KRAS silencing-induced growth inhibition, which may be crucial for improving CRC therapy.

Abstract: Colorectal cancer (CRC) responses to KRAS-targeted inhibition have been limited due to low response rates, the mechanisms of which remain unknown. Herein, we explored the cancer-associated fibroblasts (CAFs) secretome as a mediator of resistance to KRAS silencing. CRC cell lines HCT15, HCT116, and SW480 were cultured either in recommended media or in conditioned media from a normal colon fibroblast cell line (CCD-18Co) activated with rhTGF-β1 to induce a CAF-like phenotype. The expression of membrane stem cell markers was analyzed by flow cytometry. Stem cell potential was evaluated by a sphere formation assay. RNAseq was performed in KRAS-silenced HCT116 colonospheres treated with either control media or conditioned media from CAFs. Our results demonstrated that KRAS-silencing up-regulated CD24 and down-regulated CD49f and CD104 in the three cell lines, leading to a reduction in sphere-forming efficiency. However, CAF-secreted factors restored stem cell marker expression and increased stemness. RNA sequencing showed that CAF-secreted factors up-regulated genes associated with pro-tumorigenic pathways in KRAS-silenced cells, including KRAS, TGFβ, NOTCH, WNT, MYC, cell cycle progression and exit from quiescence, epithelial-mesenchymal transition, and immune regulation. Overall, our results suggest that resistance to KRAS-targeted inhibition might derive not only from cell-intrinsic causes but also from external elements, such as fibroblast-secreted factors.

Keywords: KRAS; colorectal cancer; fibroblasts; cancer stemness; epithelial-mesenchymal transition; therapy resistance

1. Introduction

KRAS mutations occur in approximately 40% of colorectal cancers (CRCs) and are associated with a poor prognosis and resistance to therapy [1,2]. For over three decades, the KRAS oncogene has been considered an undruggable target [2–4], but recent breakthroughs in developing allele-specific covalent inhibitors, such as those targeting the KRAS G12C mutation, have paved the way for a new era of KRAS-targeted therapies [4,5]. These innovative therapies, now in various stages of development, show promising results by selectively inhibiting KRAS function in cancer cells [4]. Nevertheless, a disheartening reality persists: in CRC, the therapeutic response remains limited, with only a subset of patients experiencing significant benefits [2]. Furthermore, even among initial responders, resistance to treatment rapidly develops, posing a significant challenge to the long-term efficacy of these therapies [6]. While the mechanisms of acquired resistance have been extensively studied [7–10], the mechanisms of innate resistance remain poorly understood. One possible explanation is the independence of mutant KRAS cancer cells from KRAS oncogenic signaling. This has been well demonstrated in a subset of lung, pancreatic, and colorectal cancer cell lines [11–14], which remain viable upon KRAS silencing. However, clinical data shows that the most frequent outcome of KRAS-targeted inhibition in colorectal cancer is stable disease [15,16], indicating that cancer cells are indeed sensitive to KRAS inhibition to the point of stalling their growth rate. It is, therefore, imperative to better understand the mechanisms that might be involved in resistance to KRAS-targeted therapies in CRC to unveil novel combinatorial therapies that can improve their therapeutic efficacy or identify biomarkers predictive of response.

Among the several factors that have been shown to contribute to the acquisition of resistance to therapy, the tumor microenvironment, which includes cell populations such as cancer stem cells and cancer-associated fibroblasts (CAFs), is also known to drive this process [17–19]. Cancer stem cells possess self-renewal potential and the capacity to stay quiescent for extended periods, which confers greater resistance to various forms of therapy [20]. The development and expansion of cancer stem cells are regulated by several signaling pathways, and in CRC, this process is selectively controlled by RAS isoforms, with the KRAS isoform being the most potent inducer of stemness characteristics. KRAS activation induces stemness by up-regulating pathways such as Wnt/β-catenin and the Hedgehog, and also by increasing surface markers of stemness [21].

Even with the ongoing debate on the definitive biomarkers of stemness [22], all studies show that CRC possesses a rare cell population that resembles cancer stem cells, regardless of the markers used for isolation [23–26]. Intriguingly, these CRC cancer stem cells preferentially localize in areas enriched in CAFs [27]. This association is particularly evident in consensus molecular subtype 4 tumors (CMS4), one of the four molecular subtypes identified in CRC. These tumors are enriched in CAFs and exhibit up-regulation of genes related to cancer stem cells, highlighting their interconnection [28]. CAFs are responsible for the induction and maintenance of the cancer stem cell phenotype through their secreted factors [29–32], underscoring the co-dependence of both these cell types.

In addition to regulating the cancer stem cell phenotype, CAFs can also alter the proteome profile associated with CRC cells harboring a KRAS mutation. Specifically, we demonstrated that in colorectal cancer cells exposed to fibroblast-derived factors, KRAS oncogenic signaling is predominantly governed by fibroblast-secreted factors. However, most of the fibroblast-induced signaling is independent of mutant KRAS [33]. This suggests that CAFs might be key players in independently modulating cancer cell phenotypes in the context of KRAS-targeted therapies.

Given the symbiotic relationship between KRAS, cancer stem cells, and fibroblasts, we hypothesized that CAF-derived factors might be responsible for driving cancer cell stemness in CRC cell lines, subsequently inducing resistance to KRAS-targeted inhibition. Our results confirm that both KRAS and the secreted factors from recombinant human (rh) TGF-β1 activated fibroblasts enhance CRC stem cell activity. Furthermore, such fibroblast-derived factors recover the stemness potential lost upon KRAS silencing, leading to a more mesenchymal phenotype through the up-regulation of epithelial-to-mesenchymal transition (EMT) and pro-tumorigenic pathways. These results identify a novel potential mechanism of resistance to KRAS-target therapies, mediated by the fibroblast secretome, and open new avenues to improve the efficacy of these treatments.

2. Materials and Methods

2.1. Cell Culture

Human CRC cell lines HCT116, HCT15, SW480 (Table 1) and normal human intestinal fibroblast cell line CCD-18Co were purchased from the American Type Culture Collection (ATCC). Cells were routinely maintained at 37 °C in a humidified atmosphere with 5% CO_2 in the recommended media: RPMI-1640 media (Gibco, Thermo Fisher Scientific, Waltham, MA, USA) for the CRC cell lines and DMEM (Gibco, Thermo Fisher Scientific, USA) for the fibroblasts, both supplemented with 10% heat-inactivated HyClone fetal bovine serum—FBS (Cytiva, Marlborough, MA, USA) and 1% penicillin-streptomycin—P/S (10,000 U/mL; Gibco, Thermo Fisher Scientific, USA).

Table 1. Genetic and histological characterization of KRAS mutant CRC cell lines. Adapted from [34].

Cell Line	KRAS Mutation	Disease	Derived from	Reference
HCT116	G13D	Colorectal Carcinoma	Primary tumor	[34]
HCT15	G13D	Colorectal Adenocarcinoma	DLD-1 misclassified	
SW480	G12V	Colorectal Adenocarcinoma	Primary tumor	

2.2. Production of Conditioned Media for Fibroblasts

Fibroblasts were plated into T75 culture flasks and cultured in DMEM supplemented with 10% heat-inactivated FBS and 1% P/S at 37 °C in a humidified atmosphere with 5% CO_2 until approximately 90% of confluence. After washing two times with phosphate-buffered saline (PBS), the media of the fibroblasts was changed to DMEM supplemented with 1% P/S plus 10 ng/mL recombinant human (rh)TGF-β1 (ImmunoTools, Friesoythe, Germany). The addition of rhTGF-β1 in the media leads to the activation of the fibroblasts, conferring on them a CAF-like phenotype. For the sphere formation experiments, the fibroblast-conditioned media was prepared with DMEM without phenol red (Gibco, Thermo Fisher Scientific, USA) in the same conditions. After 4 days in culture, the conditioned media was harvested, centrifuged at 1200 revolutions per minute (rpm) for 5 min, filtered through a 0.2 μm filter, and stored at −20 °C. The cells were harvested with 0.05% Trypsin-EDTA (Gibco, Thermo Fisher Scientific, USA), counted, and total protein extraction was performed. The confirmation of fibroblast activation was assessed through the evaluation of alpha-smooth muscle actin (α-SMA) expression by western blotting (Figures S1 and S2).

2.3. Cell Culture with Conditioned Media

Cancer cells were plated into a 6-well plate at a confluence of 1.5×10^5 cells per well. After 16 h the cells were transfected with siRNA. Upon 6 h of transfection, the conditioned media of activated fibroblasts (grown in DMEM + 1% P/S + 10 ng/mL rhTGF-β1) was added. After 48 h of incubation with the conditioned media (total of 72 h of transfection), the cells were harvested with 0.05% Trypsin-EDTA (Gibco, Thermo Fisher Scientific, USA), counted, and collected for flow cytometry and total protein extraction. KRAS silencing efficiency was assessed by western blotting (Figures S1 and S2).

2.4. siRNA Transfection

Knockdown of KRAS was achieved by gene silencing using a pool of 4 small interfering RNAs (siRNA) specific for KRAS (siKRAS, ON-TARGETplus SMARTpool, from Dharmacon, GE Healthcare, Lafayette, CO, USA). This approach was previously assessed by us to determine its impact on KRAS signaling and functionality [33]. The desired cell line was plated into a 6-well plate and allowed to adhere (1.5×10^5 cells per well). The following day, transfection was carried out. KRAS silencing was conducted according to manufacturer specifications using a specific ON-TARGETplus SMARTpool small interfering RNA (L-005069-00-0010; Dharmacon, GE Healthcare, USA) at a final concentration of 10 nM. As a control, a condition using non-targeting siRNA (ON-TARGETplus Non-targeting Control siRNA #1, from Dharmacon, GE Healthcare, USA) was used at the same concentration as the siRNA targeting the gene of interest. KRAS silencing efficiency was assessed by western blot after 72 h (Figures S1 and S2).

2.5. Protein Extraction and Western Blotting

Cells were lysed using RIPA lysis buffer [50 mM TrisHCl pH 7.5, 1% (*v*/*v*) IGEPAL CA-630, 150 mM NaCl, and 2 mM EDTA] supplemented with 1:7 protease inhibitor cocktail (Roche Diagnostics GmbH, Rotkreuz, Switzerland) and 1:100 phosphatase inhibitor cocktail (Sigma Aldrich, St. Louis, MO, USA). Cells were centrifuged at 14,000 rpm at 4 °C for 10 min. The supernatants were collected and stored at −20 °C. Protein concentration was determined using the Bradford assay (Bio-Rad Protein Assay kit, Bio-Rad, Hercules, CA, USA). Equal amounts of protein from each sample were dissolved in sample buffer [Laemmli with 5% (*v*/*v*) 2-β-mercaptoethanol and 5% (*v*/*v*) bromophenol blue] and denaturated for 5 min at 95 °C. Samples were separated in a 12% sodium dodecyl sulfate-polyacrylamide gel electrophoresis (SDS–PAGE), and proteins were transferred into nitrocellulose membranes (Amersham Protran Premium 0.45 µm nitrocellulose blotting membranes, Cytiva, USA). For immunostaining, membranes were blocked with 5% (*w*/*v*) non-fat dry milk in PBS containing 0.5% (*v*/*v*) Tween20 (0.5% PBS-T) (Sigma Aldrich, USA), and primary antibodies were incubated overnight at 4 °C with agitation (Table 2). After washing five times with 0.5% PBS-T for 5 min, membranes were incubated with HRP-conjugated anti-mouse secondary antibodies (GE Healthcare, USA) for 1 h at room temperature. Membranes were washed again five times with 0.5% PBS-T for 5 min. Bands were developed using ECL blotting substrate (Clarity Western ECL Substrate, Bio-Rad, USA). ImageJ software, version 1.54f, was used for protein quantification.

Table 2. List of antibodies used for western blotting.

Ab [1]	MW [2] (kDA) [3]	Blocking	Dilution	Species	Brand	Catalog no.	Storage (°C)	Secondary Dilution
α-SMA	42	5% milk in 0.5% PBS-T	1:250	Mouse	Abcam	ab7817	−20	1:3000
GAPDH	37	5% milk in 0.5% PBS-T	1:10,000	Mouse	SantaCruz	sc-47724	4	1:20,000
KRAS	21	5% milk in 0.5% PBS-T	1:4000	Mouse	LSBio	C175665	−20	1:8000

[1] Antibody. [2] Molecular weight. [3] Kilodalton.

2.6. Flow Cytometry

For flow cytometry analysis, cells were harvested using trypsin and resuspended in RPMI supplemented with 10% heat-inactivated FBS and 1% P/S to inactivate the trypsin (complete media). Cells were allowed to recover their membrane markers through an incubation of 20 min in complete media in an incubator at 37 °C in a humidified atmosphere with 5% CO_2. For the labeling, 2×10^5 cells were used per condition. Cells were washed with wash buffer (0.5% FBS in PBS) and resuspended in that same wash buffer. Single-cell suspension was labeled using a 1:100 concentration (*v*:*v*) of a single antibody (Miltenyi

Biotec, Bergisch Gladbach, Germany) in wash buffer (Table 3). As a control, a condition where no antibodies were added to the cells (unstained) was used, since it was not required to make a mix of antibodies. Fluorochrome-conjugated antibodies were incubated at room temperature, in the dark, for 15 min. Labeled cells were then rinsed in wash buffer and finally resuspended in PBS. Cells were analyzed using a FACS Canto-II (BD Biosciences) or BD Accuri C6 (BD Biosciences) flow cytometer. Data were analyzed using FlowJo version 10 cytometry analysis program. After performing a doublet exclusion gate, both the percentage of positive cells and the median fluorescence intensity (MFI) were analyzed. The gating strategy used can be found as supplementary data (Supplementary Figure S1).

Table 3. List of anti-human antibodies used for flow cytometry.

Antibody	Conjugate	Clone	Catalog no.	Brand
CD24	PE	32D12	130-098-861	Miltenyi Biotec
CD49f	APC	GoH3	130-100-147	Miltenyi Biotec
CD104	FITC	REA236	130-124-266	Miltenyi Biotec
CD44	FITC	DB105	130-113-896	Miltenyi Biotec
CD44.V6	APC	REA706	130-111-425	Miltenyi Biotec
CD133	PE	AC133	130-098-826	Miltenyi Biotec
CD166	APC	REA442	130-106-619	Miltenyi Biotec

2.7. Sphere-Forming Assay

Following siRNA transfection, cells were harvested using Trypsin and resuspended in RPMI supplemented with 10% heat inactivated FBS and 1% P/S. Cells were centrifuged at 1200 rpm for 5 min, the supernatant was removed, and the pellet was washed with PBS. Cells were again centrifuged at similar conditions and resuspended in DMEM without phenol red and 1% P/S. Single-cell suspension was achieved by physical dissociation with a 25-gauge needle. Cells were then plated at a density of 500 cells/cm^2 into 6-well plates coated with 1.2% poly(2-hydroxyethylmethacrylate) (Merck KGaA, Darmstadt, Germany) in 95% ethanol (Sigma-Aldrich, USA) to create non-adherent culture conditions. Cells were cultured in optimal conditions for 5 days in DMEM without phenol red containing 1× B27 and 1× N2 supplements (Life Technologies, Carlsbad, CA, USA), 20 ng/mL epidermal growth factor (EGF) (Sigma-Aldrich, USA), 10 ng/mL basic fibroblast growth factor (bFGF) (Life Technologies, USA), and 1% P/S in an incubator at 37 °C in a humidified atmosphere with 5% CO_2. For the conditions where fibroblast-conditioned media was used, only 1× B27 and 1× N2 supplements were added because fibroblasts produce growth factors on their own [35–37]. Sphere-forming efficiency (SFE) was calculated as the number of spheres ($\geq$50 μm) formed divided by the number of cells plated and multiplied by 100 to be expressed as a percentage: (SFE = Number of spheres formed/Number of cells plated × 100). To observe and acquire pictures of the spheres (in brightfield), the IN Cell Analyzer 2000 (GE Healthcare, USA) microscope was used. To automatically identify the spheres in the images, ilastik version 1.3.3, an interactive machine learning tool for (bio)image analysis as used, as well as the Cell Profiler^{TM} 4.0.7 cell image analysis software. Finally, each dataset was manually curated using the ImageJ software, version 1.54f.

2.8. RNA Extraction

RNA was extracted from the cells lysed with RLT+ buffer using the RNeasy Mini kit (Qiagen, Hilden, Germany) according to the manufacturer's instructions. RNA concentration and purity were measured using the UV-Vis spectrophotometer NanoDrop 1000 (Thermo Fisher Scientific, USA). RNA was stored at −80 °C until required.

2.9. Library Preparation and Transcriptome Sequencing

Spheres of KRAS-silenced cells formed in the conditioned media of activated fibroblasts were compared with their KRAS-silenced counterparts, formed in control media. Due to the low sphere-forming efficiency, the quantity of RNA extracted from each experiment was low, especially for the control condition. Therefore, a pool of 3 biological replicates was used per condition. The samples were processed using Ion Torrent technology. Libraries were prepared for each sample with the Ion AmpliSeq Transcriptome Human Gene Expression Kit targeting 20,802 genes and sequenced on a 540-chip using the Ion 540 Kit-Chef and the S5 XL instrument (IonS5XL, Thermo Fisher, USA). The average mean read length was 113 base pairs. The produced reads were aligned to hg19 AmpliSeq Transcriptome ERCC v1 reference sequence and hg19_AmpliSeq_Transcriptome_21K_v1 target regions.

2.10. *Database for Annotation, Visualization, and Integrated Discovery (DAVID)*

TAC (Transcriptome Analysis Console) software was used to generate the datasets. Differential expression was analyzed with the help of DAVID online software, version 6.8 (https://david.ncifcrf.gov/home.jsp accessed on 15 July 2021). This comprehensive approach allowed a preliminary gene annotation and visualization to explore the biological functions and signaling pathways associated with both our gene sets and create tables of functional gene enrichment.

2.11. *Gene Set Enrichment Analysis (GSEA)*

Gene expression data were analyzed for enrichment using GSEA software (Broad Institute, version 4.1.0) and Human Molecular Signatures Databases (MSigDB) v2023.1.Hs, following the described guidelines [38–40]. Several human collections from the Molecular Signatures Database were used as the gene sets of interest, allowing us to explore the enrichment of biologically relevant pathways and functional annotations. The normalized enrichment score (NES) for each gene was calculated and used for further analysis and graphical representation.

2.12. *Quantification of G0 Arrest*

The quantification of the G0 arrest in the RNAseq data was adapted from the G0 arrest score quantification described in [41]. R software was used for gene enrichment and determination of G0 arrest score (original code available at: https://github.com/secrierlab/CancerG0Arrest/tree/main/TCGA_QuiescenceEvaluation accessed on 15 October 2023). A final positive G0 arrest score indicates that cells are quiescent, while if negative, cells are cycling.

2.13. *Statistical Analysis*

Results are representative of three or more independent experiments. Quantifications are expressed as mean $\pm$ standard deviation (SD) of the biological replicates considered. Statistical analyses were performed using GraphPad Prism v8 (GraphPad Software Inc., USA). All samples were tested for normality, and significant statistical difference was considered when the p value was less than 0.05. The statistical tests performed for each analysis are indicated in the corresponding figure legend.

3. Results

3.1. *Colorectal Cancer Cells Express Variable Basal Levels of Membrane Cancer Stem Cell Markers*

To explore the role of mutant KRAS in modulating the cancer stem cell landscape in CRC, we selected three CRC cell lines (HCT116, HCT15, and SW480), all harboring a KRAS mutation (Table 1), though with different origins and genetic profiles. We began by characterizing the basal stemness potential of each cell line by flow cytometry analysis, focusing on the most commonly used membrane stem cell markers: CD24, CD133, CD166, and CD44 and its isoform CD44v6 (Table 3). Additionally, we also included CD49f (Integrin $\alpha 6$) and its binding partner CD104 (Integrin $\beta 4$) in our analysis. CD49f, in particular, has been described as a biomarker transversally present amongst stem cell populations, including the intestinal one [42], an enhancer of tumorigenesis [23], and a possible regulator of other stem cell markers, such as CD44 [43].

Our analysis revealed a heterogeneous expression of cancer stem cell markers within and across the CRC cell lines, both in the percentage of positive cells and in the level of expression per cell, denoted by the median fluorescence intensity (MFI) (Figure 1). Interestingly, only CD49f and CD104 receptors were highly and consistently expressed across the three CRC cell lines.

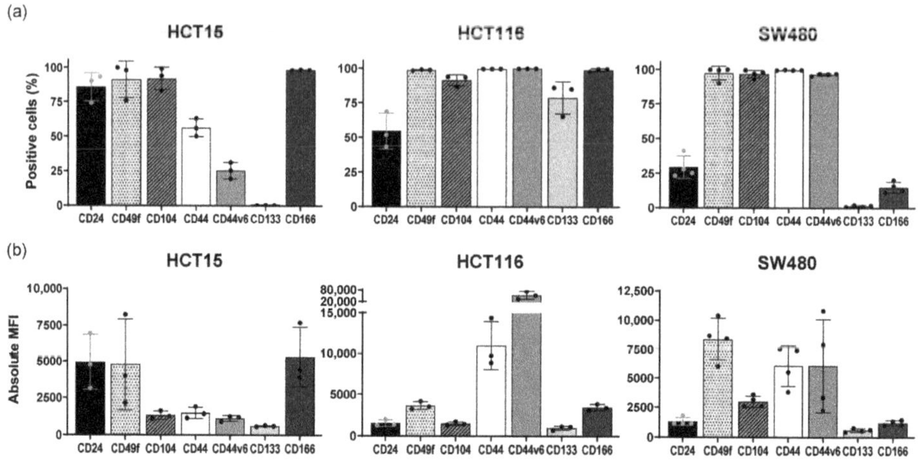

Figure 1. Characterization of the basal levels of stem cell marker expression in CRC cell lines by flow cytometry. Mean and standard deviation are represented in each bar. Each dot represents a biological replicate. (**a**) Percentage of positive cells; (**b**) Absolute median fluorescence intensity (MFI).

3.2. *KRAS Silencing Up-Regulates CD24 and Down-Regulates CD49f and CD104 Stemness Markers across Cell Lines*

Having established the basal expression levels of the stemness markers, we next investigated the role of KRAS in regulating this stemness signature. To do so, we silenced the expression of KRAS by RNAi in the three cell lines, followed by flow cytometry analysis of the abovementioned stem cell markers.

Silencing KRAS led to significant alterations in the expression of stem cell markers, predominantly reflected in changes in the MFI rather than the percentage of positive cells (Figure 2). In the HCT15 cells, KRAS silencing up-regulated CD24 and CD44 MFI, while decreasing the MFI of CD49f and CD104. CD44v6, CD133, and CD166 remained unchanged. In the HCT116 cells, KRAS silencing increased the percentage of CD24-positive cells. On the other hand, CD49f MFI decreased, along with both the MFI and percentage of positive cells, for CD104, CD44, CD44v6, and CD133. CD166 remained unchanged. In the SW480 cells, KRAS silencing up-regulated CD24 and the MFI of CD133 and CD166, whereas the MFI of CD49f, CD104, and CD44 decreased. No alterations were found in CD44v6 expression.

In summary, KRAS silencing induced significant, cell line-dependent changes in the expression of the stem cell markers. Despite this apparent cell line dependence, consistent alterations in CD24, CD49f, and CD104 expression were evident across all cell lines: after KRAS silencing, there was an up-regulation of CD24 (except the MFI in HCT116) while the MFI of CD49f and CD104 decreased. These changes strongly suggest a decrease in cancer stemness potential upon KRAS silencing.

3.3. *Fibroblast-Secreted Factors Attenuate the Capacity of KRAS Silencing to Regulate the Expression of Cancer Stem Cell Markers*

We then examined whether the factors secreted by activated fibroblasts could influence the modulatory effects of KRAS on stem cell markers. KRAS-silenced cells were cultured with conditioned media from activated fibroblasts (CCD-18Co normal colon cell line treated with rhTGF-β1). We then analyzed the abovementioned stem cell markers by flow cytometry.

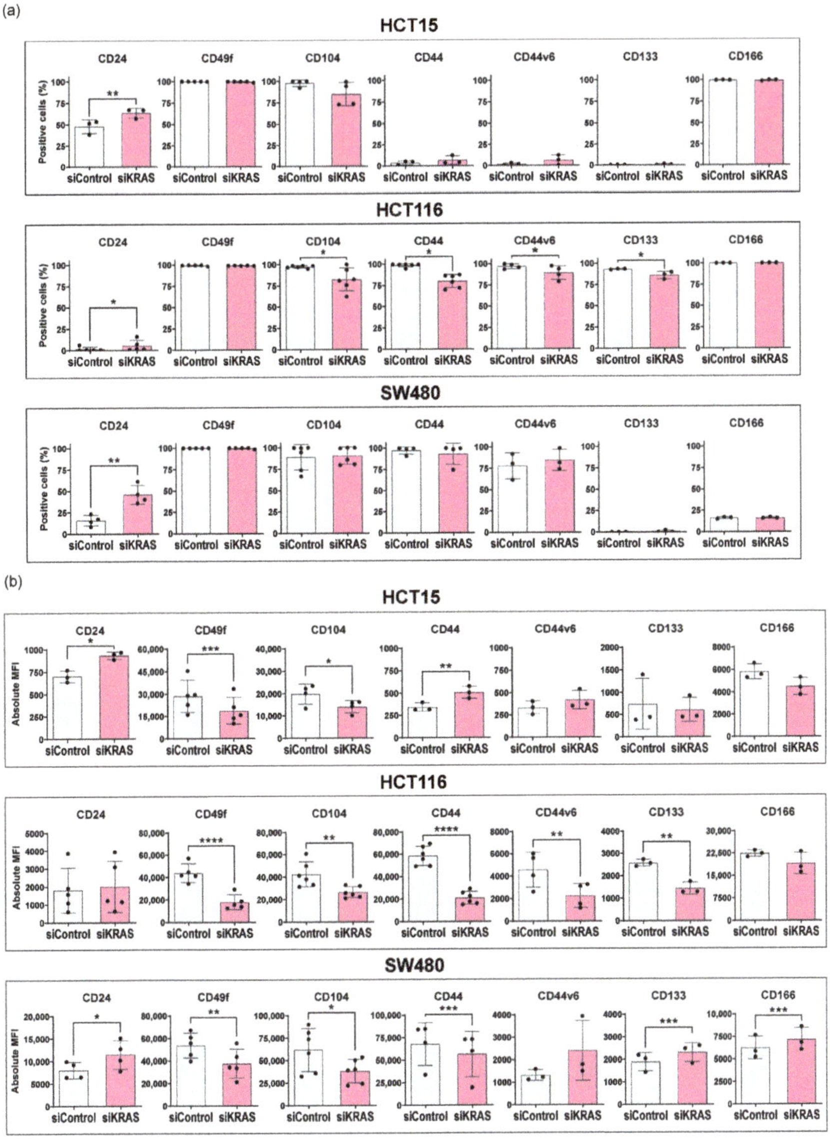

Figure 2. Stem cell marker expression by flow cytometry in CRC cell lines after KRAS silencing. For all cell lines, normality of the data was tested using Shapiro-Wilk normality test. A one-tailed paired *t*-test was performed, testing for a *p*-value < 0.05. The symbols *, **, ***, and **** were used to denote levels 0.05, 0.01, 0.001, and 0.0001 of statistical significance, respectively. For the samples that did not follow normality, a Wilcoxon matched-pairs signed rank test was used. Mean and standard deviation are represented in each bar. Each dot represents a biological replicate. (**a**) Percentage of positive cells; (**b**) Absolute median fluorescence intensity (MFI).

Strikingly, treatment with conditioned media of activated fibroblasts attenuated the differences in cancer stem cell expression previously observed between the control and

KRAS-silenced cells (Figure 3). After KRAS silencing and treatment with conditioned media in the HCT15 cells, CD24 expression increased, as seen with KRAS silencing alone, but the reduction in the MFI of CD49f and CD104 was no longer evident. Furthermore, CD44 expression decreased, contrary to KRAS silencing alone. CD44v6, CD133, and CD166 remained unchanged. In HCT116 cells, all significant differences between control and KRAS-silenced cells were lost after treatment with conditioned media. Similarly, in SW480 cells, conditioned media annulled the differences in most markers, though CD44 reduction remained. CD44v6 expression decreased, contrary to KRAS silencing alone.

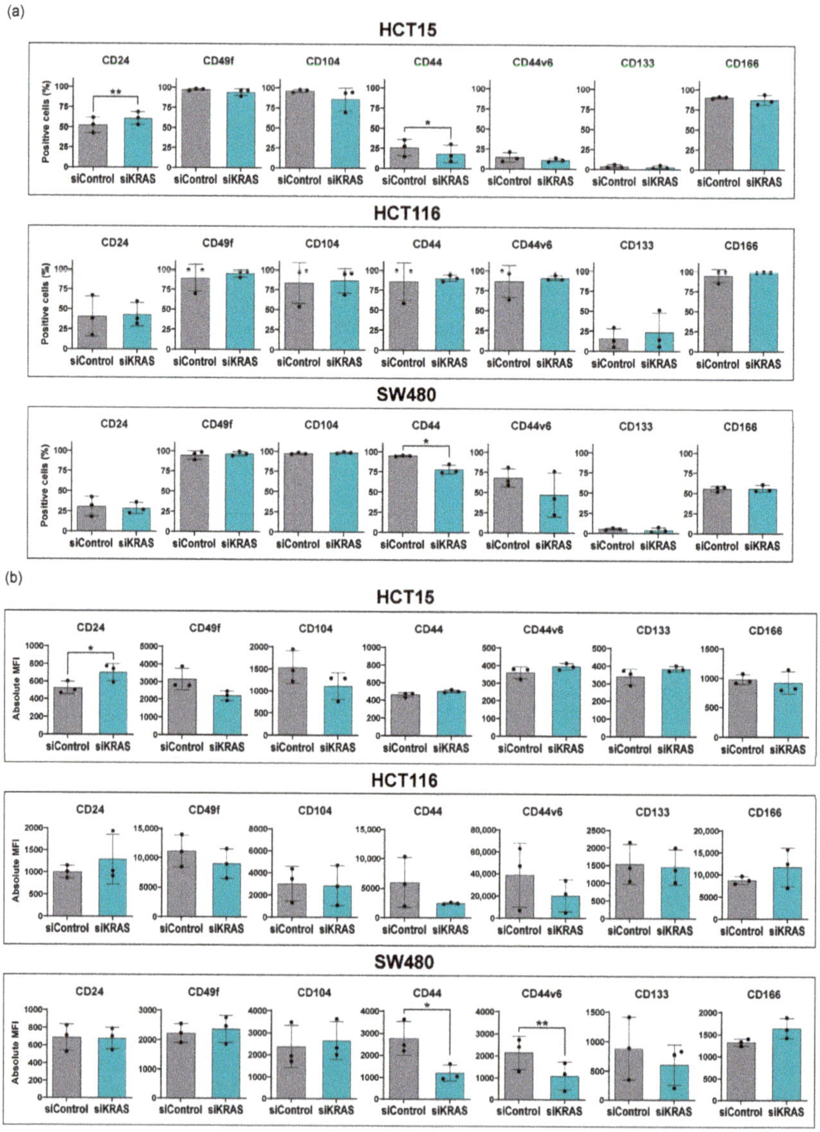

Figure 3. Stem cell marker expression by flow cytometry in CRC cell lines after KRAS silencing, plus treatment with conditioned media from fibroblasts. For all cell lines, normality of the data was tested using Shapiro-Wilk normality test. A one-tailed paired *t*-test was performed, testing for a *p*-value < 0.05.

The symbols * and ** were used to denote levels 0.05 and 0.01 of statistical significance, respectively. For the samples that did not follow normality, a Wilcoxon matched-pairs signed rank test was used. Mean and standard deviation are represented in each bar. Each dot represents a biological replicate. (**a**) Percentage of positive cells; (**b**) Absolute median fluorescence intensity.

These findings highlight the role of fibroblast-secreted factors in shaping the response of CRC cells to KRAS silencing. They suggest that fibroblast-secreted factors can mitigate the effects of KRAS silencing on stem cell marker levels in the membrane of CRC cells, thus unveiling another layer of intricacy in the interplay between KRAS signaling and fibroblasts.

3.4. Fibroblast-Secreted Factors Attenuate the Inhibitory Effect Promoted by KRAS Silencing in the Sphere Formation Assay

To further understand how KRAS modulation and fibroblast-derived signals influence the stem cell-like phenotype, we performed an in vitro sphere formation assay, a technique commonly used for assessing stem cell potential based on the capacity of cells for self-renewal. For this assay, cells were plated in anchorage-free conditions. Since resistance to anoikis is a characteristic of stem cells, the higher the number of spheres formed, the greater the stem cell potential of that cell line [44].

We started by investigating the role of KRAS silencing alone. KRAS silencing reduced sphere-forming efficiency (SFE) across all cell lines, consistent with the decreased stem cell marker expression observed earlier by flow cytometry (Figure 4a).

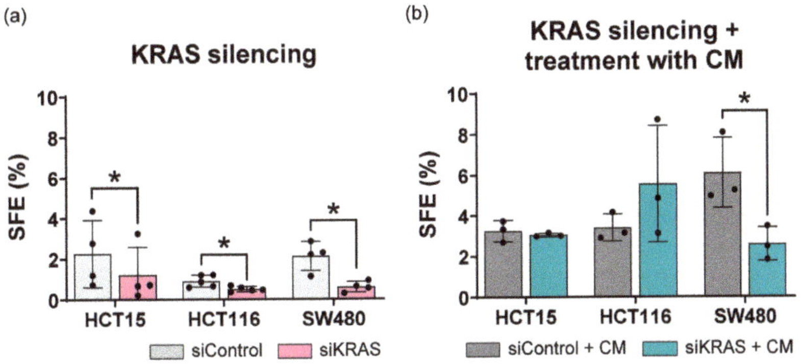

Figure 4. Sphere formation assay of CRC cell lines after silencing with siControl or siKRAS. For all cell lines, normality of the data was tested using a Shapiro-Wilk normality test, and a one-tailed Paired t-test was performed, testing for a p-value < 0.05 (symbol * denotes level 0.05 of statistical significance). Mean and standard deviation are represented in each bar. Each dot represents a biological replicate. (**a**) Sphere forming efficiency (SFE) percentage after treatment with sphere formation assay medium; (**b**) SFE percentage after treatment with conditioned media from activated fibroblasts (CM).

Subsequently, we tested the effect of fibroblast-secreted factors on the SFE of control and KRAS-silenced cells. Treatment with fibroblast-conditioned media enhanced sphere formation in all cell lines, including control and KRAS-silenced cells. Except for SW480, conditioned media of fibroblasts abolished the decrease in SFE induced by KRAS silencing (Figure 4b).

These results suggest that fibroblast-secreted factors not only enhance the self-renewal and proliferation capacity of CRC cell lines but also confer resistance to KRAS silencing.

3.5. Fibroblast-Secreted Factors Up-Regulate Proliferation, Pro-Tumorigenic, EMT, and Immune System Regulation Pathways in KRAS-Silenced Cells

To uncover the molecular mechanisms by which fibroblast-conditioned media counteract KRAS silencing, we conducted an RNASeq analysis. We focused on the HCT116 cell line since it showed the highest SFE increase in the KRAS-silenced condition after treatment with conditioned media of fibroblasts. We compared the gene expression profiles of KRAS-silenced cells treated with conditioned media versus control media to identify specific genes and pathways influenced by fibroblast-derived factors.

Analysis of the hallmark gene sets in the human molecular signature database revealed that KRAS-silenced cells treated with conditioned media exhibited a distinctive gene expression profile (Figure 5a–c and Supplementary Table S1). Treatment with conditioned media of activated fibroblasts up-regulated pathways associated with cell cycle control (E2F targets, G2-M checkpoints, MYC targets, mitotic spindle pathways), EMT, immune system regulation (IL6, IL2, interferon-gamma, complement pathways), and several pro-tumorigenic pathways (KRAS, NOTCH, TGF-β, and WNT).

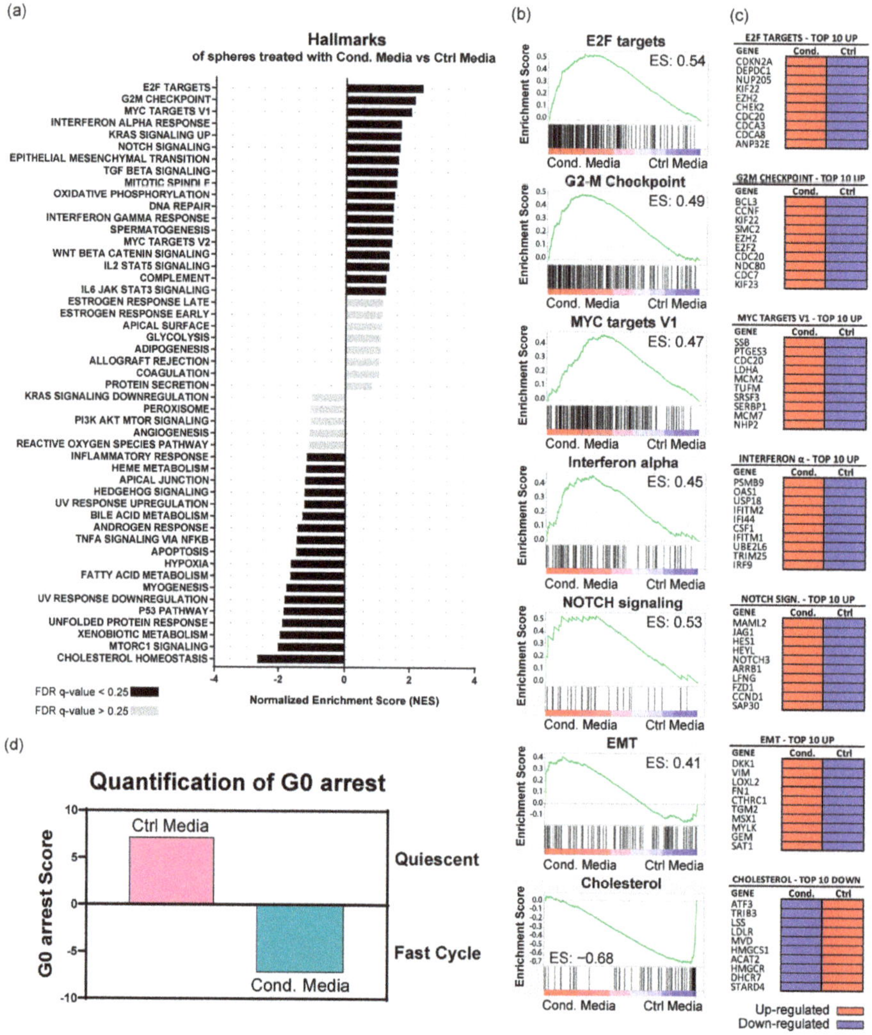

Figure 5. RNAseq analysis of KRAS-silenced HCT116 spheres treated with conditioned media of activated fibroblasts (Cond. Media) against control media (Ctrl Media). (**a**) Results of GSEA Hallmark analysis showing enriched gene sets. Black bars indicate significant enrichment at a false

discovery rate (FDR) < 25%, while gray bars represent gene sets with FDR > 25%. A positive normalized enrichment score (NES) value indicates enrichment in the cells treated with conditioned media of fibroblasts, whereas a negative NES indicates enrichment in the cells treated with control media; (**b**) Enrichment plots for the most relevant data sets enriched in GSEA Hallmark analysis, showing the profile of the running enrichment score (ES) and positions of gene set members on the rank-ordered list; (**c**) Tables showing the top 10 enriched genes in each data set. Red indicates up-regulation of the gene, while purple indicates down-regulation; (**d**) Quantification of G0 arrest. Positive values indicate that cells are in a quiescent state, whereas negative values identify cells that are in a proliferative cycle phase.

In contrast, treatment with control media up-regulated pathways related to metabolism (cholesterol, fatty acids, and xenobiotic metabolism), cellular stress (apoptosis and hypoxia, TNF-α and mTOR signaling pathways), and DNA damage response (unfolded protein response and P53 pathways).

Since we had up-regulation of pathways related to cell cycle control after treatment with conditioned media of fibroblasts, we evaluated the G0 arrest transcriptional signature. This methodology distinguishes between cells in a state of G0 arrest (encompassing quiescence, senescence, and dormancy) and those in a rapid cell cycle progression state [41]. This analysis revealed that fibroblast-conditioned media drove KRAS-silenced cells into a state of rapid cell cycling (negative G0 score), akin to stem cell self-renewal, while control media maintained cells in a quiescent state (positive G0 score) (Figure 5d).

These results indicate that fibroblast-secreted factors override the growth-inhibitory effects of KRAS silencing, enhancing proliferative potential and promoting a mesenchymal phenotype through EMT and stemness-related pathways.

4. Discussion

In an era of intensive research to develop effective therapeutic strategies against KRAS mutant tumors, our in vitro study uncovers a crucial mechanism that bypasses KRAS silencing. This mechanism operates independently of KRAS and is orchestrated by CAF-derived factors that trigger cancer stem cell activity, proliferation, EMT, modulation of immune response, and up-regulation of tumorigenic pathways.

KRAS, located downstream of many cell surface receptors, is a central regulator of intracellular signaling in response to extracellular stimuli, making it crucial for regulating diverse cancer cell activities, including stemness [2,33,45]. This role of KRAS could be exploited, since targeting cancer stem cells to prevent cancer recurrence remains one of the most significant challenges in oncology. Therefore, understanding the factors and circumstances that regulate cancer stem cell induction and maintenance—and consequently govern the development of recurrence—is essential for improving treatment efficacy [18].

We found that while the expression of stem cell markers varied in a cell line-dependent manner, KRAS silencing consistently resulted in up-regulation of CD24 and down-regulation of CD49f and CD104 across cell lines. Although CD24 up-regulation has been previously correlated with tumor progression, invasiveness, differentiation, and chemotherapy resistance [46–48], its role in modulating the CRC stem cell phenotype remains inconclusive. Some studies suggest that loss of CD24 expression is associated with poorer outcomes [49], while others suggest that CD24 is a good prognosis marker in CRC, being down-regulated in stage IV colorectal adenocarcinoma [46]. Regarding the expression of CD49f/CD104, our findings align with existing literature: both CD49f and CD104 were highly expressed at basal levels across the CRC cell lines analyzed. The CD49f/CD104 complex is crucial for cell-cell and cell-matrix interactions and is typically overexpressed in CRC. Depletion of the integrin CD49f/CD104 complex reduces the invasive and migratory capabilities of cancer cells [50], highlighting its role in promoting an aggressive phenotype. Although the mechanisms controlling its expression are still not fully known, transcription factors, such as MYC, have been implicated in this process. Our data reinforces KRAS as a regulator of CD49f/CD104 expression in CRC cells. KRAS silencing led to a reduction in the expression of these markers, indicating a decreased stemness potential, as confirmed by our functional assay that showed a reduction in SFE. However, treatment with conditioned media from fibroblasts annulled the effects of KRAS silencing. It restored the expression of CD49f and CD104, indicating an increase in stemness potential, also confirmed by the enhanced SFE. Accordingly, CD49f/CD104 can increase cell proliferation through the activation of the Wnt/β-catenin pathway, a regulator of stem cell homeostasis in the intestinal crypts and an upstream effector of MYC [51].

Our RNAseq results further reinforced the role of fibroblasts as mediators of resistance to KRAS silencing. Fibroblast-derived factors induced the up-regulation of multiple pathways linked to critical cellular processes such as cell cycle control, EMT, immune system regulation, and tumor development. Among these pathways, there were key signaling routes such as KRAS, NOTCH, TGF-β, MYC, and WNT known to play significant roles in cancer progression and therapy resistance. Specifically, E2F, MYC, and G2M pathways, often down-regulated upon KRAS inhibition [52,53], were among the top-ranked processes in KRAS-silenced cells stimulated by fibroblast-secreted factors. Up-regulation of these pathways in the presence of fibroblast-secreted factors suggests a mechanism by which fibroblasts can counteract the effects of KRAS silencing, promoting cell cycle progression and proliferation despite the absence of KRAS activity. This was further validated by a G0 arrest transcriptional signature analysis, revealing that KRAS-silenced cells treated with fibroblast-conditioned media shift from a quiescent state to active cell cycling, similar to stem cell renewal.

Interestingly, the stem cell-like phenotype and EMT are deeply intertwined, as cells undergoing EMT can acquire cancer stem cell properties, while cancer stem cells themselves can undergo EMT to facilitate metastasis [54]. One way of inducing EMT is through the activation of the canonical TGF-β pathway. TGF-β signaling has been correlated with CRC subtypes with a worse prognosis and increased relapse. Conditioned media of CAFs are enriched in TGF-β1 and, therefore, can induce EMT [55]. It also contains other factors such as fibroblast growth factor (FGF), interleukin-6 (IL-6), hepatocyte growth factor (HGF), osteopontin (OPN), and stromal-derived factor-1α (SDF1) [54]. The latter three can modulate cancer cells into a more stem cell-like phenotype, particularly through the Wnt/β-catenin pathway, increasing CD44v6 expression. IL-6 plays a critical role in immune regulation by promoting a chronic inflammatory environment and controlling NOTCH activation, a pathway responsible for intestinal stem cell self-renewal, cancer stem cell maintenance, TGF-induced EMT, and therapy resistance [54,56,57]. In our study, NOTCH3 was one of the top up-regulated genes in the NOTCH pathway, commonly aberrantly expressed in human cancers. In CRC, NOTCH3 expression increases with tumor staging and is correlated with worse prognosis, poor overall survival, and CMS4 tumors, which are CAF-enriched and also associated with TGF-β signaling [58,59]. NOTCH3 also plays a role in supporting cancer stemness and resistance to therapy [59], which can be circumvented by the use of NOTCH inhibitors since they reduce the expression of stem cell markers and improve response to chemotherapy and radiotherapy [57]. Therefore, our data opens the door for future studies addressing the effects of combining NOTCH inhibitors, such as gamma-secretase inhibitors, with KRAS-targeting therapies as a strategy to mitigate CAF-induced stemness and resistance to KRAS inhibitors.

Another way of inducing EMT by the canonical TGF-β pathway involves the activation of the Hippo pathway [60] through the regulation of several of its key effectors. One crucial interaction is between SMAD proteins, activated by TGF-β signaling, and YAP/TAZ, the primary effectors of the Hippo pathway. These proteins translocate to the nucleus, where they work together to regulate gene expression, controlling transcription of genes involved in EMT, cell proliferation, and survival [61]. KRAS-targeted inhibition (using inhibitors such as sotorasib or adagrasib) can also inadvertently activate Hippo signaling [62], which promotes the maintenance of EMT characteristics essential for cancer cells to overcome KRAS suppression and develop resistance against these inhibitors [60,62]. Our RNAseq results on KRAS-silenced cells align with the literature. Cells treated with conditioned media of fibroblasts show an up-regulation of genes in the Hippo pathway, including

YAP1, WWTR1/TAZ, and TEAD1, when compared with control treatment (Table S1). It is, therefore, plausible that in cancers harboring KRAS mutations, CAF-secreted factors can enhance the Hippo signaling pathway through the mediation of the TGF-β pathway. This interplay could further enhance EMT, facilitating tumor progression and resistance to KRAS-targeted therapies. More studies are required to further explore this putative mechanism and identify possible options to disrupt these pathways, improving therapeutic outcomes.

CAFs also play a significant role in regulating cancer metabolism, significantly impacting the metabolic landscape of the tumor microenvironment [63]. Alongside this, oncogenic KRAS also changes tumor metabolism to support quick proliferation and survival by increasing glucose uptake, enhancing glycolysis, and reprogramming lipid metabolism [64,65]. Interestingly, our RNAseq analysis revealed a down-regulation of pathways related to cholesterol, fatty acids, and xenobiotic metabolism in KRAS-silenced cells treated with fibroblast-conditioned media. Our results indicate that fibroblasts might be aiding the metabolic needs of cancer cells by providing them with substrates and signaling molecules, which can help them overcome the metabolic stress caused by KRAS inhibition. It is possible that cancer cells take up these metabolites to power oxidative phosphorylation and other growth processes. This cooperation allows cancer cells to reduce their own cholesterol and fatty acid production, relying instead on the metabolites provided by the CAFs [63]. The reduction in xenobiotic metabolism may also indicate that these cancer cells are conserving energy, focusing on growth and proliferation while leaving detoxification to the CAFs. Further studies are required to dissect the mechanisms behind this metabolic shift and identify its molecular mediators, guiding the exploration of combination therapies that target both metabolic adaptations and KRAS.

Our data pinpoints fibroblast infiltration as a potential biomarker for predicting a lack of response to KRAS-targeted treatments and lifts the veil on a fibroblast-mediated mechanism of resistance to KRAS inhibitors (Figure 6). Such insights can be particularly important in the case of CMS4 CRC, which is highly infiltrated by fibroblasts and might inherently resist KRAS inhibition. Furthermore, we previously showed that KRAS-silenced HCT116 cells promote fibroblast migration and activation [45]. When put together with our current findings, this suggests a feedback loop driving acquired resistance: fibroblast-poor CRCs, upon KRAS inhibition, recruit and activate fibroblasts into CAFs, which in turn support CRC growth independently of KRAS signaling. Our research underscores the critical need to explore these mechanisms further to improve the clinical management of KRAS mutant CRC.

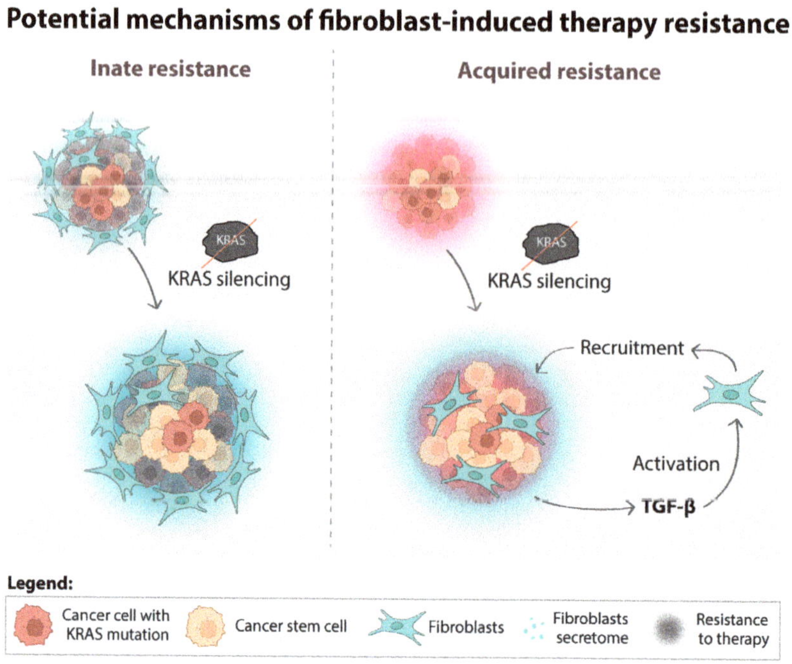

Figure 6. Fibroblast-mediated resistance to KRAS-targeted therapies: possible clinical implications. The combination of our current and previous results leads us to hypothesize two possible forms of resistance mechanisms mediated by the action of fibroblasts: innate and acquired. Innate resistance occurs in fibroblast-rich tumors, such as those from CMS4, that can inherently withstand KRAS inhibition through the support provided by resident fibroblasts. Conversely, acquired resistance may develop in initially fibroblast-poor tumors that recruit and activate fibroblasts in response to KRAS inhibition, altering the tumor microenvironment to its advantage. Both scenarios ultimately lead to resistance to KRAS-targeted therapies.

5. Conclusions

In conclusion, our findings underscore the need to rethink current therapeutic strategies for targeting KRAS-mutant CRCs. Targeting KRAS alone appears insufficient to tackle tumor growth effectively. CRC patients could benefit significantly from combined therapies that target not only KRAS but also the cancer-associated fibroblasts (CAFs) within the tumor microenvironment. Furthermore, our results highlight several potential CAF-derived molecular targets that merit further exploration as therapeutic interventions. By adopting a more comprehensive approach that addresses both cancer cells and their supportive environment, we can pave the way for more successful treatments for KRAS-mutant CRC.

Supplementary Materials: The following supporting information can be downloaded at: https://www.mdpi.com/article/10.3390/cancers16142595/s1, Figure S1: Western blot showing activation of fibroblasts, KRAS silencing in the different experiments, and gating strategy for flow cytometry; Figure S2: Original Western blots; Table S1: Differentially expressed genes with a 2.5 × fold change in the RNAseq analysis.

Author Contributions: Conceptualization: S.M.O., S.V. and M.J.O.; Formal analysis: S.M.O., P.O. and A.R.; Funding acquisition: S.V.; Investigation, S.M.O.; Methodology: S.M.O., P.D.C., A.S.-R., P.O., A.R., F.M., A.L.M. and J.C.; Project administration: S.V.; Resources: S.V.; Supervision: M.J.O. and S.V.; Visualization: S.M.O.; Writing—original draft: S.M.O.; Writing—review and editing: S.M.O. and S.V. All authors have read and agreed to the published version of the manuscript.

Funding: The following authors are supported by the Portuguese Foundation for Science and Technology (FCT): S.M.O. has a Ph.D. fellowship (10.54499/SFRH/BD/143642/2019), F.M. has a Ph.D. fellowship (10.54499/SFRH/BD/143669/2019), A.L.M. has a Ph.D. fellowship (10.54499/2020.08932.BD), P.O. is supported by FCT (10.54499/DL57/2016/CP1363/CT0013); J.C. is supported by FCT (10.54499/DL57/2016/CP1363/CT0012); S.V. is supported by FCT (10.54499/2021.01550.CEECIND/CP1663/CT0012). The work was funded by internal grants (MSI and Cancer Challenge 2022) provided by IPATIMUP and by national funds through FCT in the scope of the project 10.54499/2022.05346.PTDC. This work is part of the project "Porto Comprehensive Cancer Center Raquel Seruca (P.CCC.)"—NORTE-01-0145-FEDER-072678, supported by the Norte Portugal Regional Operational Programme (NORTE 2020), under the PORTUGAL 2020 Partnership Agreement, through the European Regional Development Fund (ERDF).

Institutional Review Board Statement: Not applicable.

Informed Consent Statement: Not applicable.

Data Availability Statement: The original data presented in the study are openly available in https://www.mdpi.com/article/10.3390/cancers16142595/s1, accessed on 29 May 2024 or article/Supplementary Material. Further inquiries can be directed to the corresponding author.

Acknowledgments: The authors acknowledge the support of the i3S Genomics Scientific Platform, in particular Ana Mafalda Rocha. The authors also acknowledge André Maia from the BioSciences Screening i3S Scientific Platform and Catarina Meireles and Emília Cardoso from the Translational Cytometry Unit (TraCy) i3S Scientific Platform.

Conflicts of Interest: The authors declare no conflict of interest. A.S.R. contributed to this work while previously affiliated with i3S and IPATIMUP. A.S.R. is currently affiliated with and employed by Novartis. Novartis did not fund or contribute to this work.

References

1. Arrington, A.K.; Heinrich, E.L.; Lee, W.; Duldulao, M.; Patel, S.; Sanchez, J.; Garcia-Aguilar, J.; Kim, J. Prognostic and Predictive Roles of KRAS Mutation in Colorectal Cancer. *Int. J. Mol. Sci.* **2012**, *13*, 12153–12168. [CrossRef] [PubMed]
2. Huang, L.; Guo, Z.; Wang, F.; Fu, L. KRAS Mutation: From Undruggable to Druggable in Cancer. *Signal Transduct. Target. Ther.* **2021**, *6*, 386. [CrossRef] [PubMed]
3. Salgia, R.; Pharaon, R.; Mambetsariev, I.; Nam, A.; Sattler, M. The Improbable Targeted Therapy: KRAS as an Emerging Target in Non-Small Cell Lung Cancer (NSCLC). *Cell Rep. Med.* **2021**, *2*, 100186. [CrossRef] [PubMed]
4. Patricelli, M.P.; Janes, M.R.; Li, L.S.; Hansen, R.; Peters, U.; Kessler, L.V.; Chen, Y.; Kucharski, J.M.; Feng, J.; Ely, T.; et al. Selective Inhibition of Oncogenic KRAS Output with Small Molecules Targeting the Inactive State. *Cancer Discov.* **2016**, *6*, 316–329. [CrossRef] [PubMed]
5. Moore, A.R.; Rosenberg, S.C.; McCormick, F.; Malek, S. RAS-Targeted Therapies: Is the Undruggable Drugged? *Nat. Rev. Drug Discov.* **2020**, *19*, 533–552. [CrossRef] [PubMed]
6. Akhave, N.S.; Biter, A.B.; Hong, D.S. Mechanisms of Resistance to Krasg12c-Targeted Therapy. *Cancer Discov.* **2021**, *11*, 1345–1352. [CrossRef]
7. Liu, J.; Kang, R.; Tang, D. The KRAS-G12C Inhibitor: Activity and Resistance. *Cancer Gene Ther.* **2022**, *29*, 875–878. [CrossRef] [PubMed]
8. Awad, M.M.; Liu, S.; Rybkin, I.I.; Arbour, K.C.; Dilly, J.; Zhu, V.W.; Johnson, M.L.; Heist, R.S.; Patil, T.; Riely, G.J.; et al. Acquired Resistance to KRAS G12C Inhibition in Cancer. *N. Engl. J. Med.* **2021**, *384*, 2382–2393. [CrossRef] [PubMed]
9. Adachi, Y.; Ito, K.; Hayashi, Y.; Kimura, R.; Tan, T.Z.; Yamaguchi, R.; Ebi, H. Epithelial-to-Mesenchymal Transition Is a Cause of Both Intrinsic and Acquired Resistance to KRAS G12C Inhibitor in KRAS G12C-Mutant Non-Small Cell Lung Cancer. *Clin. Cancer Res.* **2020**, *26*, 5962–5973. [CrossRef]
10. Tanaka, N.; Lin, J.J.; Li, C.; Ryan, M.B.; Zhang, J.; Kiedrowski, L.A.; Michel, A.G.; Syed, M.U.; Fella, K.A.; Sakhi, M.; et al. Clinical Acquired Resistance to Krasg12c Inhibition through a Novel Kras Switch-Ii Pocket Mutation and Polyclonal Alterations Converging on Ras–Mapk Reactivation. *Cancer Discov.* **2021**, *11*, 1913–1922. [CrossRef]
11. Singh, A.; Greninger, P.; Rhodes, D.; Koopman, L.; Violette, S.; Bardeesy, N.; Settleman, J. A Gene Expression Signature Associated with "K-Ras Addiction" Reveals Regulators of EMT and Tumor Cell Survival. *Cancer Cell* **2009**, *15*, 489–500. [CrossRef] [PubMed]
12. Muzumdar, M.D.; Chen, P.Y.; Dorans, K.J.; Chung, K.M.; Bhutkar, A.; Hong, E.; Noll, E.M.; Sprick, M.R.; Trumpp, A.; Jacks, T. Survival of Pancreatic Cancer Cells Lacking KRAS Function. *Nat. Commun.* **2017**, *8*, 1090. [CrossRef]
13. Scholl, C.; Fröhling, S.; Dunn, I.F.; Schinzel, A.C.; Barbie, D.A.; Kim, S.Y.; Silver, S.J.; Tamayo, P.; Wadlow, R.C.; Ramaswamy, S.; et al. Synthetic Lethal Interaction between Oncogenic KRAS Dependency and STK33 Suppression in Human Cancer Cells. *Cell* **2009**, *137*, 821–834. [CrossRef] [PubMed]

14. Singh, A.; Sweeney, M.F.; Yu, M.; Burger, A.; Greninger, P.; Benes, C.; Haber, D.A.; Settleman, J. TAK1 Inhibition Promote Apoptosis in KRAS-Dependent Colon Cancers. *Cell* **2012**, *148*, 639–650. [CrossRef] [PubMed]
15. Hong, D.S.; Fakih, M.G.; Strickler, J.H.; Desai, J.; Durm, G.A.; Shapiro, G.I.; Falchook, G.S.; Price, T.J.; Sacher, A.; Denlinge C.S.; et al. KRAS G12C Inhibition with Sotorasib in Advanced Solid Tumors. *N. Engl. J. Med.* **2020**, *383*, 1207–1217. [CrossRef [PubMed]
16. Yaeger, R.; Weiss, J.; Pelster, M.S.; Spira, A.I.; Barve, M.; Ou, S.-H.I.; Leal, T.A.; Bekaii-Saab, T.S.; Paweletz, C.P.; Heavey, G.A et al. Adagrasib with or without Cetuximab in Colorectal Cancer with Mutated KRAS G12C. *N. Engl. J. Med.* **2023**, *388*, 44–5 [CrossRef]
17. Wu, T.; Dai, Y. Tumor Microenvironment and Therapeutic Response. *Cancer Lett.* **2017**, *387*, 61–68. [CrossRef] [PubMed]
18. Prieto-Vila, M.; Takahashi, R.U.; Usuba, W.; Kohama, I.; Ochiya, T. Drug Resistance Driven by Cancer Stem Cells and Their Nich *Int. J. Mol. Sci.* **2017**, *18*, 2574. [CrossRef]
19. Hu, Y.; Yan, C.; Mu, L.; Huang, K.; Li, X.; Tao, D.; Wu, Y.; Qin, J. Fibroblast-Derived Exosomes Contribute to Chemoresistanc through Priming Cancer Stem Cells in Colorectal Cancer. *PLoS ONE* **2015**, *10*, e0125625. [CrossRef]
20. Clara, J.A.; Monge, C.; Yang, Y.; Takebe, N. Targeting Signalling Pathways and the Immune Microenvironment of Cancer Sten Cells—A Clinical Update. *Nat. Rev. Clin. Oncol.* **2020**, *17*, 204–232. [CrossRef]
21. Chippalkatti, R.; Abankwa, D. Promotion of Cancer Cell Stemness by Ras. *Biochem. Soc. Trans.* **2021**, *49*, 467–476. [CrossRef [PubMed]
22. Clevers, H. The Cancer Stem Cell: Premises, Promises and Challenges. *Nat. Med.* **2011**, *17*, 313–319. [CrossRef]
23. Haraguchi, N.; Ishii, H.; Mimori, K.; Ohta, K.; Uemura, M.; Nishimura, J.; Hata, T.; Takemasa, I.; Mizushima, T.; Yamamoto, H et al. CD49f-Positive Cell Population Efficiently Enriches Colon Cancer-Initiating Cells. *Int. J. Oncol.* **2013**, *43*, 425–430. [CrossRef [PubMed]
24. Ricci-Vitiani, L.; Lombardi, D.G.; Pilozzi, E.; Biffoni, M.; Todaro, M.; Peschle, C.; De Maria, R. Identification and Expansion o Human Colon-Cancer-Initiating Cells. *Nature* **2007**, *445*, 111–115. [CrossRef] [PubMed]
25. Dalerba, P.; Dylla, S.J.; Park, I.-K.; Liu, R.; Wang, X.; Cho, R.W.; Hoey, T.; Gurney, A.; Huang, E.H.; Simeone, D.M.; et al. Phenotypi Characterization of Human Colorectal Cancer Stem Cells. *Proc. Natl. Acad. Sci. USA* **2007**, *104*, 10158–10163. [CrossRef] [PubMed
26. Todaro, M.; Gaggianesi, M.; Catalano, V.; Benfante, A.; Iovino, F.; Biffoni, M.; Apuzzo, T.; Sperduti, I.; Volpe, S.; Cocorullo, G et al. CD44v6 Is a Marker of Constitutive and Reprogrammed Cancer Stem Cells Driving Colon Cancer Metastasis. *Cell Stem Ce* **2014**, *14*, 342–356. [CrossRef] [PubMed]
27. Lenos, K.J.; Miedema, D.M.; Lodestijn, S.C.; Nijman, L.E.; van den Bosch, T.; Romero Ros, X.; Lourenço, F.C.; Lecca, M.C.; van de Heijden, M.; van Neerven, S.M.; et al. Stem Cell Functionality Is Microenvironmentally Defined during Tumour Expansion an Therapy Response in Colon Cancer. *Nat. Cell Biol.* **2018**, *20*, 1193–1202. [CrossRef] [PubMed]
28. Guinney, J.; Dienstmann, R.; Wang, X.; De Reyniès, A.; Schlicker, A.; Soneson, C.; Marisa, L.; Roepman, P.; Nyamundanda, G Angelino, P.; et al. The Consensus Molecular Subtypes of Colorectal Cancer. *Nat. Med.* **2015**, *21*, 1350–1356. [CrossRef] [PubMed
29. Wu, F.; Yang, J.; Liu, J.; Wang, Y.; Mu, J.; Zeng, Q.; Deng, S.; Zhou, H. Signaling Pathways in Cancer-Associated Fibroblasts an Targeted Therapy for Cancer. *Signal Transduct. Target. Ther.* **2021**, *6*, 218. [CrossRef]
30. Yu, Y.; Xiao, C.H.; Tan, L.D.; Wang, Q.S.; Li, X.Q.; Feng, Y.M. Cancer-Associated Fibroblasts Induce Epithelial-Mesenchyma Transition of Breast Cancer Cells through Paracrine TGF-β Signalling. *Br. J. Cancer* **2014**, *110*, 724–732. [CrossRef]
31. Valcz, G.; Buzás, E.I.; Szállási, Z.; Kalmár, A.; Krenács, T.; Tulassay, Z.; Igaz, P.; Molnár, B. Perspective: Bidirectional Exosoma Transport between Cancer Stem Cells and Their Fibroblast-Rich Microenvironment during Metastasis Formation. *NPJ Breas Cancer* **2018**, *4*, 18. [CrossRef] [PubMed]
32. Linares, J.; Marín-Jiménez, J.A.; Badia-Ramentol, J.; Calon, A. Determinants and Functions of CAFs Secretome During Cance Progression and Therapy. *Front. Cell Dev. Biol.* **2021**, *8*, 621070. [CrossRef] [PubMed]
33. Dias Carvalho, P.; Martins, F.; Carvalho, J.; Oliveira, M.J.; Velho, S. Mutant KRAS-Associated Proteome Is Mainly Controlled b Exogenous Factors. *Cells* **2022**, *11*, 1988. [CrossRef] [PubMed]
34. Ahmed, D.; Eide, P.W.; Eilertsen, I.A.; Danielsen, S.A.; Eknæs, M.; Hektoen, M.; Lind, G.E.; Lothe, R.A. Epigenetic and Geneti Features of 24 Colon Cancer Cell Lines. *Oncogenesis* **2013**, *2*, e71. [CrossRef] [PubMed]
35. Kurobe, M.; Furukawi, S.; Hayashi, K. Synthesis and Secretion of an Epidermal Growth Factor (EGF) by Human Fibroblast Cell in Culture. *Biochem. Biophys. Res. Commun.* **1985**, *131*, 1080–1085. [CrossRef] [PubMed]
36. Powers, C.J.; Mcleskey, S.W.; Wellstein, A. Fibroblast Growth Factors, Their Receptors and Signaling. *Endocr. Relat. Cancer* **2000**, 7 165–197. [CrossRef]
37. Pratsinis, H.; Kletsas, D. Growth Factors in Fetal and Adult Wound Healing. In *Wound Healing Biomaterials*; Elsevier Inc Amsterdam, The Netherlands, 2016; Volume 1, pp. 41–68. ISBN 9780081006054.
38. Liberzon, A.; Birger, C.; Thorvaldsdóttir, H.; Ghandi, M.; Mesirov, J.P.; Tamayo, P. The Molecular Signatures Database Hallmar Gene Set Collection. *Cell Syst.* **2015**, *1*, 417–425. [CrossRef] [PubMed]
39. Mootha, V.K.; Lindgren, C.M.; Eriksson, K.-F.; Subramanian, A.; Sihag, S.; Lehar, J.; Puigserver, P.; Carlsson, E.; Ridderstråle, M Laurila, E.; et al. PGC-1α-Responsive Genes Involved in Oxidative Phosphorylation Are Coordinately Downregulated in Humar Diabetes. *Nat. Genet.* **2003**, *34*, 267–273. [CrossRef]

40. Subramanian, A.; Tamayo, P.; Mootha, V.K.; Mukherjee, S.; Ebert, B.L.; Gillette, M.A.; Paulovich, A.; Pomeroy, S.L.; Golub, T.R.; Lander, E.S.; et al. Gene Set Enrichment Analysis: A Knowledge-Based Approach for Interpreting Genome-Wide Expression Profiles. *Proc. Natl. Acad. Sci. USA* **2005**, *102*, 15545–15550. [CrossRef]
41. Wiecek, A.J.; Cutty, S.J.; Kornai, D.; Parreno-Centeno, M.; Gourmet, L.E.; Tagliazucchi, G.M.; Jacobson, D.H.; Zhang, P.; Xiong, L.; Bond, G.L.; et al. Genomic Hallmarks and Therapeutic Implications of G0 Cell Cycle Arrest in Cancer. *Genome Biol.* **2023**, *24*, 128. [CrossRef]
42. Krebsbach, P.H.; Villa-Diaz, L.G. The Role of Integrin A6 (CD49f) in Stem Cells: More than a Conserved Biomarker. *Stem Cells Dev.* **2017**, *26*, 1090–1099. [CrossRef] [PubMed]
43. Desgrosellier, J.S.; Cheresh, D.A. Integrins in Cancer: Biological Implications and Therapeutic Opportunities. *Nat. Rev. Cancer* **2010**, *10*, 9–22. [CrossRef] [PubMed]
44. Frisch, S.M.; Schaller, M.; Cieply, B. Mechanisms That Link the Oncogenic Epithelial- Mesenchymal Transition to Suppression of Anoikis. *J. Cell Sci.* **2013**, *126*, 21–29. [CrossRef] [PubMed]
45. Dias Carvalho, P.; Mendonça, S.; Martins, F.; Oliveira, M.J.; Velho, S. Modulation of Fibroblast Phenotype by Colorectal Cancer Cell-Secreted Factors Is Mostly Independent of Oncogenic KRAS. *Cells* **2022**, *11*, 2490. [CrossRef] [PubMed]
46. Yeo, M.K.; Lee, Y.M.; Seong, I.O.; Choi, S.Y.; Suh, K.S.; Song, K.S.; Lee, C.S.; Kim, J.M.; Kim, K.H. Up-Regulation of Cytoplasmic CD24 Expression Is Associated with Malignant Transformation but Favorable Prognosis of Colorectal Adenocarcinoma. *Anticancer. Res.* **2016**, *36*, 6593–6598. [CrossRef] [PubMed]
47. Paschall, A.V.; Yang, D.; Lu, C.; Redd, P.S.; Choi, J.-H.; Heaton, C.M.; Lee, J.R.; Nayak-Kapoor, A.; Liu, K. CD133 + CD24 Lo Defines a 5-Fluorouracil-Resistant Colon Cancer Stem Cell-like Phenotype. *Oncotarget* **2016**, *7*, 78698–78712. [CrossRef] [PubMed]
48. Choi, D.; Lee, H.W.; Hur, K.Y.; Kim, J.J.; Park, G.S.; Jang, S.H.; Song, Y.S.; Jang, K.S.; Paik, S.S. Cancer Stem Cell Markers CD133 and CD24 Correlate with Invasiveness and Differentiation in Colorectal Adenocarinoma. *World J. Gastroenterol.* **2009**, *15*, 2258–2264. [CrossRef] [PubMed]
49. Ahmed, M.A.H.; Al-Attar, A.; Kim, J.; Watson, N.F.S.; Scholefield, J.H.; Durrant, L.G.; Ilyas, M. CD24 Shows Early Upregulation and Nuclear Expression but Is Not a Prognostic Marker in Colorectal Cancer. *J. Clin. Pathol.* **2009**, *62*, 1117–1122. [CrossRef] [PubMed]
50. Choi, S.H.; Kim, J.K.; Chen, C.T.; Wu, C.; Marco, M.R.; Barriga, F.M.; O'Rourke, K.; Pelossof, R.; Qu, X.; Chang, Q.; et al. KRAS Mutants Upregulate Integrin B4 to Promote Invasion and Metastasis in Colorectal Cancer. *Mol. Cancer Res.* **2022**, *20*, 1305–1319. [CrossRef]
51. Beaulieu, J.F. Integrin A6β4 in Colorectal Cancer: Expression, Regulation, Functional Alterations and Use as a Biomarker. *Cancers* **2020**, *12*, 41. [CrossRef]
52. Tammaccaro, S.L.; Prigent, P.; Le Bail, J.C.; Dos-Santos, O.; Dassencourt, L.; Eskandar, M.; Buzy, A.; Venier, O.; Guillemot, J.C.; Veeranagouda, Y.; et al. TEAD Inhibitors Sensitize KRASG12C Inhibitors via Dual Cell Cycle Arrest in KRASG12C-Mutant NSCLC. *Pharmaceuticals* **2023**, *16*, 553. [CrossRef] [PubMed]
53. Hallin, J.; Engstrom, L.D.; Hargi, L.; Calinisan, A.; Aranda, R.; Briere, D.M.; Sudhakar, N.; Bowcut, V.; Baer, B.R.; Ballard, J.A.; et al. The KRASG12C Inhibitor MRTX849 Provides Insight toward Therapeutic Susceptibility of KRAS-Mutant Cancers in Mouse Models and Patients. *Cancer Discov.* **2020**, *10*, 54–71. [CrossRef] [PubMed]
54. Fiori, M.E.; Di Franco, S.; Villanova, L.; Bianca, P.; Stassi, G.; De Maria, R. Cancer-Associated Fibroblasts as Abettors of Tumor Progression at the Crossroads of EMT and Therapy Resistance. *Mol. Cancer* **2019**, *18*, 70. [CrossRef] [PubMed]
55. Zhuang, J.; Lu, Q.; Shen, B.; Huang, X.; Shen, L.; Zheng, X.; Huang, R.; Yan, J.; Guo, H. TGFβ1 Secreted by Cancer-Associated Fibroblasts Induces Epithelial-Mesenchymal Transition of Bladder Cancer Cells through LncRNA-ZEB2NAT. *Sci. Rep.* **2015**, *5*, 11924. [CrossRef] [PubMed]
56. Deshmukh, A.P.; Vasaikar, S.V.; Tomczak, K.; Tripathi, S.; den Hollander, P.; Arslan, E.; Chakraborty, P.; Soundararajan, R.; Kumar Jolly, M.; Rai, K.; et al. Identification of EMT Signaling Cross-Talk and Gene Regulatory Networks by Single-Cell RNA Sequencing. *Proc. Natl. Acad. Sci. USA* **2021**, *118*, e2102050118. [CrossRef] [PubMed]
57. Ebrahimi, N.; Afshinpour, M.; Fakhr, S.S.; Kalkhoran, P.G.; Shadman-Manesh, V.; Adelian, S.; Beiranvand, S.; Rezaei-Tazangi, F.; Khorram, R.; Hamblin, M.R.; et al. Cancer Stem Cells in Colorectal Cancer: Signaling Pathways Involved in Stemness and Therapy Resistance. *Crit. Rev. Oncol. Hematol.* **2023**, *182*, 103920. [CrossRef] [PubMed]
58. Varga, J.; Nicolas, A.; Petrocelli, V.; Pesic, M.; Mahmoud, A.; Michels, B.E.; Etlioglu, E.; Yepes, D.; Haüpl, B.; Ziegler, P.K.; et al. AKT-Dependent NOTCH3 Activation Drives Tumor Progression in a Model of Mesenchymal Colorectal Cancer. *J. Exp. Med.* **2020**, *217*, e20191515. [CrossRef]
59. Xiu, M.; Wang, Y.; Li, B.; Wang, X.; Xiao, F.; Chen, S.; Zhang, L.; Zhou, B.; Hua, F. The Role of Notch3 Signaling in Cancer Stemness and Chemoresistance: Molecular Mechanisms and Targeting Strategies. *Front. Mol. Biosci.* **2021**, *8*, 694141. [CrossRef]
60. Shao, D.D.; Xue, W.; Krall, E.B.; Bhutkar, A.; Piccioni, F.; Wang, X.; Schinzel, A.C.; Sood, S.; Rosenbluh, J.; Kim, J.W.; et al. KRAS and YAP1 Converge to Regulate EMT and Tumor Survival. *Cell* **2014**, *158*, 171–184. [CrossRef]
61. Varelas, X.; Sakuma, R.; Samavarchi-Tehrani, P.; Peerani, R.; Rao, B.M.; Dembowy, J.; Yaffe, M.B.; Zandstra, P.W.; Wrana, J.L. TAZ Controls Smad Nucleocytoplasmic Shuttling and Regulates Human Embryonic Stem-Cell Self-Renewal. *Nat. Cell Biol.* **2008**, *10*, 837–848. [CrossRef]

62. Mukhopadhyay, S.; Huang, H.Y.; Lin, Z.; Ranieri, M.; Li, S.; Sahu, S.; Liu, Y.; Ban, Y.; Guidry, K.; Hu, H.; et al. Genome-Wide CRISPR Screens Identify Multiple Synthetic Lethal Targets That Enhance KRASG12C Inhibitor Efficacy. *Cancer Res.* **2023**, *83*, 4095–4111. [CrossRef] [PubMed]
63. Karta, J.; Bossicard, Y.; Kotzamanis, K.; Dolznig, H.; Letellier, E. Mapping the Metabolic Networks of Tumor Cells and Cancer-associated Fibroblasts. *Cells* **2021**, *10*, 304. [CrossRef] [PubMed]
64. Gaglio, D.; Metallo, C.M.; Gameiro, P.A.; Hiller, K.; Danna, L.S.; Balestrieri, C.; Alberghina, L.; Stephanopoulos, G.; Chiaradonna, F. Oncogenic K-Ras Decouples Glucose and Glutamine Metabolism to Support Cancer Cell Growth. *Mol. Syst. Biol.* **2011**, *7*, 523. [CrossRef] [PubMed]
65. Pupo, E.; Avanzato, D.; Middonti, E.; Bussolino, F.; Lanzetti, L. KRAS-Driven Metabolic Rewiring Reveals Novel Actionable Targets in Cancer. *Front. Oncol.* **2019**, *9*, 848. [CrossRef] [PubMed]

cancers

MDPI

Review

Neo-RAS Wild Type or RAS Conversion in Metastatic Colorectal Cancer: A Comprehensive Narrative Review

Guido Pesola [1,†], Samantha Epistolio [2,†], Marco Cefalì [1], Elena Trevisi [1], Sara De Dosso [1,3,‡] and Milo Frattini [2,*,‡]

1 Oncology Institute of Southern Switzerland (IOSI), Ente Ospedaliero Cantonale, 6500 Bellinzona, Switzerland; guido.pesola@eoc.ch (G.P.); marco.cefali@eoc.ch (M.C.); elena.trevisi@eoc.ch (E.T.); sara.dedosso@eoc.ch (S.D.D.)
2 Laboratory of Genetics and Molecular Pathology, Istituto Cantonale di Patologia EOC, 6600 Locarno, Switzerland; samantha.epistolio@eoc.ch
3 Faculty of Biomedical Sciences, Università della Svizzera Italiana, 6900 Lugano, Switzerland
* Correspondence: milo.frattini@eoc.ch
† These authors contributed equally to this work and share co-first authorship.
‡ These authors contributed equally to this work and share co-last authorship.

Simple Summary: Patients with RAS-mutant metastatic colorectal cancer are typically treated with chemotherapy, with or without bevacizumab, as the first-line therapy. Over time, tumors in some patients may undergo plasma clearance of RAS, transitioning from being RAS-mutant to RAS-wild type, a phenomenon known as "RAS conversion" or "neo-RAS wild type". The current review focuses on this phenomenon's incidence, evaluation methodologies, and therapeutic implications, with a focus on the role that bevacizumab plays in it and its prospects.

Abstract: The management of metastatic colorectal cancer in patients harboring RAS mutations primarily involves chemotherapy, often combined with bevacizumab, as a standard first-line treatment. However, emerging evidence suggests that tumors in a subset of these patients may experience a conversion from RAS-mutant status to RAS wild type (wt) during or after chemotherapy, a process referred to as "RAS conversion" or "neo-RAS wt". Understanding the mechanisms driving the neo-RAS wt phenomenon is crucial for its application in personalized medicine. Hypotheses suggest that selective pressure from chemotherapy may lead to a decrease in the number of mutant RAS clones or an outgrowth of pre-existing RAS wt clones. Further research is needed to validate these mechanisms and understand the impact of the neo-RAS wt phenomenon on long-term outcomes, such as overall survival and progression-free survival. This review provides a comprehensive overview of the current understanding of the neo-RAS wt phenomenon, including its incidence, potential mechanisms, and clinical implications.

Keywords: RAS conversion; neo-RAS wild type; liquid biopsy; colorectal cancer; bevacizumab

1. Introduction

Mutations in RAS genes are common in patients with colorectal cancer (CRC), occurring in nearly 40% of all CRC cases, and result in resistance to treatment with epidermal growth factor receptor (EGFR) monoclonal antibodies [1,2]. Therefore, investigating the mutational status of the RAS gene is crucial for selecting the optimal treatment combination for CRC [3]. Standard first-line systemic therapy for RAS-mutant metastatic CRC (mCRC) typically involves fluoropyrimidine-based chemotherapy with the addition of the anti-angiogenic drug bevacizumab. This drug has been shown to increase overall survival (OS) (hazard ratio [HR]: 0.79; 95% confidence interval [CI]: 0.69–0.90; $p = 0.0005$), progression-free survival (PFS) (HR: 0.63; 95% CI: 0.49–0.81; $p = 0.0004$), and the response rate (RR: 1.50; 95% CI: 1.06–2.10; $p = 0.02$) in this population [4].

Even though most RAS-mutant patients treated with systemic therapy tend to remain mutated over time [5], clonal evolution can occur in some cases, leading to the disappearance of RAS-mutant clones [6–13].

Current evidence on clonal evolution primarily relies on liquid biopsy assessment, particularly plasma samples from which circulating tumor DNA (ctDNA) has been extracted. Exceptions include the studies by Epistolio et al. and Arici et al., where mutations were evaluated in resected metastases [14,15].

The liquid biopsy approach offers significant advantages: being minimally invasive, it can be repeated serially over time, and it allows real-time monitoring of tumor recurrence, metastasis, or therapeutic response, as has been widely demonstrated in recent years [6–9,11,12,16–18].

Various molecular liquid biopsy-based techniques have been considered for the treatment of CRC, such as the evaluation of circulating tumor cells, ctDNA, exosomes, and tumor-educated platelets [17]. These techniques are showing promising results in ongoing trials for choosing therapeutic strategies and predicting the recurrence or prognosis of CRC [17,19].

The disappearance of RAS clones in the plasma of patients diagnosed with RAS mutations is known as "neo-RAS wild type (wt)" or "RAS conversion". Current scientific evidence regarding this phenomenon is mixed. The concept of "neo-RAS wt" has not been unambiguously defined, partly due to the different thresholds and cut-off points applied to the methods and technologies that have been employed for liquid biopsy analysis used so far (e.g., quantitative PCR [qPCR, TaqMan methodology], Ion Torrent™ [Thermo Fisher Scientific, Waltham, MA, USA], Idylla™ [Biocartis, Melchen, Belgium], MassARRAY System® [Agena Bioscience, San Diego, CA, USA], Guardant360™ [Guardant Health, Palo Alto, CA], OncoBEAM™ [Sysmex Suisse, Horgen, Switzerland], and HapOnco CDx™ [Haplox Biotechnology, Hong Kong]). Each different methodology is based on its own limits of detection (LODs) for the RAS mutation, making it difficult to provide a global definition of the neo-RAS wt phenomenon.

In this narrative review, we aim to summarize the current literature regarding the neo-RAS wt phenomenon. Specifically, our objective is to cover this phenomenon's incidence rates, evaluation methodologies, and therapeutic implications, with a focus on the impact of bevacizumab on it and its prospects in this field.

2. General Considerations and the Role of RAS Mutations in CRC

CRC is the third most common cancer worldwide, accounting for approximately 10% of all cancer cases, and the second leading cause of cancer-related deaths. In 2022, global cancer statistics reported 1,926,118 new CRC cases [20]. When CRC is diagnosed at an advanced metastatic stage, treatment options are limited, and the five-year OS rate is lower than 20% [21]. Despite many cancer treatment advances, the therapeutic approach for stage IV CRC is generally not curative, focusing instead on increasing OS with an acceptable quality of life [3].

The introduction of specific anti-EGFR therapies, in the form of monoclonal antibodies, represented an improvement in the treatment of advanced stages of CRC. However, it soon became clear that molecular alterations in EGFR alone are insufficient to distinguish which patients would benefit from these therapies and which would not, mainly due to the molecular complexity of CRCs [22].

Analyses of not only EGFR but also all the downstream pathways have revealed how underlying markers can signify resistance mechanisms to anti-EGFR therapies.

Some of the most relevant mechanisms of resistance to anti-EGFR drugs are mutations in the RAS family genes. These mutations activate the RAS proteins and, consequently, continuously simulate downstream pathways, resulting in cell proliferation and survival. This constitutive activation can suppress the efficacy of anti-EGFR therapies [23]. KRAS mutations are present in nearly 40% of CRC cases, and the prognostic role of this gene is closely associated with the localization of these mutations and the specific mutation variants [23]. It has been demonstrated that RAS mutations are associated with a worse prognosis compared to wt (25.8 months versus 35.1 months, respectively; $p = 0.006$), especially those that occur in codon 12 when compared to codon 13 (22.4 months versus

24.8 months, respectively) [24]. The main studies that investigated the interplay between EGFR downstream alterations and the efficacy of EGFR-targeted therapies were the CRYSTAL, OPUS, and PRIME trials [25]. The first two clinical trials led to the approval of the EGFR antibody cetuximab for the treatment of mCRC patients, showing that the combination of cetuximab with chemotherapy (FOLFIRI or FOLFOX) in patients whose tumors were KRAS wt improved OS (HR: 0.81; $p = 0.0062$), PFS (HR: 0.66; $p < 0.0001$), and ORR (OR: 2.16; $p < 0.0001$) compared to chemotherapy alone [25,26].

Regarding the monoclonal antibody panitumumab, the PRIME phase III study showed that treating KRAS wt patients with this drug along with FOLFOX in the first-line treatment improved median PFS compared to FOLFOX alone (8.6 months [95% CI: 7.5–9.5 months] versus 10.0 months [95% CI: 9.3–11.4 months], respectively) [27]. Similar results were obtained in the second line with an improvement in PFS comparing panitumumab plus FOLFIRI versus FOLFIRI alone (5.9 versus 3.9 months; HR: 0.71; $p = 0.004$) [28].

This body of evidence led international agencies (FDA and EMA) to approve the administration of cetuximab and panitumumab in only KRAS wt cases.

Therefore, KRAS-mutant patients are usually excluded from treatment with EGFR-targeted therapies. However, the development of drugs specifically targeting KRAS p.G12C mutations has prompted consideration of combining anti-EGFR therapies with KRAS-specific inhibitors in patients with the KRAS p.G12C mutation. Preliminary results from clinical trials have shown that direct inhibition of KRAS p.G12C is now possible, potentially leading to the development of a novel targeted treatment for several patients with advanced CRC [29]. Indeed, the combination of a KRAS p.G12C inhibitor and an anti-EGFR treatment has been demonstrated to improve PFS versus standard care among patients with chemo-refractory mCRC harboring a KRAS p.G12C mutation in the Code Break 300 and KRYSTAL-1 trials [30,31].

Recent advances in personalized medicine for CRC have focused on refining the selection of patients who may benefit from targeted therapies. A key aspect of this approach involves hyper-selecting patients based on specific molecular alterations, as evidenced in the usage of PRESSING panels, which incorporate an expanded range of genetic markers to guide treatment decisions [32,33]. Additionally, ctDNA analysis has emerged as a valuable tool in clinical trials, such as the PARADIGM and FIRE-4 studies, where it has demonstrated the ability to further refine patient selection, specifically by identifying patients unlikely to respond to anti-EGFR therapy [34,35]. These strategies represent a significant step forward in tailoring treatments to individual patients, potentially improving outcomes in CRC therapy.

3. Neo-RAS wt: A New Reality

CRCs are highly heterogeneous tumors characterized by cells harboring different mutational profiles [36].

However, molecular concordance between biopsies of the primary tumor and the tumor during metastasis is very high at the time of diagnosis, often exceeding 90% of cases, as shown in a meta-analysis [37].

The presence of heterogeneity necessitates a methodology capable of fully assessing the different types of genetic alterations present in these tumors. It is well known that tumors can release DNA directly into the circulatory stream as ctDNA or via circulating tumor cells or exosomes. The introduction of ctDNA analysis has allowed for a comprehensive overview of cancer genetics as well as clonal evolution of CRC [34,38]. It is worth noting that 20% of CRCs do not secrete ctDNA, particularly in patients with lung metastases or peritoneal carcinosis [39,40]. Consequently, this subgroup could present non-representative results from liquid biopsy analyses. A potential advantage of liquid biopsy is its ability to provide a more comprehensive snapshot of tumor dynamics and genetic alterations over time, offering insights that a single tissue biopsy sample may not capture. This is because liquid biopsy does not suffer from the sampling bias that can affect tissue biopsy, which may capture only a subset of the tumor's heterogeneous cell population.

A concept that has gained significant relevance in recent times is the disappearance of RAS mutations observed in ctDNA from plasma over the course of the disease. This phenomenon has been referred to by various names, such as RAS reversion or RAS conversion, but it is now more commonly known as "neo-RAS wt". A biological explanation for this phenomenon has yet to be defined in the literature. Many attempts have been made to find a biological explanation for the concerns associated with the neo-RAS wt transformation, but none of them has fully explained all the aspects of this phenomenon until now.

The possibility that anti-cancer therapy can exert pressure on mutant RAS cell clones is supported by two publications on tissue analysis: the first from our group, which showed a reduction in the variant allele frequency of RAS mutations in resected liver metastases from bevacizumab-based systemic therapy-pretreated patients [15], and the second by Arici et al., which showed that 9.3% of their resected metastases had lost the RAS mutation [14]. These results suggest that it could be very useful for patients to assess the RAS mutational status after first-line treatment and that liquid biopsy represents the less invasive technique as compared to tissue biopsy, although the analysis of a tissue biopsy can also be performed if there is a need for a histological evaluation of the disease for proper clinical management.

The change in the RAS mutational status may be explained by the hypothesis that systemic therapy administered to patients affected by KRAS-mutant mCRC can have a better effect on KRAS-mutant cells even if this treatment does not include specific KRAS inhibitors. Consequently, a tumor majorly characterized by KRAS-mutant clones at diagnosis may become mostly KRAS wt. The motivation behind this selection is, to date, a matter of debate and has not been clearly described in the literature. Many biological pathways may be influenced by systemic therapy that can cause clonal selection of some molecular features, but no definitive demonstration of this has been reported so far in the literature.

The disappearance of an RAS mutation in the absence of systemic treatment could account for the percentage of patients for whom there is no concordance between primary tumor and metastasis. This discrepancy of plasma versus tissue might also be imputable to the limitations of ctDNA detection or spiral and temporal heterogeneity in RAS-mt tumor clones within the tumor issue or other aspects such as long intervals between assessments of the molecular status in tumor tissue and ctDNA, resection of the primary tumor at the time of blood draw, tumor site, and type of tissue analyzed [39].

Nonetheless, it should be noted that in the current evidence on the neo-RAS wt phenomenon, the presence of circulating mutations prior to the initiation of systemic therapy was not systematically reported. Reports from clinical trials are less affected by this bias [11,41–45].

In studies published between 2019 and 2024, the incidence of patients presenting with RAS regression is highly variable [5,6,11,13,18,21,41–47]. All authors who have explored the incidence of this phenomenon with different methodologies have highlighted this, albeit with very variable frequencies, ranging from 5.5 to 78%.

We have summarized the studies that reported having at least 50 patients in Table 1.

Table 1. Incidence of patients presenting with RAS regression and methodologies applied for identification of Neo-RAS wt from 2019 to 2024.

First Author, Year of Publication	Population (Real-World Prospective Study Versus Clinical Trial Versus Case Series)	Number of Patients	Type of ctDNA Analysis	RASmut Patients with Evaluable ctDNA Follow-Up	Neo-RAS wt Patients (n)	Neo-RAS wt Patients (%)
Henry, 2020 [41]	Real world	236	Guardant360™ (Guardant Health)	202	12	6.0%
Moati, 2020 [42]	Real world	61	Ion Torrent™ (Thermo Fisher Scientific)	36	2	5.5%
Wang, 2022 [43]	Real world	171	HapOncoCDx™ (Haplox Biotechnology)	61	26	42.6%
Sato, 2022 [11]	Real world	129	OncoBEAM™ (Sysmex)	62	27	43.5%
Sunakawa, 2022 [44]	Clinical trial	62	qPCR (TaqMan methodology)	41	32	78.0%
Nicolazzo, 2023 [18]	Case series	82	Idylla™ (Biocartis)	70	42	60.0%
Osumi, 2024 [48]	Clinical trial	478 (group A) 429 (group B)	Guardant360™ (Guardant Health)	478 (group A) 429 (group B)	91 (group A) 92 (group B)	19.8% (group A) 9.8% (group B)
Wu, 2023 [45]	Clinical trial	95	GuardantOMNI (Guardant Health)	95	6	6.3%

ctDNA: circulating tumour DNA; Neo-RAS wt: neo-RAS wild-type; RASmut: RAS mutated.

One of the most scientifically productive groups to have described this phenomenon is that of Nicolazzo and colleagues, who produced a total of four reports between 2019 and 2023 [9,18,49,50]. In their latest report in 2023, they described the presence of neo-RAS wt in 60% of the cases they analyzed, with 42 out of 70 patients given the first-line treatment followed longitudinally with Idylla (Biocartis). A high incidence was also reported by Wang and colleagues using HapOncoCDx (Haplox Biotechnology); they reported that 42.6% of 61 patients were given first-line treatment [43]. On the other hand, Osumi and colleagues reported the largest population based on the nationwide Japanese screening platform SCRUM-Japan GOZILA regarding patients with an initial diagnosis of RAS-mutant mCRC in different lines of therapies; the cfDNA test Guardant 360 highlighted a prevalence of 19.0% (91/478) of neo-RAS wt in Group A (all eligible patients) and 9.8% (42/429) in Group B, a subgroup with at least one somatic alteration detected in plasma [48]. Henry and colleagues at the MD Anderson Cancer Center found a low incidence of neo-RAS wt, between 2 and 8%, in two cohorts, analyzing a total of 236 patients in all line therapy settings [41].

Many confounding factors and biases can explain the large differences in terms of incidence, apart from the ctDNA methodology aspect, which will be discussed in the next section.

One critical aspect concerns the study population of the studies shown in the table, which is very heterogeneous.

Patients treated in the first-line setting have a higher incidence, as reported in the JACCRO CC-11 trial by Sunakawa et al., the clinical trial of Wang et al., and the series reported by Nicolazzo et al. [18,43,44]. However, the PLACOL study reported by Moati et al., which included patients in first-line treatment, reported low conversion rates [42].

In contrast, the reports by Henry, Osumi, and Wu covered more heterogeneous case histories with patients in various lines of treatment. In this case, the neo-RAS-wt rates were lower [16,41,45].

From an overall non-systematic evaluation of the studies, patients during the first line of therapy might have a greater chance of presenting a neo-RAS wt status. A possible biological rationale could be the smaller difference in terms of clonal heterogeneity and tumor burden in the first-line setting compared to heavily pre-treated patients.

On the other hand, inclusion in a clinical trial setting and the number of patients being high do not seem to be determining factors.

Furthermore, Osumi and colleagues described the characteristics of neo-RAS-wt patients with a regression multivariate analysis. They showed a higher conversion rate for patients without liver metastasis, smaller tumor diameter, and tissue RAS mutation other than KRAS exon 2 [16].

4. Methodologies for Neo-RAS wt Evaluation

Besides patient characteristics, treatment types, and the number of lines of therapies, another factor that can explain most of the differences observed in the neo-RAS wt status evaluation among the various studies published in the literature is the type of molecular tests applied to a given cohort. The reasoning behind this is that we are discussing the possibility of identifying mutations in liquid biopsies, and, to date, the only consensus among researchers regarding this is the use of a technique with "sufficient" sensitivity. Currently, there is no gold-standard method for analyzing ctDNA in plasma, nor is there a consensus on the minimum sensitivity required for the methodology. The methodologies available on the market are characterized by a wide range of sensitivities for identifying mutant alleles (not only KRAS), especially in contexts of high dilution with normal alleles and significant ctDNA degradation. The methodologies employed for investigating the presence of ctDNA are heterogeneous across all these studies, but all authors employed either qPCR (e.g., TaqMan PCR assays and Idylla™, Biocartis), NGS (e.g., Ion Torrent™, ThermoFisher Scientific; Guardant 360™ and Guardant OMNI™, Guardant Health; and HapOncoCDx™, Haplox Biotechnology), ddPCR (e.g., OncoBEAM™, Sysmex), or the

MassARRAY® System methodology (Agena Bioscience), which combines mass spectrometry with end-point PCR [5,6,9,11,16,18,21,41–47,49,51]. None of these methodologies is better or more sensible than the others because they are really different from each other regarding their strengths and limitations. The most common methodologies employ the Idylla™ (Biocartis) and Guardant™ technologies, both applied in three publications analyzing a total of six different populations, with the number of cases ranging from 23 to 82 for the IdyllaTM (Biocartis) and 95 to 478 for the Guardant™ technology [13,18,21,41,45,46,52]. The Idylla™ technology (Biocartis) has the strengths of having a high sensitivity (it can find nearly the totality of mutations analyzed—more than 73.9%) [5] and being easy to perform; however, this methodology can analyze only one marker for each experiment. A lower sensitivity (5.1–8.7%) [52] characterizes the Guardant™ technology because, being an NGS methodology, this methodology analyzes different genes in a single run. The analysis of multiple markers decreases the sensitivity of the methodology, which is also an advantage of this methodology. A disadvantage of Guardant™, when compared to Idylla™ (Biocartis), is that it requires highly specialized personnel to carry out the experiment. Comparing the qPCR methods, we can affirm that Idylla™ (Biocartis) has the same sensitivity as other TaqMan PCR methodologies but offers the advantages of being easier to apply and using less time from DNA extraction to mutation detection; it can complete this in a single step on a single cartridge. Another NGS methodology is Ion Torrent™ (Thermo Fisher Scientific). Only two studies employed the NGS Ion Torrent™ technique (Thermo Fisher Scientific) [42,47]. In one of these [42], even though the number of patients may be considered representative of a real-world population (n = 61), it detected a very low percentage of neo-RAS wt patients (3.3%). Like Guardant™, the strength of Ion Torrent™ (Thermo Fisher Scientific) is the analysis of multiple markers at one time, but this reduces the sensitivity of the methodology. Ion Torrent™ (Thermo Fisher Scientific) has lower sensitivity compared to other techniques, such as MassARRAY® (Agena Bioscience), OncoBEAM™ (Sysmex), HapOncoCDx™ (Haplox Biotechnology), the TaqMan methodology, and GuardantOMNI; studies employing these found, respectively, 56%, 20.9–83.3%, 15.2%, 51.6%, and 5.1–8.7% of neo-RAS wt patients [5,6,11,13,18,21,41–46,51]. In the paper by Moati et al., the Ion Torrent™ (Thermo Fisher Scientific) methodology presents a neo-RAS wt patient rate (3.3%) closer to that of another NGS methodology described in the literature for neo-RAS wt determination: Guardant 360™ or GuardantOMNI (5.1–8.7%) [13,41,42,45]. In the other study employing Ion Torrent™ (Thermo Fisher Scientific), the neo-RAS wt rate was higher (36.4%), but no conclusions could be drawn about neo-RAS wt incidence due to the low number of cases analyzed (n = 11) [47]. In a small number of papers, the authors applied a second confirmatory method to assess the presence of ctDNA by methylation markers and evaluate the evolution of ctDNA release in the bloodstream [5,10,12,41].

Regarding the presence of evaluable ctDNA, the Idylla™ (Biocartis), MassARRAY® (Agena Bioscience), and Guardant360™ methodologies (Guardant Health) found this marker in nearly all cases. In contrast, Ion Torrent™ (Thermo Fisher Scientific), OncoBEAM™ (Sysmex Suisse), HapOncoCDx™ (Haplox Biotechnology), and qPCR detected the presence of ctDNA in 59%, 48%, 36%, and 66% of patients, respectively [5,6,11,13,18,21,41–47,51]. To sum up the advantages and disadvantages of all the methodologies reported above, we can conclude that all the NGS methodologies (e.g., Ion Torrent™, ThermoFisher Scientific; Guardant 360™ and Guardant OMNI™, Guardant Health; and HapOncoCDx™, Haplox Biotechnology) have the advantage of providing a broader evaluation of molecular markers than the other techniques, but for the same feature, they are characterized by lower sensitivity.

The KRAS mutational rates need to be considered based on the LOD of the specific methodology applied for the evaluation of RAS mutations. The LOD can be defined as the lowest concentration of the analyte that can be detected. In liquid biopsies, the LODs for the aforementioned assays are 2–5% for Idylla™ (Biocartis), 2–10% for MassARRAY® (Agena Bioscience), 2–10% for Guardant360™ (Guardant Health), 1–2% for Ion Torrent™ (Thermo Fisher Scientific), 1% for OncoBEAM™ (Sysmex Suisse), 3–5% for HapOncoCDx™ (Haplox Biotechnology), and 1–5% for qPCR [5,6,11,13,18,21,41–47,51]. As described here, the methodologies cited in the literature could have different LODs in liquid biopsies. The different LODs can influence the interpretation of the rates and presence of RAS mutations. To address this issue, future studies must confirm rather than exclude the presence of ctDNA in a plasma sample. Two approaches are possible: tumor-informed analyses or methodologies based on normalization. The first requires knowledge of the somatic mutations present in the primary tumor tissue and the investigation of the presence of somatic mutations in the ctDNA sample through next-generation sequencing (NGS) [42,47]. The second uses cancer-specific methylated biomarkers as a "normalizer" of the quantity of ctDNA available in plasma [10].

Considering these limitations, we can speculate that an accurate diagnostic method, combined with appropriate patient selection, could favor positive results, paving the way for the use of ctDNA even in patients with an initial RAS gene mutation. Analyzing the different diagnostic methods of the studies included in our review, we observe a high variety (MassARRAY [Agena Bioscience], Guardant360^{TM} [Guardant Health] or GuardantOMNI^{TM}, OncoBEAM^{TM} [Sysmex Suisse], IonTorrent^{TM} [Thermo Fisher Scientific], Idylla^{TM} [Biocartis], and HapOncoCDx^{TM} [Haplox Biotechnology]) characterized by different sensitivities. Initially, we assumed that the different rates of neo-RAS wt ctDNA evaluable at follow-ups could be connected to the different methodologies applied and their varying sensitivities. However, by observing the individual methodologies, we found that the rate of neo-RAS wt cases detected is not comparable across the different studies (except for the Guardant360^{TM} (Guardant Health) or GuardantOMNI^{TM} assays). This suggests that it is mostly sample heterogeneity, rather than the use of a specific method, that defines the percentage of neo-RAS wt characterizable in CRC populations. Consequently, the proportion of reverted patients may not be homogeneous, with some cases being incorrectly defined as neo-RAS wt because the analysis involved only the subset of cells without mutations.

Considering sensitivity and feasibility, we can conclude that different methodologies need to be applied for specific clinical and practical scenarios. For example, if a patient is highly symptomatic and there is no possibility of obtaining the mutational landscape status in a short time, methodologies that are easier and faster to conduct are recommended for a first screening (e.g., Idylla^{TM}, Biocartis). A more comprehensive analysis by an NGS methodology might instead be more recommended in the case of a patient with mutations that cannot be monitored by a smaller panel or in case the possible mechanisms of resistance to progression have to be investigated. In general, the most recommended methodology is qPCR by TaqMan technology [44]. Other methodologies with higher sensitivity (such as next-generation sequencing and ddPCR) have more complex protocols and require greater expenses for both instruments and reagents [5,11,51]. Some authors have addressed the problem of normalizing the real quantity of RAS mutations by employing a comparison with methylation profiles, as the application of circulating methylated DNA evaluation for defining the dynamics of RAS mutation clearance in plasma can be a reliable method for defining the real quantity of ctDNA in blood at different time points [5,10,12,41,42].

5. Does Bevacizumab Increase Neo-RAS wt Likelihood?

Another aspect we want to investigate further is whether specific drugs, such as anti-angiogenic compounds, may alter the neo-RAS wt rate. Although the addition of anti-angiogenic therapy has been shown to improve the effects of chemotherapy in terms of response and survival, it is unknown whether this drug can exert selective pressure on RAS-mutated clones [4].

Research conducted by our group demonstrated that bevacizumab has the biological potential to lead to clonal selection on the RAS mutation. In our retrospective analysis, patients who received bevacizumab had a greater reduction in the variant allelic frequency of RAS mutation in resected liver metastases when compared to operated patients who received standard chemotherapy but not bevacizumab (57.1% versus 8.3%) [15].

Supporting this hypothesis, Nicolazzo highlighted that 42 out of 56 patients treated with bevacizumab experienced a reversion of the RAS in ctDNA. Only two groups of authors assessed whether a statistical difference existed between the rates of neo-RAS wt cases considering the use of the drug. Klein-Scory et al. did not conclude that bevacizumab had an additive effect, while Nicolazzo et al. found that the inclusion of this anti-angiogenic drug is beneficial [5,18,49]. The discrepancies found in these works can be mainly associated with the different methodologies applied for KRAS investigation: ddPCR and OncoBEAM™ (Sysmex) in the work by Klein-Scory et al. versus Ion Torrent™ (Thermo Fisher Scientific) in the work published by Nicolazzo et al. Moreover, the different cohort sizes could have influenced the data obtained, leaning in favor of the results of the research by Nicolazzo's group, which evaluated 72 patients, versus the study conducted by Klein-Scory et al., who examined 12 patients [5,49].

The biological explanation for the selective effect of bevacizumab on mutant RAS cells remains uncertain. Some hypotheses are focused on the role of bevacizumab in inflammation and neo-angiogenesis as preliminary findings in transgenic murine models [53–55]. Bevacizumab reducing angiogenesis could have a better effect against the clones most dependent on vascular contributions, which resulted in vitro, after stimulation of angiogenesis, as the KRAS-mutant ones. A molecular explanation was described in the work by Figueras et al., who proposed how KRAS mutations influencing the Raf-RAS-ERKs pathway activate the VEGF-A promoter, creating VEGF-A-associated vascularization [55].

Other hypotheses include the increased effect of bevacizumab-induced hypoxia on these clones or the drug's ability to increase oxidative stress on mutant RAS cells [15,18].

These data suggest a potential role for bevacizumab as a drug that can increase the chance of achieving an RAS wt time window in which EGFR monoclonal antibodies can be administered. However, most current publications on neo-RAS wt do not allow for a comparison between patients treated with bevacizumab and those who did not receive the drug. We strongly recommend that researchers and clinicians considering the neo-RAS wt phenomenon in their clinical practice evaluate the potential role of bevacizumab in this context.

6. Neo-RAS wt: Possible Therapeutic Implications and Prospects

The evidence of neo-RAS wt has led clinicians to evaluate the use of anti-EGFR therapy in patients usually precluded from targeted therapy.

One of the first cases demonstrating the effectiveness of this therapeutic strategy was reported in the article by Gazzaniga et al. in 2018 [8].

To date, we can find in the literature some encouraging results regarding the use of EGFR inhibitors in neo-RAS wt patients. As summarized in Table 2, the authors of these studies presented responses and sustained PFS of EGFR inhibitor monotherapy alone or in combination with irinotecan for neo-RAS wt patients [7,10–12,16,47,48,52].

With the limitation of small sample sizes, keeping the number of patients reported by authors to a maximum of 10, the median PFS for anti-EGFR drugs ranged from 5.5 to 14.5 months. This is comparable to, or in some cases better than, the historical standard second-line treatment with anti-angiogenics such as FOLFIRI + aflibercept or FOLFIRI/FOLFOX + bevacizumab [56,57].

Even though anti-EGFR therapies have shown promising PFS in neo-RAS-wt patients, they have not been shown to increase OS when compared to standard regimens, which is a crucial endpoint in cancer treatment. Therefore, we do not consider these therapies to be usable outside of clinical trials.

These preliminary results are a good starting point for beginning to consider the efficacy of anti-EGFR therapies in neo-RAS wt cases. Some phase II trials combining chemotherapy and anti-EGFR are ongoing (Table 3). In most of them, the major confounding factors are the lack of a control arm. We see heterogeneity within these ongoing trials, particularly in the methods used to define mutations. Three trials will evaluate RAS status via OncoBEAM (Sysmex Suisse) and two via Idylla (Biocartis).

Table 2. Survival associated to administration of EGFR inhibitors in neo-RAS wt patients.

Author, Year	Patients (n)	Best Response	Survival (Months)
Bouchada, 2021 [7]	9	1 CR 4 PR 2 SD 2 PD	PFS: 8.2 OS: 22.3
Nicolazzo, 2021 [10]	10	NA	PFS: 10
Sato, 2022 [11]	4	1 PR 2 SD 1 PD	NA
Osumi, 2021 [12]	2	2 PR	NA
Osumi, 2023 [16]	6	1 PR 2 SD 3 PD	PFS: 4.8 (0.4–11.1)
Harada, 2023 [52]	2	2 PR	PFS: 5.5 (4–7)
Gramaça, 2024 [47]	4	NA	2nd line: • PFS: 14.5 • OS: 33.6 3rd line • PFS: 3.9

NA: not available; CR: complete response; PR: partial response; SD: stable disease; PD: progressive disease; PFS: progression-free survival; OS: overall survival.

Table 3. Ongoing and closed phase II trials combining chemotherapy and anti-EGFR.

Study Name, Trial ID, Country	Method of ctDNA Analysis	Estimated Patients	Setting	Phase	Experimental Arm	Control Arm	Status
MoLiMoR, NCT04554836, Germany	OncoBEAM (Sysmex)	144	First line	II	FOLFIRI + Cetuximab	FOLFIRI + Bevacizumab	Active, not recruiting
CONVERTIX, EudraCT 2017-003242-25, Spain	OncoBEAM (Sysmex)	40	Second line	II	FOLFIRI + Panitu-mumab	NA	Closed
CETIDYL, NCT04189055, France	Idylla (Biocartis)	72	≥Third line	II	Cetuximab +/− Irinotecan	NA	Recruiting
KAIROS, EudraCT 2019-001328-36, Italy	Idylla (Biocartis)	112	Second line	II	Cetuximab + Chemo Doublet	NA	Recruiting
C-PROWESS, jRCT, s031210565, Japan	OncoBEAM (Sysmex) + Guardant360 (Guardant Health)	30	≥Second line	II	Panitumumab + irinotecan	NA	Recruiting

NA: not available.

Looking ahead, we anticipate that future prospective clinical trials will not only confirm the effectiveness of this approach but also integrate comprehensive translational research involving both tissue samples and ctDNA.

Undoubtedly, an important aspect of future research will be understanding how to harmonize data from tissue analysis with ctDNA results to elucidate their clinical significance and better characterize tumor heterogeneity. The data obtained from the tissue will need to be harmonized more closely with those obtained from ctDNA analysis to give a clear picture of the mutational status of the RAS pathway during the course of the disease and treatment. This will be feasible by implementing ctDNA analysis, improving sensitivity, and comparing results from tissue analysis with the support of a molecular tumor board.

Discrepancies in results must be taken into account, and these should always be interpreted. A comprehensive analysis must take into account many aspects, such as patient characteristics, the type of medical treatment received and in progress, mutations highlighted by consideration of aspects such as neoplastic cellularity and allele frequency, and the limitations of the methods. Therefore, we believe that molecular tumor boards are essential for interpreting results and making clinical decisions.

7. Conclusions

At present, we are far from reaching definitive conclusions on the neo-RAS wt phenomenon, as the scientific community has yet to identify specific methods and patient profiles. The complexity of tumor biology and the heterogeneity of patient responses add additional layers of difficulty to this challenge.

Given these uncertainties, further studies are required to clarify the neo-RAS wt phenomenon. These studies should focus on not only expanding our understanding of the genetic and epigenetic factors involved but also refining diagnostic techniques to improve detection sensitivity. The development of more sensitive methodologies, such as enhanced sequencing technologies and comprehensive methylation profiling in ctDNA, holds promise for yielding more accurate and reliable results in the future.

Additionally, longitudinal studies and large-scale clinical trials are essential to identify potential biomarkers that can predict the emergence of neo-RAS wt and guide treatment decisions. Collaboration across research centers and the integration of multi-omics approaches will likely be crucial in overcoming the current limitations and moving towards more personalized and effective management of patients with CRC.

In conclusion, while significant progress has been made, the path forward requires a concerted effort to unravel the complexities of the neo-RAS wt phenomenon.

With ongoing research and the development of more sophisticated tools, we can anticipate a future in which the identification and management of this condition will become more precise, ultimately leading to improved patient outcomes.

Author Contributions: G.P., S.E. and M.F.: conceptualization; G.P., S.D.D., M.F. and S.E.: writing; all authors: review and editing; S.D.D. and M.F.: supervision. All authors have read and agreed to the published version of the manuscript.

Funding: This research received no external funding.

Conflicts of Interest: The authors declare no conflicts of interest.

References

1. Amado, R.G.; Wolf, M.; Peeters, M.; Van Cutsem, E.; Siena, S.; Freeman, D.J.; Juan, T.; Sikorski, R.; Suggs, S.; Radinsky, R.; et al. Wild-Type KRAS Is Required for Panitumumab Efficacy in Patients with Metastatic Colorectal Cancer. *J. Clin. Oncol.* **2008**, *26*, 1626–1634. [CrossRef] [PubMed]
2. Karapetis, C.S.; Khambata-Ford, S.; Jonker, D.J.; O'Callaghan, C.J.; Tu, D.; Tebbutt, N.C.; Simes, R.J.; Chalchal, H.; Shapiro, J.D.; Robitaille, S.; et al. K-ras Mutations and Benefit from Cetuximab in Advanced Colorectal Cancer. *N. Engl. J. Med.* **2008**, *359*, 1757–1765. [CrossRef] [PubMed]
3. Cervantes, A.; Adam, R.; Roselló, S.; Arnold, D.; Normanno, N.; Taïeb, J.; Seligmann, J.; De Baere, T.; Osterlund, P.; Yoshino, T.; et al. Metastatic colorectal cancer: ESMO Clinical Practice Guideline for diagnosis, treatment and follow-up. *Ann. Oncol.* **2023**, *34*, 10–32. [CrossRef] [PubMed]
4. Welch, S.; Spithoff, K.; Rumble, R.B.; Maroun, J. Bevacizumab combined with chemotherapy for patients with advanced colorectal cancer: A systematic review. *Ann. Oncol.* **2010**, *21*, 1152–1162. [CrossRef]

5. Klein-Scory, S.; Wahner, I.; Maslova, M.; Al-Sewaidi, Y.; Pohl, M.; Mika, T.; Ladigan, S.; Schroers, R.; Baraniskin, A. Evolution of RAS Mutational Status in Liquid Biopsies During First-Line Chemotherapy for Metastatic Colorectal Cancer. *Front. Oncol.* **2020**, *10*, 1115. [CrossRef]
6. Bouchahda, M.; Saffroy, R.; Karaboué, A.; Hamelin, J.; Innominato, P.; Saliba, F.; Levi, F.; Bosselut, N.; Lemoine, A. Undetectable RAS -Mutant Clones in Plasma: Possible Implication for Anti-EGFR Therapy and Prognosis in Patients with RAS-Mutant Metastatic Colorectal Cancer. *JCO Precis. Oncol.* **2020**, *4*, 1070–1079. [CrossRef]
7. Bouchahda, M.; Saffroy, R.; Karaboué, A.; Hamelin, J.; Innominato, P.; Saliba, F.; Levi, F.; Bosselut, N.; Lemoine, A. Efficacy of an anti-EGFR after ctDNA conversion from mutated RAS status of metastatic colorectal cancer: Results of a pilot study. *J. Clin Oncol.* **2021**, *39* (Suppl. S15), e15574. [CrossRef]
8. Gazzaniga, P.; Raimondi, C.; Urbano, F.; Cortesi, E. EGFR Inhibitor as Second-Line Therapy in a Patient with Mutant RAS Metastatic Colorectal Cancer: Circulating Tumor DNA to Personalize Treatment. *JCO Precis. Oncol.* **2018**, *2*, 1–6. [CrossRef]
9. Raimondi, C.; Nicolazzo, C.; Belardinilli, F.; Loreni, F.; Gradilone, A.; Mahdavian, Y.; Gelibter, A.; Giannini, G.; Cortesi, E. Gazzaniga, P. Transient Disappearance of RAS Mutant Clones in Plasma: A Counterintuitive Clinical Use of EGFR Inhibitors in RAS Mutant Metastatic Colorectal Cancer. *Cancers* **2019**, *11*, 42. [CrossRef]
10. Nicolazzo, C.; Barault, L.; Caponnetto, S.; De Renzi, G.; Belardinilli, F.; Bottillo, I.; Bargiacchi, S.; Macagno, M.; Grammatico, P. Giannini, G.; et al. True conversions from RAS mutant to RAS wild-type in circulating tumor DNA from metastatic colorectal cancer patients as assessed by methylation and mutational signature. *Cancer Lett.* **2021**, *507*, 89–96. [CrossRef]
11. Sato, S.; Mikayama, Y.; Shiozawa, M.; Nukada, S.; Iguchi, K.; Okamoto, H.; Kohmura, T.; Kazama, K.; Tanaka, K.; Oshima, T.; et al Chemotherapy-induced Reversion of Mutant RAS to Wild-type RAS in Metastatic Colorectal Cancer. *Anticancer. Res.* **2022**, *42* 2625–2635. [CrossRef] [PubMed]
12. Osumi, H.; Vecchione, L.; Keilholz, U.; Vollbrecht, C.; Alig, A.H.S.; Von Einem, J.C.; Stahler, A.; Striefler, J.K.; Kurreck, A.; Kind A.; et al. NeoRAS wild-type in metastatic colorectal cancer: Myth or truth?—Case series and review of the literature. *Eur. J Cancer* **2021**, *153*, 86–95. [CrossRef] [PubMed]
13. Osumi, H.; Shinozaki, E.; Nakamura, Y.; Esaki, T.; Yasui, H.; Taniguchi, H.; Satake, H.; Sunakawa, Y.; Komatsu, Y.; Kagawa, Y. et al. Neo RAS wild-type metastatic colorectal cancer in the SCRUM-Japan GOZILA study. *J. Clin. Oncol.* **2023**, *41* (Suppl. S16) 3506. [CrossRef]
14. Arici, S.; Hamdard, J.; Sakin, A.; Sengiz Erhan, S.; Atci, M.M.; Cekin, R.; Saka, B.; Köse, E.; Saydam, T.; Geredeli, C.; et al. The conversion of RAS status in metastatic colorectal cancer patients after first-line biological agent treatment. *Color. Dis.* **2021**, *23* 206–212. [CrossRef]
15. Epistolio, S.; Cefalì, M.; Spina, P.; Molinari, F.; Movilia, A.; Cergnul, M.; Mazzucchelli, L.; Dosso, S.D.; Frattini, M.; Saletti, P Occurence of RAS reversion in metastatic colorectal cancer patients treated with bevacizumab. *Oncotarget* **2021**, *12*, 1046–1056 [CrossRef]
16. Osumi, H.; Takashima, A.; Ooki, A.; Yoshinari, Y.; Wakatsuki, T.; Hirano, H.; Nakayama, I.; Okita, N.; Sawada, R.; Ouchi, K. et al. A multi-institutional observational study evaluating the incidence and the clinicopathological characteristics of NeoRAS wild-type metastatic colorectal cancer. *Transl. Oncol.* **2023**, *35*, 101718. [CrossRef]
17. Zhou, H.; Zhu, L.; Song, J.; Wang, G.; Li, P.; Li, W.; Luo, P.; Sun, X.; Wu, J.; Liu, Y.; et al. Liquid biopsy at the frontier of detection, prognosis and progression monitoring in colorectal cancer. *Mol. Cancer* **2022**, *21*, 86. [CrossRef]
18. Nicolazzo, C.; Magri, V.; Marino, L.; Belardinilli, F.; Di Nicolantonio, F.; De Renzi, G.; Caponnetto, S.; De Meo, M.; Giannini, G., Santini, D.; et al. Genomic landscape and survival analysis of ctDNA "neo-RAS wild-type" patients with originally RAS mutant metastatic colorectal cancer. *Front. Oncol.* **2023**, *13*, 1160673. [CrossRef]
19. Sartore-Bianchi, A.; Pietrantonio, F.; Lonardi, S.; Mussolin, B.; Rua, F.; Crisafulli, G.; Bartolini, A.; Fenocchio, E.; Amatu, A.; Manca, P.; et al. Circulating tumor DNA to guide rechallenge with panitumumab in metastatic colorectal cancer: The phase 2 CHRONOS trial. *Nat. Med.* **2022**, *28*, 1612–1618. [CrossRef]
20. Bray, F.; Laversanne, M.; Sung, H.; Ferlay, J.; Siegel, R.L.; Soerjomataram, I.; Jemal, A. Global cancer statistics 2022: GLOBOCAN estimates of incidence and mortality worldwide for 36 cancers in 185 countries. *CA A Cancer J. Clin.* **2024**, *74*, 229–263. [CrossRef]
21. Albuquerque, J.; Neto Da Silva, D.; Padrão, T.; Leal-Costa, L.; Bizarro, R.; Correia, J.; Baptista, C.; Machete, M.; Prazeres, G., Margarido, I.; et al. Loss of RAS Mutations in Liquid Biopsies of Patients With Multi-Treated Metastatic Colorectal Cancer *Oncologist* **2024**, *29*, e337–e344. [CrossRef] [PubMed]
22. Kasi, P.M.; Afable, M.G.; Herting, C.; Lukanowski, M.; Jin, Z. Anti-EGFR Antibodies in the Management of Advanced Colorectal Cancer. *Oncologist* **2023**, *28*, 1034–1048. [CrossRef] [PubMed]
23. Zhu, G.; Pei, L.; Xia, H.; Tang, Q.; Bi, F. Role of oncogenic KRAS in the prognosis, diagnosis and treatment of colorectal cancer. *Mol. Cancer* **2021**, *20*, 143. [CrossRef] [PubMed]
24. Jones, R.P.; Sutton, P.A.; Evans, J.P.; Clifford, R.; McAvoy, A.; Lewis, J.; Rousseau, A.; Mountford, R.; McWhirter, D.; Malik, H.Z. Specific mutations in KRAS codon 12 are associated with worse overall survival in patients with advanced and recurrent colorectal cancer. *Br. J. Cancer* **2017**, *116*, 923–929. [CrossRef] [PubMed]
25. El Bali, M.; Bakkach, J.; Bennani Mechita, M. Colorectal Cancer: From Genetic Landscape to Targeted Therapy Yuchi A, editor. *J. Oncol.* **2021**, *2021*, 9918116. [CrossRef]
26. Woo, J.; Palmisiano, N.; Tester, W.; Leighton, J.C. Controversies in antiepidermal growth factor receptor therapy in metastatic colorectal cancer. *Cancer* **2013**, *119*, 1941–1950. [CrossRef]

27. Douillard, J.Y.; Siena, S.; Cassidy, J.; Tabernero, J.; Burkes, R.; Barugel, M.; Humblet, Y.; Bodoky, G.; Cunningham, D.; Jassem, J.; et al. Final results from PRIME: Randomized phase III study of panitumumab with FOLFOX4 for first-line treatment of metastatic colorectal cancer. *Ann. Oncol.* **2014**, *25*, 1346–1355. [CrossRef]
28. Peeters, M.; Price, T.J.; Cervantes, A.; Sobrero, A.F.; Ducreux, M.; Hotko, Y.; André, T.; Chan, E.; Lordick, F.; Punt, C.J.A.; et al. Randomized Phase III Study of Panitumumab with Fluorouracil, Leucovorin, and Irinotecan (FOLFIRI) Compared with FOLFIRI Alone As Second-Line Treatment in Patients with Metastatic Colorectal Cancer. *J. Clin. Oncol.* **2010**, *28*, 4706–4713. [CrossRef]
29. Hong, D.S.; Fakih, M.G.; Strickler, J.H.; Desai, J.; Durm, G.A.; Shapiro, G.I.; Falchook, G.S.; Price, T.J.; Sacher, A.; Denlinger, C.S.; et al. KRAS G12C Inhibition with Sotorasib in Advanced Solid Tumors. *N. Engl. J. Med.* **2020**, *383*, 1207–1217. [CrossRef]
30. Yaeger, R.; Weiss, J.; Pelster, M.S.; Spira, A.I.; Barve, M.; Ou, S.-H.I.; Leal, T.A.; Bekaii-Saab, T.S.; Paweletz, C.P.; Heavey, G.A.; et al. Adagrasib with or without Cetuximab in Colorectal Cancer with Mutated KRAS G12C. *N. Engl. J. Med.* **2023**, *388*, 44–54. [CrossRef]
31. Fakih, M.G.; Salvatore, L.; Esaki, T.; Modest, D.P.; Lopez-Bravo, D.P.; Taieb, J.; Karamouzis, M.V.; Ruiz-Garcia, E.; Kim, T.-W.; Kuboki, Y.; et al. Sotorasib plus Panitumumab in Refractory Colorectal Cancer with Mutated KRAS G12C. *N. Engl. J. Med.* **2023**, *389*, 2125–2139. [CrossRef] [PubMed]
32. Morano, F.; Corallo, S.; Lonardi, S.; Raimondi, A.; Cremolini, C.; Rimassa, L.; Murialdo, R.; Zaniboni, A.; Sartore-Bianchi, A.; Tomasello, G.; et al. Negative Hyperselection of Patients with RAS and BRAF Wild-Type Metastatic Colorectal Cancer Who Received Panitumumab-Based Maintenance Therapy. *J. Clin. Oncol.* **2019**, *37*, 3099–3110. [CrossRef] [PubMed]
33. Randon, G.; Maddalena, G.; Germani, M.M.; Pircher, C.C.; Manca, P.; Bergamo, F.; Giordano, M.; Sposetti, C.; Montagna, A.; Vetere, G.; et al. Negative Ultraselection of Patients with RAS/BRAF Wild-Type, Microsatellite-Stable Metastatic Colorectal Cancer Receiving Anti–EGFR-Based Therapy. *JCO Precis. Oncol.* **2022**, *6*, e2200037. [CrossRef] [PubMed]
34. Shitara, K.; Muro, K.; Watanabe, J.; Yamazaki, K.; Ohori, H.; Shiozawa, M.; Takashima, A.; Yokota, M.; Makiyama, A.; Akazawa, N.; et al. Baseline ctDNA gene alterations as a biomarker of survival after panitumumab and chemotherapy in metastatic colorectal cancer. *Nat. Med.* **2024**, *30*, 730–739. [CrossRef]
35. Stintzing, S.; Heinemann, V.; Fischer Von Weikersthal, L.; Fuchs, M.; Kaiser, F.; Heinrich, K.; Modest, D.P.; Hofheinz, R.D.; Decker, T.; Gerger, A.; et al. Phase III FIRE-4 study (AIO KRK-0114): Influence of baseline liquid biopsy results in first-line treatment efficacy of FOLFIRI/cetuximab in patients with tissue RAS-WT mCRC. *J. Clin. Oncol.* **2023**, *41* (Suppl. S16), 3507. [CrossRef]
36. Zygulska, A.L.; Pierzchalski, P. Novel Diagnostic Biomarkers in Colorectal Cancer. *Int. J. Mol. Sci.* **2022**, *23*, 852. [CrossRef]
37. Bhullar, D.S.; Barriuso, J.; Mullamitha, S.; Saunders, M.P.; O'Dwyer, S.T.; Aziz, O. Biomarker concordance between primary colorectal cancer and its metastases. *EBioMedicine* **2019**, *40*, 363–374. [CrossRef]
38. Siravegna, G.; Mussolin, B.; Buscarino, M.; Corti, G.; Cassingena, A.; Crisafulli, G.; Ponzetti, A.; Cremolini, C.; Amatu, A.; Lauricella, C.; et al. Clonal evolution and resistance to EGFR blockade in the blood of colorectal cancer patients. *Nat. Med.* **2015**, *21*, 795–801. [CrossRef]
39. Vidal, J.; Muinelo, L.; Dalmases, A.; Jones, F.; Edelstein, D.; Iglesias, M.; Orrillo, M.; Abalo, A.; Rodríguez, C.; Brozos, E.; et al. Plasma ctDNA RAS mutation analysis for the diagnosis and treatment monitoring of metastatic colorectal cancer patients. *Ann. Oncol.* **2017**, *28*, 1325–1332. [CrossRef]
40. Bachet, J.B.; Bouché, O.; Taieb, J.; Dubreuil, O.; Garcia, M.L.; Meurisse, A.; Normand, C.; Gornet, J.M.; Artru, P.; Louafi, S.; et al. RAS mutation analysis in circulating tumor DNA from patients with metastatic colorectal cancer: The AGEO RASANC prospective multicenter study. *Ann. Oncol.* **2018**, *29*, 1211–1219. [CrossRef]
41. Henry, J.; Willis, J.; Parseghian, C.M.; Raghav, K.P.S.; Johnson, B.; Dasari, A.; Stone, D.; Jeyakumar, N.; Coker, O.; Raymond, V.M.; et al. NeoRAS: Incidence of RAS reversion from RAS mutated to RAS wild type. *J. Clin. Oncol.* **2020**, *38* (Suppl. S4), 180. [CrossRef]
42. Moati, E.; Blons, H.; Taly, V.; Garlan, F.; Wang-Renault, S.; Pietrasz, D.; Didelot, A.; Garrigou, S.; Saint, A.; Pernot, S.; et al. Plasma clearance of RAS mutation under therapeutic pressure is a rare event in metastatic colorectal cancer. *Int. J. Cancer* **2020**, *147*, 1185–1189. [CrossRef] [PubMed]
43. Wang, F.; Huang, Y.-S.; Wu, H.-X.; Wang, Z.-X.; Jin, Y.; Yao, Y.-C.; Chen, Y.-X.; Zhao, Q.; Chen, S.; He, M.-M.; et al. Genomic temporal heterogeneity of circulating tumour DNA in unresectable metastatic colorectal cancer under first-line treatment. *Gut* **2022**, *71*, 1340–1349. [CrossRef] [PubMed]
44. Sunakawa, Y.; Satake, H.; Usher, J.; Jaimes, Y.; Miyamoto, Y.; Nakamura, M.; Kataoka, M.; Shiozawa, M.; Takagane, A.; Terazawa, T.; et al. Dynamic changes in RAS gene status in circulating tumour DNA: A phase II trial of first-line FOLFOXIRI plus bevacizumab for RAS-mutant metastatic colorectal cancer (JACCRO CC-11). *ESMO Open* **2022**, *7*, 100512. [CrossRef]
45. Wu, F.T.H.; Topham, J.T.; O'Callaghan, C.J.; Feilotter, H.; Kennecke, H.F.; Banks, K.; Renouf, D.J.; Jonker, D.J.; Tu, D.; Chen, E.X.; et al. Reversion of RAS mutations in metastatic colorectal cancer in the CCTG CO.26 clinical trial. *J. Clin. Oncol.* **2023**, *41* (Suppl. S16), 3567. [CrossRef]
46. De Santiago, B.G.; López-Gómez, M.; Delgado-López, P.D.; Gordo, A.J.; Neria, F.; Thuissard-Vasallo, I.J.; Gómez-Raposo, C.; Tevar, F.Z.; Moreno-Rubio, J.; Hernández, A.M.; et al. RAS Mutational Status in Advanced Colorectal Adenocarcinoma Treated With Anti-angiogenics: Preliminary Experience With Liquid Biopsy. *Vivo* **2021**, *35*, 2841–2844. [CrossRef]
47. Gramaça, J.; Fernandes, I.G.; Trabulo, C.; Gonçalves, J.; Dos Santos, R.G.; Baptista, A.; Pina, I. Emerging role of liquid biopsy in rat sarcoma virus mutated metastatic colorectal cancer: A case report. *World J. Gastrointest. Oncol.* **2024**, *16*, 234–243. [CrossRef]

48. Osumi, H.; Shinozaki, E.; Nakamura, Y.; Esaki, T.; Yasui, H.; Taniguchi, H.; Satake, H.; Sunakawa, Y.; Komatsu, Y.; Kagawa, Y. et al. Clinical features associated with NeoRAS wild-type metastatic colorectal cancer A SCRUM-Japan GOZILA substudy. *Na Commun.* **2024**, *15*, 5885. [CrossRef]
49. Nicolazzo, C.; Belardinilli, F.; Vestri, A.; Magri, V.; De Renzi, G.; De Meo, M.; Caponnetto, S.; Di Nicolantonio, F.; Cortesi, E Giannini, G.; et al. RAS Mutation Conversion in Bevacizumab-Treated Metastatic Colorectal Cancer Patients: A Liquid Biops Based Study. *Cancers* **2022**, *14*, 802. [CrossRef]
50. Nicolazzo, C.; Belardinilli, F.; Caponnetto, S.; Gradilone, A.; Cortesi, E.; Giannini, G.; Gazzaniga, P. Why the Therapeutic Impac of RAS Mutation Clearance in Plasma ctDNA Deserves to Be Further Explored in Metastatic Colorectal Cancer. *Front. Onco* **2019**, *9*, 1414. [CrossRef]
51. Huiyan, L.; Wang, Z.; Lin, W.; Wang, Z.; Xu, R. 478P RAS mutation clearance in patients with metastatic colorectal cancer. *Ann Oncol.* **2020**, *31*, S444. [CrossRef]
52. Harada, K.; Yuki, S.; Kawamoto, Y.; Nakamura, T.; Kaneko, S.; Ishida, K.; Sakamoto, N.; Komatsu, Y. Anti-epidermal growth factor receptor treatment for patients with Neo RAS wild-type metastatic colorectal cancer: A case report of two cases. *Ther. Ad Med. Oncol.* **2023**, *15*, 17588359231216090. [CrossRef]
53. Kranenburg, O.; Gebbink, M.F.B.G.; Voest, E.E. Stimulation of angiogenesis by Ras proteins. *Biochim. Biophys. Acta (BBA)-Re Cancer* **2004**, *1654*, 23–37. [CrossRef]
54. Tang, Y.; Kim, M.; Carrasco, D.; Kung, A.L.; Chin, L.; Weissleder, R. In vivo Assessment of RAS-Dependent Maintenance of Tumo Angiogenesis by Real-time Magnetic Resonance Imaging. *Cancer Res.* **2005**, *65*, 8324–8330. [CrossRef]
55. Figueras, A.; Arbos, M.A.; Quiles, M.T.; Viñals, F.; Germà, J.R.; Capellà, G. The impact of KRAS mutations on VEGF-A productio and tumour vascular network. *BMC Cancer* **2013**, *13*, 125. [CrossRef]
56. Van Cutsem, E.; Tabernero, J.; Lakomy, R.; Prenen, H.; Prausová, J.; Macarulla, T.; Ruff, P.; Van Hazel, G.A.; Moiseyenko, V.; Ferry D.; et al. Addition of Aflibercept to Fluorouracil, Leucovorin, and Irinotecan Improves Survival in a Phase III Randomized Tria in Patients With Metastatic Colorectal Cancer Previously Treated With an Oxaliplatin-Based Regimen. *J. Clin. Oncol.* **2012**, *30* 3499–3506. [CrossRef]
57. Bennouna, J.; Sastre, J.; Arnold, D.; Österlund, P.; Greil, R.; Van Cutsem, E.; Von Moos, R.; Viéitez, J.M.; Bouché, O.; Borg, C.; et a Continuation of bevacizumab after first progression in metastatic colorectal cancer (ML18147): A randomised phase 3 trial. *Lance Oncol.* **2013**, *14*, 29–37. [CrossRef]

cancers

MDPI

Review

A Systematic Analysis of Expression and Function of RAS GTPase-Activating Proteins (RASGAPs) in Urological Cancers: A Mini-Review

Hao Song [1], Guojing Wang [1], Guoqiang Gao [1], Huayu Xia [1], Lianying Jiao [2] and Kaijie Wu [1,*]

[1] Department of Urology, The First Affiliated Hospital of Xi'an Jiaotong University, Xi'an 710061, China; song_hao@stu.xjtu.edu.cn (H.S.); 11306699@stu.xjtu.edu.cn (G.W.); gaoguoqiang@stu.xjtu.edu.cn (G.G.); huayv.xia@gmail.com (H.X.)

[2] Department of Biochemistry and Molecular Biology, School of Basic Medical Sciences, Xi'an Jiaotong University Health Science Center, Xi'an 710061, China; jiaoly@xjtu.edu.cn

* Correspondence: kaijie_wu@163.com

Simple Summary: The RAS signaling pathway, a key communication system in tumors, is often disrupted in urological cancers, such as prostate, bladder, and kidney cancers. Normally, RASGAPs act as "off switches" for this pathway by breaking down a molecule (RAS-GTP) that keeps the pathway active. When RASGAPs are lost or dysfunctional, the RAS pathway becomes overactive, fueling cancer growth and spread. Research shows that dysregulation of RASGAPs may also affect the function of treatments for urological cancers, such as chemotherapy, radiation, or targeted drugs. This suggests that restoring or targeting RASGAPs might improve outcomes for patients. While the mechanism behind this improvement is complex, advances in personalized precision medicine could unlock new therapies that exploit RASGAPs to fight these cancers more effectively. In short, understanding RASGAPs' function in the RAS signaling pathway offers hope for smarter, more precise treatments in the future.

Abstract: The RAS signaling pathway is one of the most commonly dysregulated pathways in urological cancers. This pathway can be regulated by RASGAPs, which catalyze the hydrolysis of RAS-GTP to RAS-GDP. As such, the loss of RASGAPs can promote the activation of the RAS signaling pathway. Dysregulation of RASGAPs significantly contributes to the progression of urological cancers, including prostate cancer, bladder cancer, and renal cell carcinoma. Furthermore, alterations in RASGAP expression may influence sensitivity to chemotherapy, radiotherapy, and targeted therapies, suggesting their potential as therapeutic targets. Despite the challenges involved, a deeper understanding of the complexity of the RAS signaling network, along with the evolution of personalized medicine, holds promise for delivering more precise and effective treatment options targeting RASGAPs in urological cancers.

Keywords: RASGAPs; RAS signaling pathway; cancer therapy; urological cancers

1. Introduction

Urological cancers, including prostate cancer (PCa), bladder cancer (BCa), and renal cell carcinoma (RCC), represent significant global health challenges due to their prevalence and impact on patients' quality of life [1]. These malignancies often share common molecular features, such as aberrant signaling through the RAS pathway, a critical regulator of cell growth, differentiation, and survival [2,3]. Upon activation, the RAS pathway triggers downstream signaling cascades, such as the MAPK and PI3K-Akt pathways, which are involved in driving tumor initiation and progression. The RAS gene family includes three main members (KRAS, NRAS, and HRAS), and they encode small GTPase proteins that act as molecular switches, cycling between an active GTP-bound state and an inactive GDP-bound state [4,5]. The balance of RAS activity is tightly controlled by two classes of regulatory proteins, namely guanine nucleotide exchange factors (GEFs) and RAS GTPase-activating proteins (RASGAPs). GEFs, such as SOS (son of sevenless) typically promote RAS activation, but RASGAPs, such as Disabled-2 Interacting Protein (DAB2IP), enhance the intrinsic GTPase activity of RAS, leading to its inactivation [6,7] (Figure 1).

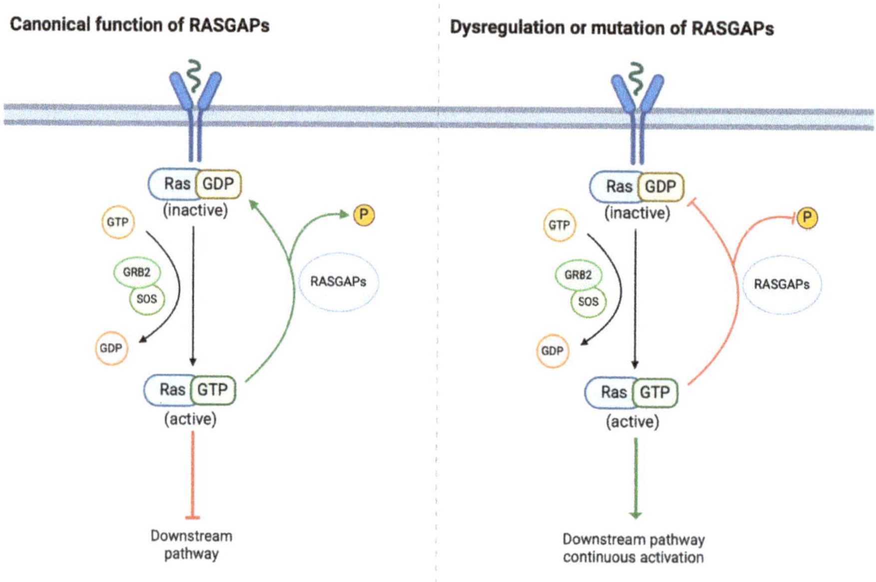

Figure 1. Summary of the function of RASGAPs in the RAS signaling pathway. In the pathway, RASGAPs are a class of important proteins that play a crucial negative regulatory role in the RAS signaling pathway. RASGAPs accelerate the hydrolysis of GTP on RAS proteins, converting them from an active state (GTP-bound state) to an inactive state (GDP-bound state), thereby shutting off the transmission of RAS signals.

Emerging research has shown that RASGAPs play crucial roles in controlling cell differentiation, proliferation, and survival [8]. This review focuses on the expression and function of RASGAPs in urological cancers, highlighting their functional significance, their dysregulation in tumor progression, and their potential as therapeutic targets.

2. Structure and Physiological Roles of RASGAPs

2.1. Structure of RASGAPs

The RAS signaling pathway is central to many oncogenic processes. Early experiments have demonstrated that RAS GTPase activity could be accelerated by specific cytosolic proteins, leading to the discovery of RASGAPs [9]. These proteins are shown to bind RAS and enhance its intrinsic GTPase activity, converting RAS-GTP to the inactive RAS-GDP form, thereby acting as critical negative regulators of RAS activity [10]. Aberrations in RASGAPs usually lead to dysregulated RAS signaling, causing excessive cell proliferation and tumorigenesis [11]. Thus, RASGAPs provide critical insights into the regulation of RAS signaling, highlighting their importance in cellular homeostasis.

Since the identification of the first RASGAP in the late 1980s, 14 types of RASGAPs have been discovered, with each exhibiting tissue-specific expression and distinct regulatory mechanisms [12]. The 14 main members include Neurofibromin 1 (NF1), DAB2-interacting protein (DAB2IP), SYNaptic GTPase Activating Protein 1 (SYNGAP1), Ras activator-like protein 1/2/3 (RASAL1/2/3), GTPase-activating protein 1/2/3/4/5 (RASA1/2/3/4/5), and IQ motif containing GTPase-activating protein homologue 1/2/3 (IQGAP1/2/3). All 14 of these RASGAPs share a conserved GAP functional domain, which binds to RAS proteins and directly modulates the RAS activity. Additional domains mediate protein–protein or protein–lipid interactions [13,14]. In the case of H-RAS inactivation, the GAP domain engages a key catalytic residue E61 in catalysis by extending an arginine side chain to the Ras active site [15,16]. However, mutations in critical residues, such as E61 or G12/13, prevent GTP hydrolysis, locking RAS in its active state to perpetuate oncogenic signaling [17,18].

Furthermore, research has also revealed that different RASGAP proteins exhibit variations in their distribution and function. Some RASGAPs regulate not only the RAS pathway but also participate in the modulation of other RAS-independent pathways [7].

2.2. *Physiological Roles of RASGAPs*

RASGAPs play crucial roles in cellular signaling, particularly in the regulation of RAS proteins. The tumor suppressor function of certain RASGAPs is primarily linked to their GAP domain's GTPase enzymatic activity and their ability to negatively regulate RAS activity, which is central to control cell proliferation, differentiation, and survival [5].

The tumor suppressor function of RASGAPs reveals their importance in cellular homeostasis and highlights them as potential targets for therapeutic intervention in cancers where RAS signaling is aberrantly activated. However, it is important to note that not all RASGAPs necessarily act as tumor suppressors. Some studies suggest that their specific roles may depend on the context and particular cellular environment. For example, certain cancers may exploit RASGAP-mediated signaling for cancer cell survival or invasion [19,20]. In the IQGAP family, the expression of IQGAP2 is reduced and acts as a tumor suppressor in most solid cancer types, while IQGAP3 is overexpressed and functions as an oncogene [21]. The dual roles of RASGAPs complicate their therapeutic targeting, highlighting the importance of understanding the specific cancer context in which RASGAPs are dysregulated.

Moreover, it has been observed that certain RASGAP proteins possess broader regulatory roles beyond the RAS pathway. They may influence multiple signaling networks, contributing to diverse biological outcomes. For example, some RASGAPs have been shown to engage in pathways related to cell adhesion, cytoskeletal dynamics, and even vesicle trafficking, independent of their canonical function in deactivating RAS [22–24]. The multiple functions of RASGAPs reveal the complexity of cellular regulation and underscore their importance in the full spectrum of activities associated with individual RASGAP proteins.

3. Dysregulation of RASGAPs Promotes Urological Cancers

The RAS pathway is highly conserved and regulates a variety of cellular processes. Dysregulation of this pathway, especially through mutations in RAS genes, can lead to uncontrolled cellular proliferation and cancer development [25]. In urological cancers, specific mutations in HRAS, KRAS, and NRAS have been linked to tumor formation and aggressive cancer phenotypes [26,27]. While mutations in RAS genes are less frequent in urological cancers, dysregulation of RASGAPs plays a critical role in the development and progression of these cancers. Members of the RASGAP family exhibit tumor-suppressive roles in urological tumors, providing new insights into their treatment. However, among the RASGAP family members, only some of them have been closely associated with the occurrence, development, and prognosis of urological tumors, with certain specific mechanisms remaining unclear. Therefore, future research must continue to explore the relationships between other RASGAP family members and urological tumors to offer more options for precise treatment. The three major urological cancers (PCa, BCa, and RCC) can exhibit dysregulation of the RAS pathway. However, the specific RASGAPs involved (e.g., DAB2IP in PCa, NF1 in BCa, RASAL2 in RCC) and their functional roles may vary. Although these cancers share common RASGAP pathway components, such as DAB2IP, the extent of pathway involvement and its functional significance can differ due to variations in molecular context, tumor biology, and genetic alterations (Figure 2).

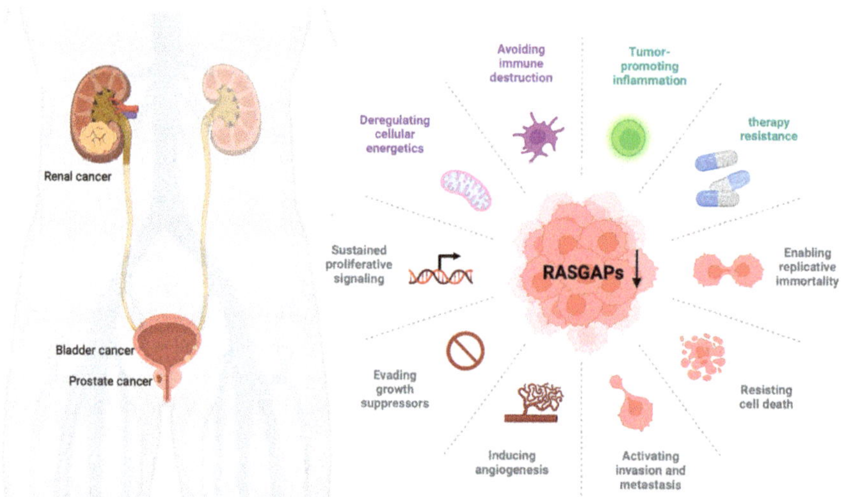

Figure 2. The oncogenic effects of RASGAP downregulation in urological cancer. RASGAP downregulation activates the RAS signaling pathway by inhibiting RAS from an active state (GTP-bound state) to an inactive state (GDP-bound state), thereby promoting urological cancers' malignancy. The interplay between RASGAP dysregulation and oncogenic processes highlights their therapeutic relevance in urological malignancies.

3.1. RASGAPs and Prostate Cancer

Prostate cancer (PCa) is one of the most frequently diagnosed urological cancers and one of the most frequent causes of cancer deaths in males [1]. Patients with PCa have benefitted from androgen deprivation therapies (ADTs) and small molecule inhibitors targeting the androgen receptor (AR). However, 30% of patients exhibit primary resistance to both forms of treatment, and the majority of patients progress from androgen-dependent prostate cancer (ADPC) to castration-resistant prostate cancer (CRPC) primarily due to the emergence of AR splice variant-7 (AR-V7) [28,29]. The AR remains a key driver of CRPC through aberrant activation in a low-androgen microenvironment. Understanding the molecular mechanisms of PCa has led to advancements in prognostic, diagnostic, and therapeutic approaches.

RAS signaling plays a significant role in PCa progression. While RAS mutations are rare in PCa, dysregulation of the RAS signaling pathway via alterations in upstream or downstream components, especially RASGAPs, is frequently observed. In PCa, several RASGAPs, such as RASAL3, have been found to be dysregulated. This dysregulation of RASGAPs ultimately leads to the activation of Ras signaling in carcinoma-associated fibroblasts [30].

Dysregulation of RAS signaling through alterations in RASGAPs significantly contributes to PCa progression [31]. RASGAPs, such as DAB2IP, can be dysregulated in PCa through several mechanisms, including genetic mutations, epigenetic modifications, and post-translational changes. H3K27 hypermethylation of the DAB2IP promoter, mediated by EZH2's histone methyltransferase activity, is a common epigenetic alteration in advanced PCa, leading to its transcriptional downregulation. Loss of DAB2IP not only negatively regulates RAS activity but also GSK3β/β-catenin and NF-κB activity [32–35]. Increased activation of NF-κB in DAB2IP-deficient PCa promotes epithelial-to-mesenchymal transition (EMT) and metastasis both in vitro and in vivo. Other studies have also shown that DAB2IP knockdown promotes EMT and metastasis through targeting PROX1/HIF1α either in LAPC-4 or RWPE-1 PCa cell lines [36,37]. This evidence forecasts that DAB2IP could inhibit EMT and metastasis in PCa. As a key regulator, the loss of DAB2IP expression results in the activation of the PI3K-AKT pathway and inactivation of the ASK1-JNK pathway, leading to accelerated PCa growth in vivo [38]. Additionally, the loss of DAB2IP function may increase AR signaling, both in vitro and in vivo, and contribute to the development of CRPC [39,40]. The tumor suppressor p53 has also been found to be associated with the RASGAP pathway in PCa. PCa with mutant p53 proteins (mut p53) responds to insulin signaling by increasing cell proliferation and invasiveness in vitro. This response is mainly due to the fact that mutant p53 enhances insulin-induced AKT1 activation by binding and inhibiting DAB2IP [41]. Furthermore, other studies have demonstrated that downregulation of DAB2IP gene expression impacts PCa's resistance to ionizing radiation (IR) [42]. DAB2IP re-expression sensitizes PCa cells to radiation and chemotherapy, underscoring the therapeutic potential of targeting RASGAPs in this malignancy. Epothilone B (EpoB), an anticancer drug, has been found to significantly increase cellular radiosensitivity in DAB2IP-deficient PCa cells [43]. Reactivating DAB2IP or inhibiting the PI3K-AKT pathway in DAB2IP-deficient PCa models has shown promise in preclinical studies [7,44]. Thus, restoring the function or expression of DAB2IP or inhibiting downstream effectors of RAS activation may provide alternative therapeutic strategies.

Moon et al. found that hypercholesterolemia was able to promote PCa metastasis by increasing IQGAP1 both in vitro and in vivo [45]. The adhesion of cancer cells to endothelial cells requires β1-integrin, and IQGAP1 regulates the transcription and expression of β1-integrin. Mechanistically, IQGAP1 functions downstream of Cdc42 to enhance β1-integrin expression through ERK/focal adhesion kinase (FAK) signaling at the protein level and by promoting myocardin-related transcription factor (MRTF)/serum response factor (SRF)-mediated transcriptional activity. This cascade ultimately upregulates β1-integrin expression, facilitating cancer cell–endothelial adhesion [46]. Xiong et al. also found that IQGAP1 is additionally associated with chemoresistance in PCa. The exocrine factor ANGPTL4 is primarily expressed in cancer-associated fibroblasts (CAFs) of PCa. Upon binding of ANGPTL4 to IQGAP1 on the membrane of PCa cells, it activates the Raf-MEK-ERK-PGC1α axis, driving mitochondrial biogenesis and oxidative phosphorylation (OXPHOS) metabolism. This mechanism promotes PCa tumor growth and confers chemoresistance [47]. IQGAP2 has been regarded as a tumor suppressor of PCa. Its expression is elevated in low-grade PCa (from prostatic intraepithelial neoplasia to Gleason 3 tumors). However, IQGAP2 is downregulated in high-grade PCa (Gleason 4-5) [48,49], and its downregulation has been shown to be positively associated with recurrence and metastasis in PCa through the activation of AKT signaling [50]. Meanwhile, IQGAP3 has been found to be positively correlated with the infiltration of B cells, macrophages, and dendritic cells, indicating its potential role as a tumor-specific antigen. Its overexpression is associated with worse overall survival rates [51].

RASA1, another RASGAP associated with aggressive PCa, has been linked to Gleason score [52]. However, the function and mechanisms of RASA1 in PCa remain incompletely understood. Meanwhile, RASAL1 has been shown to inhibit the tumorigenicity of human primary cells in both PCa and BCa [53]. Recently, RASAL2 was found to be upregulated in PCa tumors and metastatic lymph node tissues. Overexpression of RASAL2 was associated with higher PCa tumor stage, Gleason score, and poorer prognosis. Mechanistically, RASAL2 functions as an oncogene by promoting cancer cell proliferation through activation of the PI3K/AKT/cyclin D1 pathway [54]. Additionally, Tailor et al. found that RASAL2 could also function as a tumor suppressor by inhibiting cell proliferation and invasion, as well as by inducing an S phase and G2/M phase cell cycle arrest through the downregulation of TNFα [55]. RASAL3 promotes lethal PCa progression and resistance to ADT. Mechanistically, in prostatic CAFs, RASAL3 facilitates activated Ras signaling to drive macropinocytosis-mediated glutamine synthesis, fueling tumor metabolic demands. ADT further induces epigenetic silencing of RASAL3, which increases glutamine secretion from CAFs. This adaptive metabolic reprogramming sustains PCa survival and proliferation under androgen-deprived conditions, fostering therapeutic resistance and aggressive tumor growth [56].

While studies on the function of other RASGAPs in PCa are not extensive, Kachroo et al. found that downregulation of SPRED2, a RAS-regulator, is associated with a high Gleason score in PCa. SPRED2 regulates the RAS signaling pathway by modulating RASGAP's GTPase activity through direct binding. Overexpression of SPRED2 reduced ERK phosphorylation and inhibited PCa cell proliferation and migration [57]. Meanwhile, loss-function mutations of SPRED2 have been observed in human cancers, leading to tumor progression [58].

Oligophrenin 1 (OPHN1), an indirect RASGAP containing a RhoGAP domain that enhances GTPase activity [59], is located in the same region as the AR gene, which can be amplified by ADTs. Consequently, PCa undergoing ADTs may amplify both AR and OPHN1. Liu et al. found that the overexpression of OPHN1 contributes to cell viability and enhances migration in LNCaP, 22RV1, and PC3 cells [60]. Furthermore, due to the proximity of these two genes' location, long non-coding RNAs (lncRNAs), such as lnc-OPHN1-5, can increase sensitivity to enzalutamide (Enz) by interfering with AR mRNA expression. This finding could help to develop novel therapies to increase Enz treatment sensitivity [61].

RASGAPs have also been found to interact with phosphoprotein associated with glycosphingolipid microdomains 1 (PAG), a negative regulator of immune signaling in T lymphocytes. In this way, RASGAPs are enriched on the cell membrane to inhibit RAS activity, ultimately suppressing the ERK1/2 pathway and cyclin D1 expression [62].

3.2. RASGAPs and Bladder Cancer

Bladder cancer (BCa) is a common malignancy, with most cases classified as non-muscle-invasive bladder cancer (NMIBC). However, some BCa cases are characterized by high recurrence rates and progression to muscle-invasive forms (MIBC). Despite advancements in clinical treatment methods, the prognosis for BCa remains poor, particularly in the advanced stages [63,64]. RASGAPs, such as DAB2IP and NF1, have been implicated in regulating cellular growth and survival in BCa [65–68].

In addition to its essential roles in PCa, DAB2IP is a well-characterized RASGAP that regulates multiple signaling pathways in BCa. Loss of DAB2IP expression has been linked to aggressive tumor behavior in BCa [69]. In BCa, DAB2IP acts as a tumor suppressor by inhibiting the RAS and PI3K/Akt pathways. Its downregulation, mainly due to promoter hypermethylation, leads to increased cell proliferation, migration, invasion, and resistance to apoptosis, contributing to tumor growth and metastasis. Additionally, the caveolin-1 gene plays a vital role in promoter methylation of the DAB2IP gene in the progression of urinary bladder transitional cell carcinoma (TCC) from low to high potential for malignancy [70,71]. miRNA-mediated repression is another mechanism leading to DAB2IP loss in BCa. Ou et al. found that estrogen receptor beta (ERβ) induces the expression of miR-92a by binding to the estrogen-response element (ERE) in the 5′ promoter region of its host gene C13orf25. miR-92a then represses DAB2IP expression by binding to its 3′ UTR [72]. In another study, miR-92b was shown to specifically downregulate DAB2IP, promoting EMT, migration, and invasion in BCa, though it had no effect on cell proliferation [73]. Similarly, overexpression of miR-556-3p in BCa was found to downregulate DAB2IP and increase ERK1/2 phosphorylation levels [74].

As a promising biomarker, DAB2IP deficiency can promote chemoresistance and tumor recurrence in NMIBC after bladder-preserving surgery. In one study, Wu et al. observed significant downregulation of DAB2IP expression in high-grade and recurrent NMIBC specimens. This loss of DAB2IP was inversely correlated with elevated Twist1 expression and predicted poorer recurrence-free survival in patients. Mechanistically, DAB2IP deficiency in BCa cells promotes STAT3 phosphorylation and transactivation, which drives the upregulation of Twist1 and its downstream target P-glycoprotein (P-gp). This STAT3-Twist1/P-gp axis is essential for pirarubicin chemoresistance and tumor regrowth [75]. DAB2IP is also typically downregulated in BCa with a radioresistant phenotype. He et al. found that overexpression of ataxia-telangiectasia mutated (ATM), which is negatively regulated by DAB2IP, plays a central role in BCa resistance to ionizing radiation (IR). Knockdown of ATM may activate MAPK and NF-κB signaling pathways [76]. This finding suggests that ATM may be an effective target in DAB2IP-deficient BCa with an IR-resistant phenotype.

Additionally, NF1 is another important RASGAP that negatively regulates RAS signaling in BCa. NF1 deficiency has been observed in higher-grade urinary bladder transitional cell carcinoma (TCC), making it a potential biomarker [77]. High expression of NF1 could be regulated by the knockdown of Heterogeneous Nuclear Ribonucleoprotein U (HNRNPU), which enhances chemosensitivity in BCa. In T24 cancer cells with high HNRNPU expression, the knockout of HNRNPU inhibited cell proliferation, invasion, and migration. Moreover, the loss of HNRNPU promoted apoptosis and S-phase arrest in T24 cells treated with cisplatin [78]. RASAL2 expression is significantly downregulated in BCa specimens and inversely correlates with pathological grade and clinical stage. Wu et al. demonstrated that RASAL2 suppresses BCa stemness and EMT. Mechanistically, MAPK/SOX2 signaling is critical for maintaining stem-like and mesenchymal properties in RASAL2-deficient BCa cells, as inhibition of ERK activity or SOX2 knockdown reverses these phenotypes. Furthermore, RASAL2 inhibits BCa tumorigenesis and distant metastasis in vivo [79]. RASAL2 knockdown has also been shown to promote angiogenesis in BCa. Mechanistically, RASAL2 deficiency enhances AKT phosphorylation, which drives the transcriptional upregulation of ETS1 and VEGFA. This RASAL2-AKT-ETS1/VEGFA signaling axis orchestrates proangiogenic reprogramming, thereby potentiating tumor vascularization and progression in BCa [80].

Hensel et al. found that IQGAP1, a membrane of the RASGAP family, was downregulated in BCa. Its deficiency was shown to increase TGFβ signaling in BCa, suggesting that IQGAP1 has the potential to serve as both a clinical biomarker and a cancer growth suppressor [81]. Compared to normal controls, IQGAP3 was found to be highly expressed in urine samples from BCa patients. The IQGAP3/BMP4 ratio in urinary cell-free DNA has been demonstrated to be a diagnostic marker for BCa, showing high sensitivity and specificity, with a specificity for hematuria reaching 90.3% [82,83]. CDC42 and IQGAP3 were co-upregulated in both BCa tissues and cell lines. Mechanistically, CDC42 silencing reduced IQGAP3 expression and suppressed RAS/ERK signaling while concurrently inducing apoptosis and inhibiting BCa cell proliferation. Notably, IQGAP3 overexpression abolished CDC42 silencing-mediated proliferation arrest and apoptotic induction. Collectively, these findings demonstrate that CDC42 drives RAS/ERK signaling through IQGAP3 to enhance proliferation and suppress apoptosis, thereby contributing to BCa pathogenesis [84]. In contrast, IQGAP2, which is downregulated in BCa, may function to inhibit tumor proliferation, migration, and invasion by regulating the MAPK/ERK signaling pathway and cytokines [85].

Additionally, ARHGAP family genes, which share a homology with RASGAPs in their catalytic domains, have been associated with a tumor-promoting immune microenvironment in BCa. This microenvironment is characterized by a lower Th1/Th2 cell ratio, higher dendritic cell (DC) infiltration, increased Treg cell infiltration, and a T cell exhaustion phenotype [86].

3.3. *RASGAPs and Renal Cancer*

Renal cell carcinoma (RCC) accounts for over 90% of all kidney cancers and poses significant challenges due to its resistance to chemotherapy and radiation [87,88]. Currently, surgical intervention remains the primary treatment strategy for early-stage kidney tumors, while targeted therapy and immunotherapy are the main treatments for advanced stages. RCC is characterized by complex molecular mechanisms, with RAS signaling being implicated in its progression [89,90]. Understanding the molecular drivers of RCC has led to the development of targeted therapies, such as VEGF inhibitors and mTOR inhibitors, though resistance to these treatments is still a hurdle.

Emerging evidence suggests that dysregulation of RAS signaling contributes to RCC tumorigenesis. RASGAPs, particularly through the RAS/RAF/MEK/ERK and PI3K/Akt/mTOR pathways, play a central role in RCC progression [91,92]. First, DAB2IP is reported to be a key RASGAP involved in the regulation of multiple signaling pathways in RCC. Downregulation of DAB2IP in RCC has been associated with increased tumor growth and metastasis. Mechanistically, the loss of DAB2IP leads to activation of the PI3K/Akt pathway, enhancing RCC cell survival and proliferation [93]. In one study, high DAB2IP mRNA expression correlated with smaller tumor volume and better survival outcomes compared to patients with low expression. Zhou et al. found that the proline-rich domain in the C terminal (CPR) of DAB2IP could suppress AKT phosphorylation and p27 expression [93]. DAB2IP is frequently epigenetically silenced in RCC, with its loss correlating with poor overall survival. RCC cells with DAB2IP downregulation exhibit enhanced sensitivity to growth factor stimulation and resistance to small-molecule inhibitors, such as mTOR inhibitors. Mechanistically, DAB2IP deficiency leads to simultaneous activation of the ERK/ribosomal S6 kinase 1 (RSK1) and PI3K/mTOR pathways, which synergistically induce HIF-2α expression. Elevated HIF-2α subsequently represses p21/WAF1 transcription, a critical mediator of mTOR inhibitor resistance. These findings position DAB2IP as both a prognostic biomarker and a predictive marker for therapy response in RCC [94].

Moreover, the loss of DAB2IP results in elevated PARP-1 protein levels, which are associated with IR resistance in RCC, providing a new targeting strategy to improve the efficacy of radiotherapy of RCC. Mechanistically, DAB2IP acts as a scaffold to assemble a ternary complex with PARP-1 and E3 ubiquitin ligases, facilitating PARP-1 ubiquitination and proteasomal degradation, offering a new targeting strategy to enhance radiotherapy efficacy in RCC [95]. Primary cilia are microtubule-based organelles that function as sensors for physical and biochemical cues, regulating physiological and developmental processes.

Their loss has been implicated in multiple cancers, including RCC. DAB2IP has been identified as a novel interactor of kinesin family member 3A (KIF3A), a core component essential for primary cilia assembly. DAB2IP stabilizes KIF3A within the axoneme, thereby maintaining primary cilia integrity. Notably, KIF3A deficiency promotes RCC tumorigenesis, establishing the DAB2IP-KIF3A complex as a critical homeostatic regulator in normal renal epithelia. Mechanistically, KIF3A interacts with the N-terminal pleckstrin homology (PH) domain of DAB2IP, which extends KIF3A's half-life and facilitates ciliogenesis. KIF3A loss-induced ciliary disassembly drives renal tumorigenesis, highlighting the tumor-suppressive role of primary cilia stability mediated by this complex [96].

RCC, characterized by hypervascularity, is clinically managed using targeted therapies against the VEGFA/VEGFR2 signaling axis. Zhu et al. identified the lncRNA DMDRMR as a molecular sponge for miR-378a-5p, which upregulates EZH2 and SMURF1 expression. This DMDRMR/miR-378a-5p axis promotes the dual suppression of DAB2IP through EZH2-mediated transcriptional silencing and SMURF1-dependent proteasomal degradation. Consequently, DAB2IP loss activates VEGFA/VEGFR2 signaling, driving angiogenesis and conferring sunitinib resistance in ccRCC [97]. Yun et al. demonstrated that DAB2IP enhances miR-138 expression, which suppresses stem-like phenotypes in RCC by directly targeting ATP-binding cassette subfamily A member 13 (ABCA13) and EZH2. However, miR-138 downregulation in RCC is driven by DNA methyltransferase 1 (DNMT1)-mediated epigenetic silencing via promoter hypermethylation. DAB2IP may interact with DNMT1 to facilitate promoter methylation, amplifying ABCA13/EZH2-dependent pro-tumorigenic signaling [98]. Yeh et al. demonstrated that tumor-infiltrating T lymphocytes enhance RCC invasiveness by upregulating estrogen receptor β (Erβ) expression, which suppresses DAB2IP-dependent tumor-suppressive signaling. This T cell/ERβ/DAB2IP axis drives a pro-invasive cascade, suggesting that therapeutically targeting this pathway may disrupt RCC progression [99].

Second, RASAL2, another RASGAP in RCC, has been shown to inhibit angiogenesis both in vitro and in vivo by decreasing the expression of vascular endothelial growth factor A (VEGFA) through the p-GSK3β/c-FOS pathway. This study provides new insights into preventing RCC resistance to anti-vascular therapy [100].

In addition, aldehyde dehydrogenases 9 family A1 (ALDH9A1) deficiency has been linked to increased tumor proliferation, invasion, migration, and lipid accumulation in RCC. Mechanistically, reduced ALDH9A1 levels impair its cytoplasmic sequestration of nucleophosmin 1 (NPM1), leading to suppressed transcription of IQGAP2. This transcriptional repression subsequently activates the AKT-mTOR signaling pathway, which drives tumor progression and dysregulates lipid metabolism in RCC [101].

Mutations in RASGAPs also play crucial roles in regulating the RAS signaling pathway. Vanli et al. found that a point mutation in RASGAP prevented its cleavage by caspase-3, thereby enhancing RAS signaling [102]. In non-small cell lung cancer (NSCLC), NF1 co-mutation with RASA1 has been described [103]. Similarly, in breast cancer, mutations in RASAL2 have been reported to promote tumor growth, progression, and metastasis in mouse models [104]. Furthermore, loss-of-function mutations of RASA2 have been linked to increased RAS activation, melanoma cell growth, and migration in melanomas [105]. Inactivation mutations of RASGAPs promote tumorigenesis by disrupting the negative regulation of the Ras signaling pathway, potentially holding significant pathological implications in urological malignancies. However, direct evidence of RASGAP mutations in urological tumors remains limited, with most conclusions extrapolated from general mechanisms and other tumor types. Future research should focus on comprehensive genomic sequencing and functional assays to delineate the specific mutation profiles of RASGAPs in urological cancers. This would clarify their precise mechanistic roles and therapeutic potential in urological cancers. To summarize the functions of RASGAPs discussed above, Table 1 presents their expression patterns, biological effects, and mechanisms in urological cancers.

Table 1. The expression and function of RASGAPs in PCa, BCa, and RCC. The protein expression level of RASGAPs in urological organs (data adapted from the Human Protein Atlas database: https://www.proteinatlas.org/, accessed on 16 February 2025). The expression levels are categorized as +, ++, or +++. Note that this is a generalized representation, and specific expression levels may vary depending on the tissue and cell type.

Tumor Type	RASGAPs	Exp.	Biological Effect	Mechanism	Ref.
PCa	DAB2IP	+++	DAB2IP loss promoted PCa EMT and metastasis	Tumor suppressor: targeted GSK3β/β-catenin; targeted NF-κB signaling; targeted PROX1/HIF1α	[35–37]
			DAB2IP loss accelerated PCa growth in vivo	Tumor suppressor: targeted PI3K-AKT and ASK1-JNK pathway	[38]
			DAB2IP loss contributed to the development of CRPC	Tumor suppressor: targeted testosterone synthesis and AR signaling	[39,40]
			DAB2IP loss increased cell proliferation and invasiveness	Tumor suppressor: mut p53 enhanced insulin-induced AKT1 activation by binding and inhibiting DAB2IP	[41]
			DAB2IP loss promoted resistance to ionizing radiation	Tumor suppressor: enhanced DSB repair, robust G(2)-M checkpoint control, and resistance to apoptosis and	[42–44]
	IQGAP1	+++	IQGAP1 promoted cancer cell dissemination and metastasis	Oncogene: regulated β1-integrin via FAK and MRTF/SRF	[46]
			IQGAP1 promoted PCa tumor growth and increased chemoresistance	Oncogene: activated by ANGPTL4 in CAFs to activate Raf-MEK-ERK-PGC1α axis and drive mitochondrial biogenesis and OXPHOS metabolism	[47]
	IQGAP2	+++	IQGAP2 downregulation was associated with high Gleason score, recurrence and metastasis	Tumor suppressor: activated AKT signaling	[50]
	IQGAP3	+	IQGAP3 was positively correlated with infiltration of B cells, macrophages and dendritic cells	-	[51]
	RASA1	++	RASA1 was positively associated with aggressive PCa and Gleason score	-	[52]
	RASAL1	+	RASAL1 inhibited tumorigenicity of human primary cells	-	[53]
	RASAL2	+	RASAL2 promoted tumor cell proliferation, the transition from G1 to S phase in vitro and tumor growth in vivo	Oncogene: activated PI3K/AKT/cyclin D1 pathway	[54]
			RASAL2 overexpression inhibited cell proliferation and invasion and induced an S phase plus G2/M phase cell cycle arrest	Tumor suppressor: downregulated TNFα expression	[55]
	RASAL3	+	Epigenetic silencing of RASAL3 promoted lethal PCa growth and the development of resistance to ADT	Tumor suppressor: expressed in prostatic CAFS; activated Ras signaling and drived macropinocytosis-mediated glutamine synthesis; ADT promoted RASAL3 epigenetic silencing and glutamine secretion	[56]

Table 1. *Cont.*

Tumor Type	RASGAPs	Exp.	Biological Effect	Mechanism	Ref.
BCa	DAB2IP	+++	DAB2IP-deficient BCa cells promoted chemoresistance and tumor recurrence in NMIBC	Tumor suppressor: targeted STAT3 phosphorylation and transactivation; elevated Twist1 and P-glycoprotein expression	[75]
			DAB2IP-knockdown induced BCa resistance to IR	Tumor suppressor: elevated expression of ATM; inhibited MAPK and NF-κB signaling pathways	[76]
	NF1	++	Knockout of HNRNPU enhanced cisplatin sensitivity by regulating NF1 expression	-	[78]
	RASAL1	+	RASAL1 inhibited the tumorigenicity of human primary cells	-	[53]
	RASAL2	+	RASAL2 BCa tumorigenesis and distant metastasis in vivo	Tumor suppressor: inhibited MAPK/SOX2 signaling and BCa stemness and EMT	[79]
			Knockdown of RASAL2 promoted angiogenesis in BCa	Tumor suppressor: targeted RASAL2-AKT-ETS1/VEGFA signaling axis	[80]
	IQGAP1	+++	IQGAP1 inhibited cancer growth	Tumor suppressor: regulated TGFβ signaling	[81]
	IQGAP2	+	Reduced IQGAP2 promoted tumor proliferation, migration, invasion and EMT	Tumor suppressor: regulated MAPK/ERK pathway and reduced cytokines	[85]
	IQGAP3	++	IQGAP3 inhibited apoptosis and promoted BCa cells proliferation	Oncogene: activated RAS/ERK signaling	[84]
RCC	DAB2IP	+++	DAB2IP knockdown increased cell proliferation, promoted cell cycle progression in G1/S phase	Tumor suppressor: regulated the phosphorylation level of AKT and p27	[93]
			Loss of DAB2IP enhanced RCC sensitivity to growth factor stimulation and resistance to mTOR inhibitors	Tumor suppressor: targeted ERK/RSK1 and PI3K/mTOR pathways; induced HIF-2α expression; repressed p21/WAF1 transcription	[94]
			Loss of DAB2IP elevated PARP-1 protein levels; RCC acquired IR-resistance	Tumor suppressor: DAB2IP acted as a scaffold to assemble a ternary complex with PARP-1 and E3 ubiquitin ligases, facilitating PARP-1 ubiquitination and subsequent proteasomal degradation	[95]
			DAB2IP-KIF3A complex suppressed renal tumorigenesis	Tumor suppressor: KIF3A interacted with the N-terminal PH domain of DAB2IP; extended KIF3A's half-life and facilitates ciliogenesis	[96]
			DAB2IP loss drived angiogenesis and conferring sunitinib resistance in RCC	Tumor suppressor: DMDRMR/miR-378a-5p/DAB2IP axis; targeted VEGFA/VEGFR2 signaling	[97]
			DAB2IP-mediated miR-138 in modulating RCC stem-like phenotypes	Tumor suppressor: DAB2IP loss interact with DNMT1 to facilitate promoter methylation of miR-138; miR-138 could suppress ABCA13 and EZH2	[98]
			DAB2IP inhibition promoted RCC cell invasion	Tumor suppressor: targetd infiltrating T cells/ERβ/DAB2IP signals	[99]
	RASAL2	++	RASAL2 inhibited angiogenesis	Tumor suppressor: decreased the expression of VEGFA through p-GSK3β/c-FOS pathway	[100]
	IQGAP2	++	ALDH9A1 deficiency promoted tumor proliferation, invasion, migration, and lipid ac-cumulation in RCC through downregulating IQGAP2	Tumor suppressor; involved in ALDH9A1-NPM1-IQGAP2-AKT axis	[101]

4. Targeting the RASGAPs Is a Significant Strategy in Cancer Therapy

The therapeutic targeting of RAS and its regulators has long been a challenge due to the "undruggable" nature of RAS [106]. RAS proteins were historically considered difficult due to their smooth surface and the high affinity of GTP binding [107]. Despite extensive research, directly inhibiting RAS has remained challenging [108]. However, recent advances in the development of direct RAS inhibitors, as well as therapies targeting downstream effectors of RAS, offer promising avenues for cancer treatment [109].

RASGAPs, which regulate multiple aspects of RAS signaling, represent attractive targets for therapeutic intervention. Restoring RASGAP function or inhibiting compensatory pathways activated by RASGAP loss may provide novel treatment strategies for urological cancers (Figure 3). One approach involves restoring the expression of downregulated RASGAPs, such as DAB2IP, using gene therapy or small-molecule drugs. Studies have shown that targeting inhibitory miRNAs can restore endogenous DAB2IP protein levels and function. However, a major limitation of RNA-based strategies in clinical applications is efficient delivery to tumor cells [110–112]. Significant research is currently underway to develop both lipid and non-lipid nanocarriers for improved RNA delivery [113].

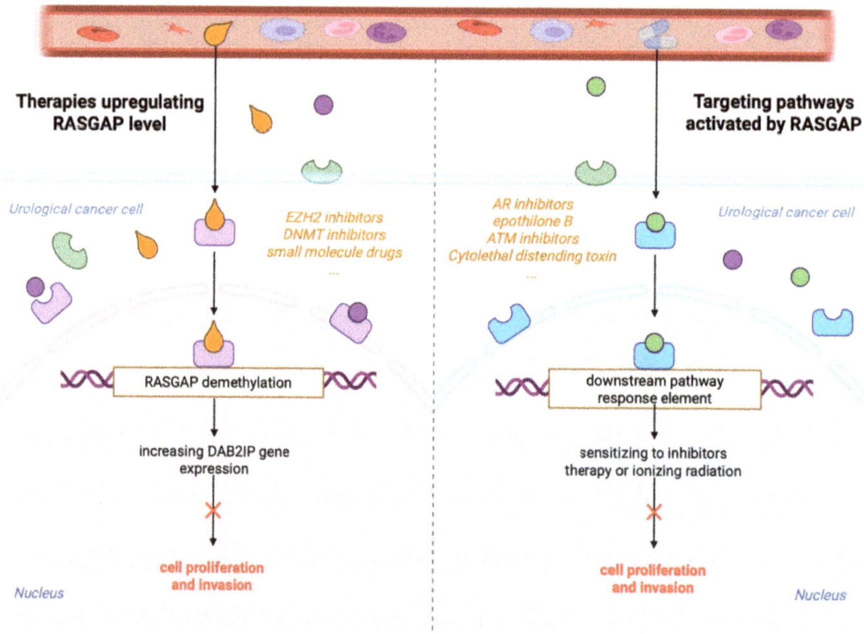

Figure 3. Potential therapies targeting RASGAPs in urological cancers. Therapies aimed at increasing RASGAP levels result in increased RASGAP gene expression by contributing to the demethylation of RASGAPs. Therapies targeting pathways activated by RASGAPs are shown. By targeting these pathways, the aim is to sensitize cancer cells to inhibitors, therapy, or ionizing radiation, thereby reducing cell proliferation and invasion. Consequently, these strategies seek to mitigate the effects of RAS signaling in urological cancer cells, either by enhancing the expression of RASGAPs or by directly interfering with downstream signaling pathways.

Furthermore, RASGAP expression is regulated at epigenetic levels, including promoter DNA and histone hypermethylation. Since DAB2IP transcription is repressed by the epigenetic factor EZH2, targeting EZH2 with its inhibitor GSK126 could restore DAB2IP expression [114]. Similarly, inhibitors of DNA methyltransferase 3A (DNMT3A) have been proven to reduce the proliferation and survival of colorectal cancer cells by upregulating the DAB2IP protein level [115].

Alternatively, targeting pathways activated by RASGAP loss may be an effective therapeutic strategy. The negative feedback regulation between AR and PI3K/AKT signaling networks has been previously demonstrated [116]. For instance, PCa with enhanced

PI3K/AKT/mTOR signaling due to DAB2IP downregulation may be more sensitive to combined treatment with PI3K inhibitors and AR pathway inhibitors, such as enzalutamide and abiraterone [117,118]. Additionally, several genotoxic drugs, including cytolethal distending toxin (CDT), epothilone B, and ATM inhibitors, have been found to resensitize DAB2IP-deficient prostate or bladder cancer cells to IR [43,119,120].

Given the complexity of RAS signaling, combination therapies targeting both RAS and its regulatory pathways may be required. Combining RASGAP modulators with inhibitors of downstream effectors, such as MEK or PI3K inhibitors, may enhance therapeutic efficacy [61]. In PCa, AR signaling supports tumor growth and survival, while PI3K/AKT activation, due to PTEN loss, drives proliferation and resistance to AR-targeted therapies. The reciprocal feedback between these pathways, where inhibition of one often leads to compensatory activation of the other, contributes to treatment resistance. These findings highlight the complexity of PCa biology and underscore the need for integrated, precision-based approaches to target interconnected pathways and improve patient outcomes [117,118].

Additionally, immunotherapies that exploit the immune-modulatory effects of RASGAPs in the tumor microenvironment are a promising avenue for future research [121]. Urological tumors with RASGAP deficiencies may possess a distinct immune landscape. Combining RAS pathway inhibitors with immune checkpoint inhibitors (e.g., anti-PD-1/PD-L1 agents) could potentially enhance antitumor immune responses and overcome therapeutic resistance.

Furthermore, the development of small molecules or peptides that mimic RASGAP function may provide a direct approach to suppressing RAS signaling. However, RAS-driven tumors are highly heterogeneous, with diverse genetic alterations influencing their response to therapy [122]. This heterogeneity complicates treatment strategies, as not all tumors will respond to the same therapeutic approach. Personalized medicine, which tailors therapies based on the specific genetic profile of each tumor, holds great promise for overcoming these challenges and optimizing therapeutic outcomes.

5. Perspectives on the Future of RASGAPs

While significant progress has been made in targeting the RAS signaling pathway and RASGAPs, several challenges remain. These include the development of resistance to targeted therapies [123], the heterogeneity of RAS mutations across different urological cancer types [108], and the intricate complexity of RAS signaling networks [106]. Since RASGAP-related pathways play crucial roles in normal cells, therapies targeting these pathways may induce toxicity in normal tissues. Moreover, urological tumors exhibit high heterogeneity, and RASGAP-deficient tumors may display diverse molecular characteristics and clinical behaviors. A deeper understanding of the interaction between RAS signaling and other oncogenic pathways could lead to the identification of novel therapeutic targets, particularly in PCa [124].

Another important tumor suppressor, p53, plays a vital role in preventing tumor initiation by regulating the cell cycle and DNA repair. p53 mutations, especially missense mutations, are prevalent across various cancers, including urological cancers. Mutant p53 (mutp53) can regulate key oncogenic pathways, such as PI3K/AKT/mTOR and RAS/MAPK [125]. Giulio et al. found that mutp53 enhances NF-κB activation while inhibiting ASK1/JNK via TNFα, a function that depends on its binding to and inhibition of DAB2IP in the cytoplasm [126]. As previously discussed, Elena et al. found that mutp53 promotes cell proliferation and invasiveness by interacting with DAB2IP in PCa [41]. In BCa, p53 mutations occur with high incidence [127]. However, the extensive function of

DAB2IP in PCa and other urological cancers remains insufficiently understood. Further research is needed to clarify its specific mechanism.

The comprehensive review of RASGAPs in urological cancers underscores their important roles in tumor progression. While significant advances have been made in elucidating their mechanisms, several critical gaps and contradictions limit their translational application. The paradoxical roles of RASGAPs across tumor types highlight the complexity of their regulatory networks. For instance, IQGAP1 acts as an oncogene in PCa by activating Raf-MEK-ERK signaling but functions as a tumor suppressor by inhibiting TGFβ signaling in BCa. Similarly, RASAL2 inhibits MAPK/SOX2-driven stemness in BCa yet promotes PI3K/AKT activation in PCa. These discrepancies may arise from tissue-specific interactomes or post-translational modifications, but the lack of systematic comparative studies across malignancies hampers the development of unified mechanistic models.

Although DAB2IP's tumor-suppressive functions are well documented, its application in therapy strategies of CRPC remains poorly integrated into clinical paradigms. While preclinical studies suggest DAB2IP restoration sensitizes tumors to therapy, clinical validation is still lacking. Additionally, much of the current research heavily relies on cell line studies (e.g., PC-3, T24) and xenograft models, which inadequately recapitulate human tumor heterogeneity and stromal interactions. For example, while RASAL3's role in CAF-mediated glutamine synthesis in CRPC is mechanistically compelling, it lacks validation in patient-derived organoids or clinical cohorts. Similarly, proposed therapeutic strategies—such as using EZH2 inhibitors to reactivate DAB2IP or PI3K/AR inhibitor combinations—are supported by preclinical data but face challenges regarding delivery specificity and toxicity. The absence of clinical trials directly targeting RASGAPs represents a critical gap in the field. Moreover, while this review highlights RASGAPs' interplay with diverse pathways (e.g., NF-κB, HIF-2α, STAT3), the hierarchical and temporal regulation of these interactions remains unclear. For instance, while DAB2IP's suppression of PI3K-AKT and activation of ASK1-JNK in PCa is well documented, how these opposing signals are balanced in vivo remains unexplored.

To address these gaps, future efforts should focus on identifying biomarkers that predict responses to RAS-targeted therapies, understanding the resistance mechanisms, and developing more effective combination therapies [128]. Emerging technologies, such as single-cell RNA sequencing (scRNA-seq) and spatial transcriptomics, may provide deeper insights into RASGAP heterogeneity within tumor subpopulations. Additionally, prospective clinical trials correlating RASGAP expression and mutation status (e.g., DAB2IP promoter methylation) with therapeutic responses to PARP or MEK inhibitors could aid in clinical biomarker validation. Furthermore, rational combination strategies, such as pairing RASGAP modulators (e.g., EZH2 inhibitors) with immune checkpoint blockers, may enhance treatment efficacy. By integrating these approaches, we can improve our understanding of the initiation and development of urological cancers, ultimately laying the groundwork for new targeted therapies. Advances in precision medicine, coupled with deeper insights into RASGAP biology, are expected to pave the way for more personalized and effective treatments for RAS-driven cancers.

6. Conclusions

Urological cancers, including PCa, BCa, and RCC, demonstrate a significant reliance on aberrant RAS signaling, underscoring the importance of the RAS pathway as a therapeutic target. The RAS family of small GTPases plays an important role in regulating cellular processes, such as growth, differentiation, and survival, primarily through downstream signaling cascades, including the MAPK and PI3K-Akt pathways. Dysregulation of the RAS signaling pathway is commonly mediated by the interaction of RAS with GEFs and

RASGAPs. RASGAPs are essential for inactivating RAS by accelerating the conversion of RAS-GTP to RAS-GDP, thus acting as critical regulatory checkpoints.

RASGAPs are closely associated with urological cancers, and their function in these cancers is complicated. Therefore, future research should focus on elucidating the mechanisms through which RASGAPs contribute to the development and progression of urological cancers. Such research could provide new insights into targeted therapies in urological cancers.

The distinct molecular features of RASGAPs in urological cancers provide a basis for exploring the targeting of the RAS signaling pathway therapies. Continued research into the molecular mechanisms by which RASGAPs govern RAS activation and inactivation, as well as into the specific roles of RASGAPs in various cancer contexts, will enhance our understanding of tumor biology and offer promising avenues for therapeutic intervention. However, the complexity of the RAS signaling pathway necessitates a nuanced approach, as therapeutic targeting must consider the broader signaling network and potential compensatory mechanisms within cancer cells.

Author Contributions: H.S. conceived and wrote the first draft of both the text and figures. H.S., G.W., G.G., H.X., L.J. and K.W. conceptualized, wrote, revised and edited the manuscript. All authors have read and agreed to the published version of the manuscript.

Funding: This research was funded by the Urologic Oncological Project from Bethune Charitable Foundation (mnzl202010 to K.W.) and the Key research and development program of Shaanxi Province (No. 2022SF-307 to K.W.).

Acknowledgments: The authors acknowledge the contributions of all researchers to this article. Figures 1–3 were created with BioRender.com (accessed on 12 April 2025). We have obtained all confirmation of publication and licensing rights for open access. The agreement numbers for these three figures are RU28503327, PE28504X7K, and GX28505L1K.

Conflicts of Interest: The authors declare that they have no conflicts of interest.

Abbreviations

ADPC	Androgen-dependent prostate cancer
ADTs	Androgen deprivation therapies
AKT	Protein Kinase B (also known as PKB)
AR	Androgen receptor
ARHGAP	Rho GTPase-Activating Protein
ASK1-JNK	Apoptosis Signal-regulating Kinase 1–c-Jun N-terminal Kinase
ATM	Ataxia-Telangiectasia Mutated
BCa	Bladder cancer
CAFs	Cancer-associated fibroblasts
CDT	Cytolethal distending toxin
CRPC	Castration-resistant prostate cancer
CPR	Proline-rich domain in C terminal
DAB2IP	Disabled-2 Interacting Protein
DC	Dendritic cell
DNMT1/3A	DNA Methyltransferase 1/3A
EMT	Epithelial-to-mesenchymal transition
Enz	Enzalutamide
ERE	Estrogen-response element
ERβ	Estrogen receptor beta

ERK1/2	Extracellular Signal-Regulated Kinase 1/2
GEFs	Guanine nucleotide exchange factors
GSK3β	Glycogen Synthase Kinase 3 Beta
GSK126	EZH2 inhibitor (specific compound)
HIF1α	Hypoxia-Inducible Factor 1-alpha
HIF2α	Hypoxia-Inducible Factor 2-alpha
HNRNPU	Heterogeneous Nuclear Ribonucleoprotein U
IQGAP1	IQ motif-containing GTPase-Activating Protein 1
IR	Ionizing radiation
KIF3A	Kinesin family member 3A
lncRNAs	Long non-coding RNAs
MAPK	Mitogen-Activated Protein Kinase
MEK	Mitogen-Activated Protein Kinase Kinase
MIBC	Muscle-invasive bladder cancer
mTOR	Mammalian target of rapamycin
mutp53	Mutant p53 proteins
NF-κB	Nuclear factor kappa B
NF1	Neurofibromin 1
NMIBC	Non-muscle-invasive bladder cancer
NSCLC	Non-small-cell lung cancer
OPHN1	Oligophrenin 1
PAG	Phosphoprotein Associated with Glycosphingolipid Microdomains 1
PARP-1	Poly (ADP-ribose) Polymerase 1
PCa	Prostate cancer
PD-L1	Programmed Death-Ligand 1
PI3K	Phosphoinositide 3-Kinase
PROX1	Prospero Homeobox 1
RASA1/2/3/4	Ras GTPase-Activating Protein family members
RASAL1/2/3	Ras Activator-like proteins
RASGAPs	RAS GTPase-Activating Proteins
RCC	Renal cell carcinoma
RSK1	Ribosomal S6 kinase 1
SPRED2	Sprouty-Related EVH1 Domain-Containing Protein 2
SYNGAP1	SYNaptic GTPase Activating Protein 1
TCC	Transitional cell carcinoma
TGFβ	Transforming growth factor beta
Th1/Th2	T-helper 1/T-helper 2 cells
Treg	Regulatory T cell
UTR	Untranslated region
VEGF	Vascular endothelial growth factor

References

1. Siegel, R.L.; Giaquinto, A.N.; Jemal, A. Cancer statistics, 2024. *CA Cancer J. Clin.* **2024**, *74*, 12–49. [CrossRef]
2. Malumbres, M.; Barbacid, M. RAS oncogenes: The first 30 years. *Nat. Rev. Cancer* **2003**, *3*, 459–465. [CrossRef] [PubMed]
3. Yang, X.; Wu, H. RAS signaling in carcinogenesis, cancer therapy and resistance mechanisms. *J. Hematol. Oncol.* **2024**, *17*, 10 [CrossRef] [PubMed]
4. Vigil, D.; Cherfils, J.; Rossman, K.L.; Der, C.J. Ras superfamily GEFs and GAPs: Validated and tractable targets for cancer therapy *Nat. Rev. Cancer* **2010**, *10*, 842–857. [CrossRef] [PubMed]
5. Simanshu, D.K.; Nissley, D.V.; McCormick, F. RAS Proteins and Their Regulators in Human Disease. *Cell* **2017**, *170*, 17–3 [CrossRef]
6. Liu, L.; Xu, C.; Hsieh, J.T.; Gong, J.; Xie, D. DAB2IP in cancer. *Oncotarget* **2016**, *7*, 3766–3776. [CrossRef]
7. Bellazzo, A.; Di Minin, G.; Collavin, L. Block one, unleash a hundred. Mechanisms of DAB2IP inactivation in cancer. *Cell Deat Differ.* **2017**, *24*, 15–25. [CrossRef]

8. Stites, E.C.; Trampont, P.C.; Ma, Z.; Ravichandran, K.S. Network analysis of oncogenic Ras activation in cancer. *Science* **2007**, *318*, 463–467. [CrossRef]
9. Trahey, M.; McCormick, F. A cytoplasmic protein stimulates normal N-ras p21 GTPase, but does not affect oncogenic mutants. *Science* **1987**, *238*, 542–545. [CrossRef]
10. Martin, G.A.; Viskochil, D.; Bollag, G.; McCabe, P.C.; Crosier, W.J.; Haubruck, H.; Conroy, L.; Clark, R.; O'Connell, P.; Cawthon, R.M.; et al. The GAP-related domain of the neurofibromatosis type 1 gene product interacts with ras p21. *Cell* **1990**, *63*, 843–849. [CrossRef]
11. Cox, A.D.; Der, C.J. The dark side of Ras: Regulation of apoptosis. *Oncogene* **2003**, *22*, 8999–9006. [CrossRef] [PubMed]
12. Bernards, A. GAPs galore! A survey of putative Ras superfamily GTPase activating proteins in man and Drosophila. *Biochim. Biophys. Acta* **2003**, *1603*, 47–82. [CrossRef]
13. Bernards, A.; Settleman, J. GAP control: Regulating the regulators of small GTPases. *Trends Cell Biol.* **2004**, *14*, 377–385. [CrossRef] [PubMed]
14. Bos, J.L.; Rehmann, H.; Wittinghofer, A. GEFs and GAPs: Critical elements in the control of small G proteins. *Cell* **2007**, *129*, 865–877. [CrossRef]
15. Ahmadian, M.R.; Kiel, C.; Stege, P.; Scheffzek, K. Structural fingerprints of the Ras-GTPase activating proteins neurofibromin and p120GAP. *J. Mol. Biol.* **2003**, *329*, 699–710. [CrossRef] [PubMed]
16. Scheffzek, K.; Ahmadian, M.R.; Kabsch, W.; Wiesmüller, L.; Lautwein, A.; Schmitz, F.; Wittinghofer, A. The Ras-RasGAP complex: Structural basis for GTPase activation and its loss in oncogenic Ras mutants. *Science* **1997**, *277*, 333–338. [CrossRef]
17. Pylayeva-Gupta, Y.; Grabocka, E.; Bar-Sagi, D. RAS oncogenes: Weaving a tumorigenic web. *Nat. Rev. Cancer* **2011**, *11*, 761–774. [CrossRef]
18. Prior, I.A.; Lewis, P.D.; Mattos, C. A comprehensive survey of Ras mutations in cancer. *Cancer Res.* **2012**, *72*, 2457–2467. [CrossRef]
19. Feng, M.; Bao, Y.; Li, Z.; Li, J.; Gong, M.; Lam, S.; Wang, J.; Marzese, D.M.; Donovan, N.; Tan, E.Y.; et al. RASAL2 activates RAC1 to promote triple-negative breast cancer progression. *J. Clin. Investig.* **2014**, *124*, 5291–5304. [CrossRef]
20. Lin, Y.; Pal, D.S.; Banerjee, P.; Banerjee, T.; Qin, G.; Deng, Y.; Borleis, J.; Iglesias, P.A.; Devreotes, P.N. Ras suppression potentiates rear actomyosin contractility-driven cell polarization and migration. *Nat. Cell Biol.* **2024**, *26*, 1062–1076. [CrossRef]
21. Song, F.; Dai, Q.; Grimm, M.O.; Steinbach, D. The Antithetic Roles of IQGAP2 and IQGAP3 in Cancers. *Cancers* **2023**, *15*, 1115. [CrossRef] [PubMed]
22. Fincham, V.J.; Chudleigh, A.; Frame, M.C. Regulation of p190 Rho-GAP by v-Src is linked to cytoskeletal disruption during transformation. *J. Cell Sci.* **1999**, *112 Pt 6*, 947–956. [CrossRef] [PubMed]
23. Lodhi, I.J.; Chiang, S.H.; Chang, L.; Vollenweider, D.; Watson, R.T.; Inoue, M.; Pessin, J.E.; Saltiel, A.R. Gapex-5, a Rab31 guanine nucleotide exchange factor that regulates Glut4 trafficking in adipocytes. *Cell Metab.* **2007**, *5*, 59–72. [CrossRef]
24. Noguchi, T.; Matozaki, T.; Inagaki, K.; Tsuda, M.; Fukunaga, K.; Kitamura, Y.; Kitamura, T.; Shii, K.; Yamanashi, Y.; Kasuga, M. Tyrosine phosphorylation of p62(Dok) induced by cell adhesion and insulin: Possible role in cell migration. *EMBO J.* **1999**, *18*, 1748–1760. [CrossRef]
25. Hobbs, G.A.; Der, C.J.; Rossman, K.L. RAS isoforms and mutations in cancer at a glance. *J. Cell Sci.* **2016**, *129*, 1287–1292. [CrossRef] [PubMed]
26. Schubbert, S.; Shannon, K.; Bollag, G. Hyperactive Ras in developmental disorders and cancer. *Nat. Rev. Cancer* **2007**, *7*, 295–308. [CrossRef]
27. Mustachio, L.M.; Chelariu-Raicu, A.; Szekvolgyi, L.; Roszik, J. Targeting KRAS in Cancer: Promising Therapeutic Strategies. *Cancers* **2021**, *13*, 1204. [CrossRef]
28. Zheng, Z.; Li, J.; Liu, Y.; Shi, Z.; Xuan, Z.; Yang, K.; Xu, C.; Bai, Y.; Fu, M.; Xiao, Q.; et al. The Crucial Role of AR-V7 in Enzalutamide-Resistance of Castration-Resistant Prostate Cancer. *Cancers* **2022**, *14*, 4877. [CrossRef]
29. Niu, Y.; Guo, C.; Wen, S.; Tian, J.; Luo, J.; Wang, K.; Tian, H.; Yeh, S.; Chang, C. ADT with antiandrogens in prostate cancer induces adverse effect of increasing resistance, neuroendocrine differentiation and tumor metastasis. *Cancer Lett.* **2018**, *439*, 47–55. [CrossRef]
30. Mishra, R.; Haldar, S.; Suchanti, S.; Bhowmick, N.A. Epigenetic changes in fibroblasts drive cancer metabolism and differentiation. *Endocr. Relat. Cancer* **2019**, *26*, R673–R688. [CrossRef]
31. Chen, H.; Toyooka, S.; Gazdar, A.F.; Hsieh, J.T. Epigenetic regulation of a novel tumor suppressor gene (hDAB2IP) in prostate cancer cell lines. *J. Biol. Chem.* **2003**, *278*, 3121–3130. [CrossRef] [PubMed]
32. Chen, H.; Tu, S.W.; Hsieh, J.T. Down-regulation of human DAB2IP gene expression mediated by polycomb Ezh2 complex and histone deacetylase in prostate cancer. *J. Biol. Chem.* **2005**, *280*, 22437–22444. [CrossRef] [PubMed]
33. Duggan, D.; Zheng, S.L.; Knowlton, M.; Benitez, D.; Dimitrov, L.; Wiklund, F.; Robbins, C.; Isaacs, S.D.; Cheng, Y.; Li, G.; et al. Two genome-wide association studies of aggressive prostate cancer implicate putative prostate tumor suppressor gene DAB2IP. *J. Natl. Cancer Inst.* **2007**, *99*, 1836–1844. [CrossRef]

34. Wu, K.; Xie, D.; Zou, Y.; Zhang, T.; Pong, R.C.; Xiao, G.; Fazli, L.; Gleave, M.; He, D.; Boothman, D.A.; et al. The mechanism o DAB2IP in chemoresistance of prostate cancer cells. *Clin. Cancer Res.* **2013**, *19*, 4740–4749. [CrossRef]
35. Min, J.; Zaslavsky, A.; Fedele, G.; McLaughlin, S.K.; Reczek, E.E.; De Raedt, T.; Guney, I.; Strochlic, D.E.; Macconaill, L.E. Beroukhim, R.; et al. An oncogene-tumor suppressor cascade drives metastatic prostate cancer by coordinately activating Ras and nuclear factor-kappaB. *Nat. Med.* **2010**, *16*, 286–294. [CrossRef]
36. Wang, B.; Huang, J.; Zhou, J.; Hui, K.; Xu, S.; Fan, J.; Li, L.; Wang, X.; Hsieh, J.T.; He, D.; et al. DAB2IP regulates EMT and metastasis of prostate cancer through targeting PROX1 transcription and destabilizing HIF1α protein. *Cell. Signal.* **2016**, *28* 1623–1630. [CrossRef]
37. Xie, D.; Gore, C.; Liu, J.; Pong, R.C.; Mason, R.; Hao, G.; Long, M.; Kabbani, W.; Yu, L.; Zhang, H.; et al. Role of DAB2IP in modulating epithelial-to-mesenchymal transition and prostate cancer metastasis. *Proc. Natl. Acad. Sci. USA* **2010**, *107*, 2485–2490 [CrossRef]
38. Xie, D.; Gore, C.; Zhou, J.; Pong, R.C.; Zhang, H.; Yu, L.; Vessella, R.L.; Min, W.; Hsieh, J.T. DAB2IP coordinates both PI3K-Akt and ASK1 pathways for cell survival and apoptosis. *Proc. Natl. Acad. Sci. USA* **2009**, *106*, 19878–19883. [CrossRef]
39. Wu, K.; Liu, J.; Tseng, S.F.; Gore, C.; Ning, Z.; Sharifi, N.; Fazli, L.; Gleave, M.; Kapur, P.; Xiao, G.; et al. The role of DAB2IP in androgen receptor activation during prostate cancer progression. *Oncogene* **2014**, *33*, 1954–1963. [CrossRef] [PubMed]
40. Gu, Y.; Wu, S.; Chong, Y.; Guan, B.; Li, L.; He, D.; Wang, X.; Wang, B.; Wu, K. DAB2IP regulates intratumoral testosterone synthesis and CRPC tumor growth by ETS1/AKR1C3 signaling. *Cell. Signal.* **2022**, *95*, 110336. [CrossRef]
41. Valentino, E.; Bellazzo, A.; Di Minin, G.; Sicari, D.; Apollonio, M.; Scognamiglio, G.; Di Bonito, M.; Botti, G.; Del Sal, G.; Collavin L. Mutant p53 potentiates the oncogenic effects of insulin by inhibiting the tumor suppressor DAB2IP. *Proc. Natl. Acad. Sci. USA* **2017**, *114*, 7623–7628. [CrossRef]
42. Kong, Z.; Xie, D.; Boike, T.; Raghavan, P.; Burma, S.; Chen, D.J.; Habib, A.A.; Chakraborty, A.; Hsieh, J.T.; Saha, D. Downregulation of human DAB2IP gene expression in prostate cancer cells results in resistance to ionizing radiation. *Cancer Res.* **2010**, *70* 2829–2839. [CrossRef]
43. Kong, Z.; Raghavan, P.; Xie, D.; Boike, T.; Burma, S.; Chen, D.; Chakraborty, A.; Hsieh, J.T.; Saha, D. Epothilone B confers radiation dose enhancement in DAB2IP gene knock-down radioresistant prostate cancer cells. *Int. J. Radiat. Oncol. Biol. Phys.* **2010**, *78* 1210–1218. [CrossRef]
44. De Florian Fania, R.; Bellazzo, A.; Collavin, L. An update on the tumor-suppressive functions of the RasGAP protein DAB2IP with focus on therapeutic implications. *Cell Death Differ.* **2024**, *31*, 844–854. [CrossRef]
45. Moon, H.; Ruelcke, J.E.; Choi, E.; Sharpe, L.J.; Nassar, Z.D.; Bielefeldt-Ohmann, H.; Parat, M.O.; Shah, A.; Francois, M.; Inder, K.L. et al. Diet-induced hypercholesterolemia promotes androgen-independent prostate cancer metastasis via IQGAP1 and caveolin-1 *Oncotarget* **2015**, *6*, 7438–7453. [CrossRef]
46. Cerutti, C.; Lucotti, S.; Menendez, S.T.; Reymond, N.; Garg, R.; Romero, I.A.; Muschel, R.; Ridley, A.J. IQGAP1 and NWASP promote human cancer cell dissemination and metastasis by regulating β1-integrin via FAK and MRTF/SRF. *Cell Rep.* **2024**, *43*, 113989. [CrossRef]
47. Xiong, Z.; Zhuang, R.L.; Yu, S.L.; Xie, Z.X.; Peng, S.R.; Li, Z.A.; Li, B.H.; Xie, J.J.; Li, Y.N.; Li, K.W.; et al. Cancer-associated fibroblasts regulate mitochondrial metabolism and inhibit chemosensitivity via ANGPTL4-IQGAP1 axis in prostate cancer. *J. Adv. Res.* **2024**. [CrossRef]
48. Ernst, T.; Hergenhahn, M.; Kenzelmann, M.; Cohen, C.D.; Bonrouhi, M.; Weninger, A.; Klären, R.; Gröne, E.F.; Wiesel, M., Güdemann, C.; et al. Decrease and gain of gene expression are equally discriminatory markers for prostate carcinoma: A gene expression analysis on total and microdissected prostate tissue. *Am. J. Pathol.* **2002**, *160*, 2169–2180. [CrossRef]
49. Xie, Y.; Yan, J.; Cutz, J.C.; Rybak, A.P.; He, L.; Wei, F.; Kapoor, A.; Schmidt, V.A.; Tao, L.; Tang, D. IQGAP2, A candidate tumour suppressor of prostate tumorigenesis. *Biochim. Biophys. Acta* **2012**, *1822*, 875–884. [CrossRef]
50. Xie, Y.; Zheng, L.; Tao, L. Downregulation of IQGAP2 Correlates with Prostate Cancer Recurrence and Metastasis. *Transl. Oncol.* **2019**, *12*, 236–244. [CrossRef]
51. Zheng, X.; Xu, H.; Yi, X.; Zhang, T.; Wei, Q.; Li, H.; Ai, J. Tumor-antigens and immune landscapes identification for prostate adenocarcinoma mRNA vaccine. *Mol. Cancer* **2021**, *20*, 160. [CrossRef] [PubMed]
52. Berndt, S.I.; Wang, Z.; Yeager, M.; Alavanja, M.C.; Albanes, D.; Amundadottir, L.; Andriole, G.; Beane Freeman, L.; Campa, D.; Cancel-Tassin, G.; et al. Two susceptibility loci identified for prostate cancer aggressiveness. *Nat. Commun.* **2015**, *6*, 6889. [CrossRef] [PubMed]
53. Kolfschoten, I.G.; van Leeuwen, B.; Berns, K.; Mullenders, J.; Beijersbergen, R.L.; Bernards, R.; Voorhoeve, P.M.; Agami, R. A genetic screen identifies PITX1 as a suppressor of RAS activity and tumorigenicity. *Cell* **2005**, *121*, 849–858. [CrossRef]
54. Wang, Q.; Wu, S.; Gu, Y.; Liang, H.; He, F.; Wang, X.; He, D.; Wu, K. RASAL2 regulates the cell cycle and cyclin D1 expression through PI3K/AKT signalling in prostate tumorigenesis. *Cell Death Discov.* **2022**, *8*, 275. [CrossRef]
55. Tailor, K.; Paul, J.; Ghosh, S.; Kumari, N.; Kwabi-Addo, B. RASAL2 suppresses the proliferative and invasive ability of PC3 prostate cancer cells. *Oncotarget* **2021**, *12*, 2489–2499. [CrossRef]

6. Mishra, R.; Haldar, S.; Placencio, V.; Madhav, A.; Rohena-Rivera, K.; Agarwal, P.; Duong, F.; Angara, B.; Tripathi, M.; Liu, Z.; et al. Stromal epigenetic alterations drive metabolic and neuroendocrine prostate cancer reprogramming. *J. Clin. Investig.* **2018**, *128*, 4472–4484. [CrossRef]
7. Kachroo, N.; Valencia, T.; Warren, A.Y.; Gnanapragasam, V.J. Evidence for downregulation of the negative regulator SPRED2 in clinical prostate cancer. *Br. J. Cancer* **2013**, *108*, 597–601. [CrossRef]
8. Lorenzo, C.; McCormick, F. SPRED proteins and their roles in signal transduction, development, and malignancy. *Genes. Dev.* **2020**, *34*, 1410–1421. [CrossRef] [PubMed]
9. Billuart, P.; Bienvenu, T.; Ronce, N.; des Portes, V.; Vinet, M.C.; Zemni, R.; Roest Crollius, H.; Carrié, A.; Fauchereau, F.; Cherry, M.; et al. Oligophrenin-1 encodes a rhoGAP protein involved in X-linked mental retardation. *Nature* **1998**, *392*, 923–926. [CrossRef]
0. Liu, J.; Zhang, Y.; Li, S.; Sun, F.; Wang, G.; Wei, D.; Yang, T.; Gu, S. Androgen deprivation-induced OPHN1 amplification promotes castration-resistant prostate cancer. *Oncol. Rep.* **2022**, *47*, 3. [CrossRef]
1. Hua, H.; Zhang, H.; Chen, J.; Wang, J.; Liu, J.; Jiang, Y. Targeting Akt in cancer for precision therapy. *J. Hematol. Oncol.* **2021**, *14*, 128. [CrossRef] [PubMed]
2. Yu, W.; Wang, Y.; Gong, M.; Pei, F.; Zheng, J. Phosphoprotein associated with glycosphingolipid microdomains 1 inhibits the proliferation and invasion of human prostate cancer cells in vitro through suppression of Ras activation. *Oncol. Rep.* **2012**, *28*, 606–614. [CrossRef]
3. Siegel, R.L.; Miller, K.D.; Jemal, A. Cancer statistics, 2020. *CA Cancer J. Clin.* **2020**, *70*, 7–30. [CrossRef]
4. Lv, M.; Shang, S.; Liu, K.; Wang, Y.; Xu, P.; Song, H.; Zhang, J.; Sun, Z.; Yan, Y.; Zhu, Z.; et al. Revitalizing Bacillus Calmette-Guérin Immunotherapy for Bladder Cancer: Nanotechnology and Bioengineering Approaches. *Pharmaceutics* **2024**, *16*, 1067. [CrossRef]
5. Weinstein, J.N.; Akbani, R.; Broom, B.M.; Wang, W.; Verhaak, R.G.W.; McConkey, D.; Lerner, S.; Morgan, M.; Creighton, C.J.; Smith, C.; et al. Comprehensive molecular characterization of urothelial bladder carcinoma. *Nature* **2014**, *507*, 315–322. [CrossRef]
6. Goebell, P.J.; Knowles, M.A. Bladder cancer or bladder cancers? Genetically distinct malignant conditions of the urothelium. *Urol. Oncol.* **2010**, *28*, 409–428. [CrossRef] [PubMed]
7. Beukers, W.; Hercegovac, A.; Zwarthoff, E.C. HRAS mutations in bladder cancer at an early age and the possible association with the Costello Syndrome. *Eur. J. Hum. Genet.* **2014**, *22*, 837–839. [CrossRef]
8. Cerami, E.; Gao, J.; Dogrusoz, U.; Gross, B.E.; Sumer, S.O.; Aksoy, B.A.; Jacobsen, A.; Byrne, C.J.; Heuer, M.L.; Larsson, E.; et al. The cBio cancer genomics portal: An open platform for exploring multidimensional cancer genomics data. *Cancer Discov.* **2012**, *2*, 401–404. [CrossRef]
9. Jou, Y.C.; Tsai, Y.S.; Chen, S.Y.; Hsieh, H.Y.; Tsai, H.T.; Tzai, T.S. Loss of DAB2IP expression in human urothelial carcinoma is associated with poorer recurrence-free survival. *Virchows Arch.* **2016**, *468*, 733–740. [CrossRef]
0. Shen, Y.J.; Kong, Z.L.; Wan, F.N.; Wang, H.K.; Bian, X.J.; Gan, H.L.; Wang, C.F.; Ye, D.W. Downregulation of DAB2IP results in cell proliferation and invasion and contributes to unfavorable outcomes in bladder cancer. *Cancer Sci.* **2014**, *105*, 704–712. [CrossRef]
1. Kunze, E.; Von Bonin, F.; Werner, C.; Wendt, M.; Schlott, T. Transitional cell carcinomas and nonurothelial carcinomas of the urinary bladder differ in the promoter methylation status of the caveolin-1, hDAB2IP and p53 genes, but not in the global methylation of Alu elements. *Int. J. Mol. Med.* **2006**, *17*, 3–13. [CrossRef]
2. Ou, Z.; Wang, Y.; Chen, J.; Tao, L.; Zuo, L.; Sahasrabudhe, D.; Joseph, J.; Wang, L.; Yeh, S. Estrogen receptor β promotes bladder cancer growth and invasion via alteration of miR-92a/DAB2IP signals. *Exp. Mol. Med.* **2018**, *50*, 1–11. [CrossRef] [PubMed]
3. Huang, J.; Wang, B.; Hui, K.; Zeng, J.; Fan, J.; Wang, X.; Hsieh, J.-T.; He, D.; Wu, K. miR-92b targets DAB2IP to promote EMT in bladder cancer migration and invasion. *Oncol. Rep.* **2016**, *36*, 1693–1701. [CrossRef] [PubMed]
4. Feng, C.; Sun, P.; Hu, J.; Feng, H.; Li, M.; Liu, G.; Pan, Y.; Feng, Y.; Xu, Y.; Feng, K.; et al. miRNA-556-3p promotes human bladder cancer proliferation, migration and invasion by negatively regulating DAB2IP expression. *Int. J. Oncol.* **2017**, *50*, 2101–2112. [CrossRef] [PubMed]
5. Wu, K.; Wang, B.; Chen, Y.; Zhou, J.; Huang, J.; Hui, K.; Zeng, J.; Zhu, J.; Zhang, K.; Li, L.; et al. DAB2IP regulates the chemoresistance to pirarubicin and tumor recurrence of non-muscle invasive bladder cancer through STAT3/Twist1/P-glycoprotein signaling. *Cell. Signal.* **2015**, *27*, 2515–2523. [CrossRef]
6. He, H.; Chang, R.; Zhang, T.; Yang, C.; Kong, Z. ATM mediates DAB2IP-deficient bladder cancer cell resistance to ionizing radiation through the p38MAPK and NF-κB signaling pathway. *Mol. Med. Rep.* **2017**, *16*, 1216–1222. [CrossRef]
7. Aaltonen, V.; Boström, P.J.; Söderström, K.O.; Hirvonen, O.; Tuukkanen, J.; Nurmi, M.; Laato, M.; Peltonen, J. Urinary bladder transitional cell carcinogenesis is associated with down-regulation of NF1 tumor suppressor gene in vivo and in vitro. *Am. J. Pathol.* **1999**, *154*, 755–765. [CrossRef]
8. Shi, Z.D.; Hao, L.; Han, X.X.; Wu, Z.X.; Pang, K.; Dong, Y.; Qin, J.X.; Wang, G.Y.; Zhang, X.M.; Xia, T.; et al. Targeting HNRNPU to overcome cisplatin resistance in bladder cancer. *Mol. Cancer* **2022**, *21*, 37. [CrossRef]
9. Hui, K.; Gao, Y.; Huang, J.; Xu, S.; Wang, B.; Zeng, J.; Fan, J.; Wang, X.; Yue, Y.; Wu, S.; et al. RASAL2, a RAS GTPase-activating protein, inhibits stemness and epithelial-mesenchymal transition via MAPK/SOX2 pathway in bladder cancer. *Cell Death Dis.* **2017**, *8*, e2600. [CrossRef]

80. Hui, K.; Wu, S.; Yue, Y.; Gu, Y.; Guan, B.; Wang, X.; Hsieh, J.T.; Chang, L.S.; He, D.; Wu, K. RASAL2 inhibits tumor angiogenesi via p-AKT/ETS1 signaling in bladder cancer. *Cell. Signal.* **2018**, *48*, 38–44. [CrossRef]
81. Hensel, J.; Duex, J.E.; Owens, C.; Dancik, G.M.; Edwards, M.G.; Frierson, H.F.; Theodorescu, D. Patient Mutation Directed shRN Screen Uncovers Novel Bladder Tumor Growth Suppressors. *Mol. Cancer Res.* **2015**, *13*, 1306–1315. [CrossRef] [PubMed]
82. Xu, Y.; Kim, Y.H.; Jeong, P.; Piao, X.M.; Byun, Y.J.; Kang, H.W.; Kim, W.T.; Lee, J.Y.; Kim, I.Y.; Moon, S.K.; et al. Diagnostic value c combined IQGAP3/BMP4 and IQGAP3/FAM107A expression ratios in urinary cell-free DNA for discriminating bladder cance from hematuria. *Urol. Oncol.* **2019**, *37*, 86–96. [CrossRef] [PubMed]
83. Xu, Y.; Kim, Y.H.; Jeong, P.; Piao, X.M.; Byun, Y.J.; Seo, S.P.; Kang, H.W.; Kim, W.T.; Lee, J.Y.; Ryu, D.H.; et al. Urinary Cell-Fre DNA IQGAP3/BMP4 Ratio as a Prognostic Marker for Non-Muscle-Invasive Bladder Cancer. *Clin. Genitourin. Cancer* **2019**, *1* e704–e711. [CrossRef]
84. Li, G.; Wang, Y.; Guo, X.B.; Zhao, B. CDC42 Regulates Cell Proliferation and Apoptosis in Bladder Cancer via the IQGAP Mediated Ras/ERK Pathway. *Biochem. Genet.* **2022**, *60*, 2383–2398. [CrossRef] [PubMed]
85. Song, F.; Kotolloshi, R.; Gajda, M.; Hölzer, M.; Grimm, M.O.; Steinbach, D. Reduced IQGAP2 Promotes Bladder Cancer throug Regulation of MAPK/ERK Pathway and Cytokines. *Int. J. Mol. Sci.* **2022**, *23*, 13508. [CrossRef]
86. Yang, C.; Wu, S.; Mou, Z.; Zhou, Q.; Zhang, Z.; Chen, Y.; Ou, Y.; Chen, X.; Dai, X.; Xu, C.; et al. Transcriptomic Analysi Identified ARHGAP Family as a Novel Biomarker Associated With Tumor-Promoting Immune Infiltration and Nanomechanica Characteristics in Bladder Cancer. *Front. Cell Dev. Biol.* **2021**, *9*, 657219. [CrossRef]
87. Li, M.; Li, L.; Zheng, J.; Li, Z.; Li, S.; Wang, K.; Chen, X. Liquid biopsy at the frontier in renal cell carcinoma: Recent analysis c techniques and clinical application. *Mol. Cancer* **2023**, *22*, 37. [CrossRef]
88. Gray, R.E.; Harris, G.T. Renal Cell Carcinoma: Diagnosis and Management. *Am. Fam. Physician* **2019**, *99*, 179–184.
89. Hsieh, J.J.; Purdue, M.P.; Signoretti, S.; Swanton, C.; Albiges, L.; Schmidinger, M.; Heng, D.Y.; Larkin, J.; Ficarra, V. Renal ce carcinoma. *Nat. Rev. Dis. Primers* **2017**, *3*, 17009. [CrossRef]
90. Linehan, W.M.; Ricketts, C.J. The metabolic basis of kidney cancer. *Semin. Cancer Biol.* **2013**, *23*, 46–55. [CrossRef]
91. Roberts, P.J.; Der, C.J. Targeting the Raf-MEK-ERK mitogen-activated protein kinase cascade for the treatment of cancer. *Oncogen* **2007**, *26*, 3291–3310. [CrossRef]
92. Bahar, M.E.; Kim, H.J.; Kim, D.R. Targeting the RAS/RAF/MAPK pathway for cancer therapy: From mechanism to clinica studies. *Signal Transduct. Target. Ther.* **2023**, *8*, 455. [CrossRef]
93. Zhou, J.; Deng, Z.; Pei, X.; Lai, J.; Qu, W. DAB2IP stabilizes p27(Kip1) via suppressing PI3K/AKT signaling in clear cell renal ce carcinoma. *Funct. Integr. Genom.* **2023**, *23*, 326. [CrossRef]
94. Zhou, J.; Luo, J.; Wu, K.; Yun, E.J.; Kapur, P.; Pong, R.C.; Du, Y.; Wang, B.; Authement, C.; Hernandez, E.; et al. Loss of DAB2IP i RCC cells enhances their growth and resistance to mTOR-targeted therapies. *Oncogene* **2016**, *35*, 4663–4674. [CrossRef] [PubMed
95. Yun, E.J.; Lin, C.J.; Dang, A.; Hernandez, E.; Guo, J.; Chen, W.M.; Allison, J.; Kim, N.; Kapur, P.; Brugarolas, J.; et al. Downregula tion of Human DAB2IP Gene Expression in Renal Cell Carcinoma Results in Resistance to Ionizing Radiation. *Clin. Cancer Re* **2019**, *25*, 4542–4551. [CrossRef] [PubMed]
96. Lin, C.J.; Dang, A.; Hernandez, E.; Hsieh, J.T. DAB2IP modulates primary cilia formation associated with renal tumorigenesi *Neoplasia* **2021**, *23*, 169–180. [CrossRef]
97. Zhu, Y.; Liu, X.; Wang, Y.; Pan, Y.; Han, X.; Peng, B.; Zhang, X.; Niu, S.; Wang, H.; Ye, Q.; et al. DMDRMR promotes angiogenesi via antagonizing DAB2IP in clear cell renal cell carcinoma. *Cell Death Dis.* **2022**, *13*, 456. [CrossRef] [PubMed]
98. Yun, E.J.; Zhou, J.; Lin, C.J.; Xu, S.; Santoyo, J.; Hernandez, E.; Lai, C.H.; Lin, H.; He, D.; Hsieh, J.T. The network of DAB2IP-miF 138 in regulating drug resistance of renal cell carcinoma associated with stem-like phenotypes. *Oncotarget* **2017**, *8*, 66975–6698 [CrossRef]
99. Yeh, C.R.; Ou, Z.Y.; Xiao, G.Q.; Guancial, E.; Yeh, S. Infiltrating T cells promote renal cell carcinoma (RCC) progression via alterin the estrogen receptor β-DAB2IP signals. *Oncotarget* **2015**, *6*, 44346–44359. [CrossRef]
100. Hui, K.; Yue, Y.; Wu, S.; Gu, Y.; Guan, B.; Wang, X.; Hsieh, J.T.; Chang, L.S.; He, D.; Wu, K. The expression and function of RASAL in renal cell carcinoma angiogenesis. *Cell Death Dis.* **2018**, *9*, 881. [CrossRef]
101. Tan, D.; Miao, D.; Zhao, C.; Shi, J.; Lv, Q.; Lu, F.; Ruan, H.; Xiong, Z.; Zhang, X. N6-methyladenosine-modified ALDH9A modulates lipid accumulation and tumor progression in clear cell renal cell carcinoma through the NPM1/IQGAP2/AK signaling pathway. *Cell Death Dis.* **2024**, *15*, 520. [CrossRef]
102. Vanli, G.; Sempoux, C.; Widmann, C. The caspase-3/p120 RasGAP stress-sensing module reduces liver cancer incidence but doe not affect overall survival in gamma-irradiated and carcinogen-treated mice. *Mol. Carcinog.* **2017**, *56*, 1680–1684. [CrossRef]
103. Kitajima, S.; Barbie, D.A. RASA1/NF1-Mutant Lung Cancer: Racing to the Clinic? *Clin. Cancer Res.* **2018**, *24*, 1243–124 [CrossRef] [PubMed]
104. McLaughlin, S.K.; Olsen, S.N.; Dake, B.; De Raedt, T.; Lim, E.; Bronson, R.T.; Beroukhim, R.; Polyak, K.; Brown, M.; Kuperwasse C.; et al. The RasGAP gene, RASAL2, is a tumor and metastasis suppressor. *Cancer Cell* **2013**, *24*, 365–378. [CrossRef] [PubMed

105. Arafeh, R.; Qutob, N.; Emmanuel, R.; Keren-Paz, A.; Madore, J.; Elkahloun, A.; Wilmott, J.S.; Gartner, J.J.; Di Pizio, A.; Winograd-Katz, S.; et al. Recurrent inactivating RASA2 mutations in melanoma. *Nat. Genet.* **2015**, *47*, 1408–1410. [CrossRef] [PubMed]
106. Cox, A.D.; Fesik, S.W.; Kimmelman, A.C.; Luo, J.; Der, C.J. Drugging the undruggable RAS: Mission possible? *Nat. Rev. Drug Discov.* **2014**, *13*, 828–851. [CrossRef]
107. Papke, B.; Der, C.J. Drugging RAS: Know the enemy. *Science* **2017**, *355*, 1158–1163. [CrossRef]
108. Ostrem, J.M.; Shokat, K.M. Direct small-molecule inhibitors of KRAS: From structural insights to mechanism-based design. *Nat. Rev. Drug Discov.* **2016**, *15*, 771–785. [CrossRef]
109. Ostrem, J.M.; Peters, U.; Sos, M.L.; Wells, J.A.; Shokat, K.M. K-Ras(G12C) inhibitors allosterically control GTP affinity and effector interactions. *Nature* **2013**, *503*, 548–551. [CrossRef]
110. Wang, J.; Liu, Y.; Wang, X.; Li, J.; Wei, J.; Wang, Y.; Song, W.; Zhang, Z. MiR-1266 promotes cell proliferation, migration and invasion in cervical cancer by targeting DAB2IP. *Biochim. Biophys. Acta Mol. Basis Dis.* **2018**, *1864*, 3623–3630. [CrossRef]
111. Li, X.; Zhang, X.; Zhang, Q.; Lin, R. miR-182 contributes to cell proliferation, invasion and tumor growth in colorectal cancer by targeting DAB2IP. *Int. J. Biochem. Cell Biol.* **2019**, *111*, 27–36. [CrossRef] [PubMed]
112. Chen, S.; Liu, R.; Wang, H.; Liu, Q. Hypoxia-driven miR-1307-3p promotes hepatocellular carcinoma cell proliferation and invasion by modulating DAB2 interacting protein. *Pathol. Res. Pract.* **2022**, *237*, 154066. [CrossRef]
113. Zhu, Y.; Zhu, L.; Wang, X.; Jin, H. RNA-based therapeutics: An overview and prospectus. *Cell Death Dis.* **2022**, *13*, 644. [CrossRef]
114. Schade, A.E.; Kuzmickas, R.; Rodriguez, C.L.; Mattioli, K.; Enos, M.; Gardner, A.; Cichowski, K. Combating castration-resistant prostate cancer by co-targeting the epigenetic regulators EZH2 and HDAC. *PLoS Biol.* **2023**, *21*, e3002038. [CrossRef]
115. Zhou, Y.; Yang, Z.; Zhang, H.; Li, H.; Zhang, M.; Wang, H.; Zhang, M.; Qiu, P.; Zhang, R.; Liu, J. DNMT3A facilitates colorectal cancer progression via regulating DAB2IP mediated MEK/ERK activation. *Biochim. Biophys. Acta Mol. Basis Dis.* **2022**, *1868*, 166353. [CrossRef]
116. Chandarlapaty, S. Negative feedback and adaptive resistance to the targeted therapy of cancer. *Cancer Discov.* **2012**, *2*, 311–319. [CrossRef]
117. Ku, S.Y.; Gleave, M.E.; Beltran, H. Towards precision oncology in advanced prostate cancer. *Nat. Rev. Urol.* **2019**, *16*, 645–654. [CrossRef]
118. Tortorella, E.; Giantulli, S.; Sciarra, A.; Silvestri, I. AR and PI3K/AKT in Prostate Cancer: A Tale of Two Interconnected Pathways. *Int. J. Mol. Sci.* **2023**, *24*, 2046. [CrossRef]
119. Lai, C.H.; Chang, C.S.; Liu, H.H.; Tsai, Y.S.; Hsu, F.M.; Yu, Y.L.; Lai, C.K.; Gandee, L.; Pong, R.C.; Hsu, H.W.; et al. Sensitization of radio-resistant prostate cancer cells with a unique cytolethal distending toxin. *Oncotarget* **2014**, *5*, 5523–5534. [CrossRef] [PubMed]
120. Zhang, T.; Shen, Y.; Chen, Y.; Hsieh, J.T.; Kong, Z. The ATM inhibitor KU55933 sensitizes radioresistant bladder cancer cells with DAB2IP gene defect. *Int. J. Radiat. Biol.* **2015**, *91*, 368–378. [CrossRef] [PubMed]
121. Carnevale, J.; Shifrut, E.; Kale, N.; Nyberg, W.A.; Blaeschke, F.; Chen, Y.Y.; Li, Z.; Bapat, S.P.; Diolaiti, M.E.; O'Leary, P.; et al. RASA2 ablation in T cells boosts antigen sensitivity and long-term function. *Nature* **2022**, *609*, 174–182. [CrossRef] [PubMed]
122. Ruscetti, M.; Leibold, J.; Bott, M.J.; Fennell, M.; Kulick, A.; Salgado, N.R.; Chen, C.C.; Ho, Y.J.; Sanchez-Rivera, F.J.; Feucht, J.; et al. NK cell-mediated cytotoxicity contributes to tumor control by a cytostatic drug combination. *Science* **2018**, *362*, 1416–1422. [CrossRef] [PubMed]
123. Yaeger, R.; Corcoran, R.B. Targeting Alterations in the RAF-MEK Pathway. *Cancer Discov.* **2019**, *9*, 329–341. [CrossRef]
124. Downward, J. RAS Synthetic Lethal Screens Revisited: Still Seeking the Elusive Prize? *Clin. Cancer Res.* **2015**, *21*, 1802–1809. [CrossRef]
125. Yue, X.; Wu, F.; Li, Y.; Liu, J.; Boateng, M.; Mandava, K.; Zhang, C.; Feng, Z.; Gao, J.; Hu, W. Gain of function mutant p53 protein activates AKT through the Rac1 signaling to promote tumorigenesis. *Cell Cycle* **2020**, *19*, 1338–1351. [CrossRef]
126. Di Minin, G.; Bellazzo, A.; Dal Ferro, M.; Chiaruttini, G.; Nuzzo, S.; Bicciato, S.; Piazza, S.; Rami, D.; Bulla, R.; Sommaggio, R.; et al. Mutant p53 reprograms TNF signaling in cancer cells through interaction with the tumor suppressor DAB2IP. *Mol. Cell* **2014**, *56*, 617–629. [CrossRef]
127. Sidransky, D.; Von Eschenbach, A.; Tsai, Y.C.; Jones, P.; Summerhayes, I.; Marshall, F.; Paul, M.; Green, P.; Hamilton, S.R.; Frost, P.; et al. Identification of p53 gene mutations in bladder cancers and urine samples. *Science* **1991**, *252*, 706–709. [CrossRef] [PubMed]
128. Prior, I.A.; Hood, F.E.; Hartley, J.L. The Frequency of Ras Mutations in Cancer. *Cancer Res.* **2020**, *80*, 2969–2974. [CrossRef]

cancers

Review

Unlocking the Genetic Secrets of Pancreatic Cancer: *KRAS* Allelic Imbalances in Tumor Evolution

Vasiliki Liaki [1,*,†], Blanca Rosas-Perez [1,†] and Carmen Guerra [1,2]

1 Molecular Oncology Program, Spanish National Cancer Research Center (CNIO), 28029 Madrid, Spain; brosas@cnio.es (B.R.-P.); mcguerra@cnio.es (C.G.)
2 Centro de Investigación Biomédica en Red de Cáncer (CIBERONC), Instituto de Salud Carlos III, 28029 Madrid, Spain
* Correspondence: lvasiliki@cnio.es
† These authors contributed equally to this work.

Simple Summary: Pancreatic cancer is one of the most heterogeneous and highly aggressive malignancies. Traditionally, the gradual evolution model of pancreatic tumors has been based on the progressive acquisition of key genetic alterations. However, a complementary model based on rapid genetic evolution has also emerged, according to which driver gene inactivations can occur simultaneously. In both models, an activating mutation in the KRAS oncogene is the main initiating event in acinar pancreatic cells. However, in advanced stages of the disease, additional amplifications and major imbalances in the mutant KRAS allele are often detected. Here, we discuss how KRAS allelic imbalances can arise during tumor evolution. We also analyze the importance of KRAS allelic imbalances in pancreatic cancer biology and their possible therapeutic implications.

Abstract: Pancreatic Ductal Adenocarcinoma (PDAC) belongs to the types of cancer with the highest lethality. It is also remarkably chemoresistant to the few available cytotoxic therapeutic options. PDAC is characterized by limited mutational heterogeneity of the known driver genes, KRAS, CDKN2A, TP53, and SMAD4, observed in both early-stage and advanced tumors. In this review, we summarize the two proposed models of genetic evolution of pancreatic cancer. The gradual or stepwise accumulated mutations model has been widely studied. On the contrary, less evidence exists on the more recent simultaneous model, according to which rapid tumor evolution is driven by the concurrent accumulation of genetic alterations. In both models, oncogenic KRAS mutations are the main initiating event. Here, we analyze the emerging topic of KRAS allelic imbalances and how it arises during tumor evolution, as it is often detected in advanced and metastatic PDAC. We also summarize recent evidence on how it affects tumor biology, metastasis, and response to therapy. To this extent, we highlight the necessity to include studies of KRAS allelic frequencies in the design of future therapeutic strategies against pancreatic cancer.

Keywords: pancreatic ductal adenocarcinoma (PDAC); KRAS; allelic imbalance; wild-type allele; loss of heterozygosity; tumor evolution; genomic rearrangements; metastasis

1. Introduction

Pancreatic cancer is one of the most aggressive malignancies and has become the third leading cause of cancer-related deaths worldwide [1]. With a 5-year survival rate of only 5%, the median survival is less than 6 months [2]. This elevated mortality is mainly attributed to late diagnosis, as locally advanced disease is largely asymptomatic.

Unfortunately, this dismal prognosis has remained unchanged over the years, as minimal improvements have been made in the field of treatment [3]. Currently, surgical resection remains the only curative option for early-stage local disease, which corresponds to 20-30% of patients. Nevertheless, the recurrence of pancreatic tumors occurs in up to 85% of these patients [4]. So far, adjuvant treatment with 5-fluorouracil, leucovorin, irinotecan, and oxaliplatin (FOLFIRINOX) has been demonstrated to provide the longest median overall survival (OS) of 54 months in patients with resectable disease [5].

Approximately 70–80% of patients do not benefit from surgery, because at the time of diagnosis they display advanced disease with metastasis. Thus far, cytotoxic chemotherapy is the standard treatment for late-stage patients, with overall survival in the range of weeks to a few months [3]. More specifically, the standard of care consists of modern combinations of highly cytotoxic agents, such as nab-paclitaxel plus gemcitabine or FOLFIRINOX, that only provide modest improvements in OS, in the range of weeks to months [6,7]. Undoubtedly, there is urgent need for novel targeted therapies with reduced toxicities. At present, there are numerous ongoing clinical trials investigating the efficacy of inhibitors against druggable mutations, such as KRAS, PARP, and SHP2 [8–10]. Yet, only few targeted therapies or immunotherapeutic approaches have been approved for clinical use [11].

Pancreatic Ductal Adenocarcinoma (PDAC) accounts for more than 90% of all pancreatic tumors. The majority are non-hereditary PDACs and are caused by the accumulation of somatic mutations in oncogenes and tumor suppressor genes. All large-scale genomic studies have confirmed the most common alterations in the KRAS oncogene, and the tumor suppressor genes, CDKN2A, TP53, and SMAD4 [12–15]. More specifically, activating mutations in KRAS are detected in 95% of all pancreatic tumors [14,15]. CDKN2A, TP53, and SMAD4 are commonly inactivated in 50–80% of pancreatic tumors by methylation or homozygous deletions [16]. Other mutated genes in ~10% of tumors include TGFBR1, ARID1A, GLI3, and TGFBR2 [16,17].

Additionally, other types of genomic changes, such as copy number alterations, that involve high numbers of amplified or lost genes, are also often detected [17,18]. Waddell et al. have separated pancreatic tumors based on the number and location of structural variants. Subtypes designated as locally rearranged or unstable seem to be defined by a large (>200) number of events or structural variants, respectively [13,16]. These genomic rearrangements occur frequently during tumor evolution, explaining the limited heterogeneity observed among the known driver genes between primary and metastatic PDAC [19].

Here, we present a brief overview of the proposed models of PDAC evolution, in correlation with these genomic alterations. We also focus on the main driver oncogene, KRAS, and discuss how different allelic frequencies arise during tumor progression and how they affect tumor biology.

2. Tumor Evolution

Two models of tumor evolution have been proposed for the genomic events that occur during PDAC evolution. According to the classical or gradual progression model, precursor lesions accumulate genetic alterations over many years and eventually lead to malignancy [20–22]. Alternatively, the more recent model of simultaneous progression proposes that PDAC driver gene activations and inactivations can occur concomitantly through complex chromosomal rearrangements, leading to rapid tumor development and dissemination [23,24].

2.1. Gradual Progression Model

Most PDACs emerge from non-cystic precursor lesions, such as pancreatic intraepithelial neoplasias (PanINs) [25], although a small minority consists of cystic lesions, such as intraductal papillary mucinous neoplasm (IPMN) and mucinous cystic neoplasm (MCN) [26]. PanINs are microscopic, non-invasive epithelial neoplasias confined within pancreatic ducts, generated from acinar cells through a reprogramming process called acinar-to-ductal metaplasia (ADM) [27,28]. In normal pancreas, ADM is a temporary and reversible process However, persistent ADM can lead to PanIN development and, eventually, PDAC [29]. The transition from ADM to PanINs has been well-documented in both mouse models and humans [25,28,30–32]. It is characterized by the induction of ductal cell identity and the suppression of acinar cell characteristics, both at gene expression and morphology levels [33,34]. Interestingly, PanINs are surrounded by unique microenvironment, are markedly different from ADM, and are transcriptionally distinct from ducts [35]. The earliest genetic event that initiates alterations is the oncogenic activation of KRAS that forces acinar cells into a constant ADM state, which can quickly lead to the formation of PanINs [36,37] (Figure 1).

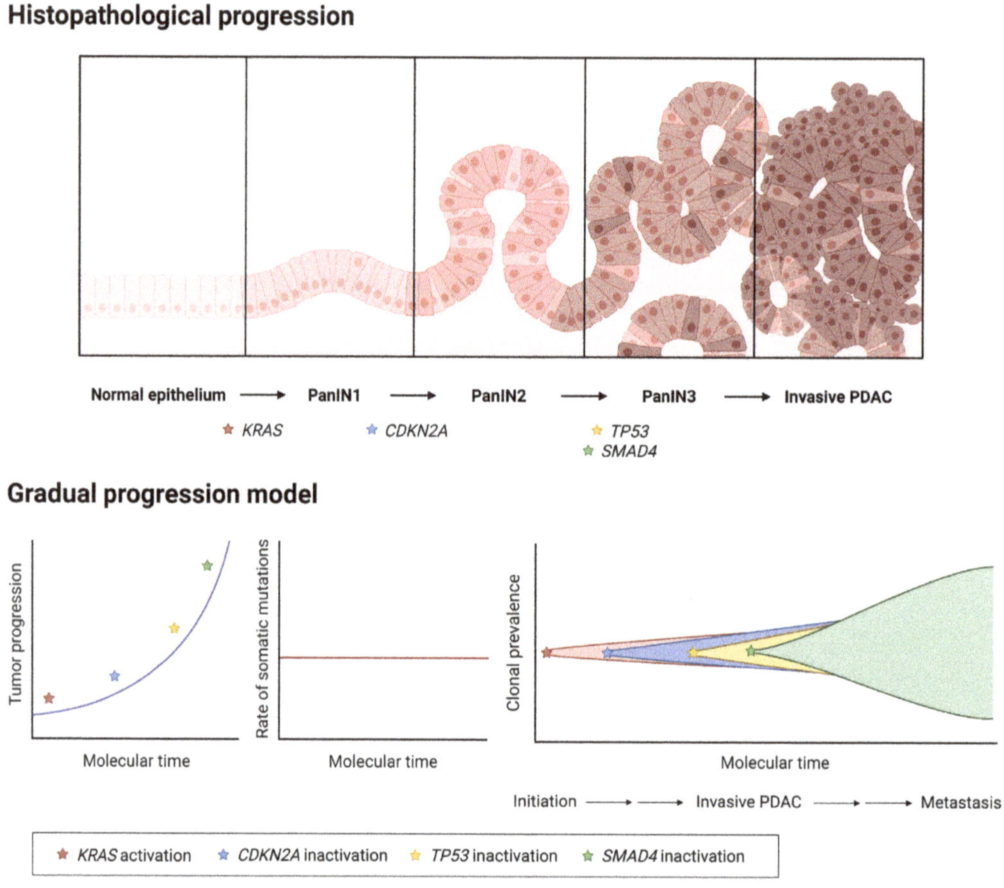

Figure 1. Schematic representation of the gradual progression model of PDAC evolution. Histopathological progression, accumulation rate of somatic mutations, and clonal prevalence over molecular time, are represented.

PanIN lesions are divided into three grades of dysplasia, according to the grade of cytological and architectural atypia. PanIN1A and PanIN1B are classified as low grade, although they are subdivided based on the extent of the structural abnormalities (Figure 1).

On the other hand, PanIN2 and PanIN3 present high-grade dysplasia, and PanIN3 are considered in situ carcinomas [28,38,39]. Progression from low-grade to high-grade and, eventually, to locally invasive PDAC is caused by the accumulation of different genetic and non-genetic alterations (Figure 1).

The gradual progression model posits that histopathological lesions are driven by the stepwise accumulation of specific known mutations [21]. As mentioned before, the main driver genes of PanINs and PDAC include KRAS, TP53, CDKN2A, and SMAD4 [14,21,40]. KRAS mutations are present in low-grade PanIN, indicating that it is the earliest alteration that drives ADM [41,42]. More importantly, they are also detected in healthy pancreatic tissue, thus confirming its role as the initiating oncogenic event [35]. Mutations in the tumor suppressor CDKN2A, are also often found early in the tumorigenesis process [43]. Further accumulation of inactivating mutations in the tumor suppressor genes TP53 and SMAD4, leads to progression to PDAC (Figure 1) [44–46]. The accumulation of genetic and morphological alterations is proposed to occur during one to two decades until invasive carcinoma is developed [44]. However, although it appears to be a gradual process, not all mutations are present in all patients. The inactivation of different tumor suppressors, their type of alteration, and the combination of both, results in wide tumoral heterogeneity. These evolutionary trajectories give rise to a large number of inter- and intra-tumoral differences that define tumor behavior [44,47,48].

2.2. Simultaneous Progression Model

Traditionally, pancreatic tumor evolution is considered a gradual process. However, recent studies on the simultaneous progression model propose that genetic alterations can also occur in a short period of time or even simultaneously (Figure 2) [23,24,49,50]. Such alterations include the synchronous inactivation of tumor suppressor genes and are associated with complex genomic rearrangements (Figure 2) [51–53]. Therefore, they can induce accelerated tumor progression and set off invasive cancer, as they are commonly detected in advanced stages of pancreatic tumors (Figure 2) [50,53]. Mechanistically, the processes involving such genomic events include mitotic errors, chromothripsis, and polyploidization [13,49,51]. However, even though these molecular events have been previously described, our understanding of their impact on tumor evolution remains incomplete.

It is also worth mentioning that periods of latency between morphological states can follow a different order to that of the classical gradual progression model (Figure 2). Considering the lesions found in samples from patients and healthy individuals, it seems that there is a relatively long latency period between PanIN1 and -2, but rapid evolution between PanIN2 and PDAC [23]. In this scenario, PanIN3 may contain all the molecular information for this rapid progression, as in some cases they display higher somatic mutation burden [50,54].

In the simultaneous progression model, two key steps are identified: the cancer-initiating event and the cancer-transforming event [49,50,55]. KRAS mutations act as the initial trigger of pancreatic cancer. Using this as the reference point in the concept of the gradual progression model, it is widely considered that metastasis occurs late in the genomic timeline [44,56]. However, metastasis may not always be a late event, but can arise concurrently with primary tumor development. According to the simultaneous model, this is induced by cancer-transforming events (Figure 2). Such events either induce or are necessary for transformation, and endow cells with both invasive and metastatic abilities, resulting in a short latency period between local expansion and metastasis. Indeed, it was demonstrated that PDAC cells with high metastatic capacity can also appear simultaneously or even earlier than the formation of primary tumors [57–59]. More specifically, lineage-labelling studies involving genetically engineered mouse models have revealed the early dissemination of invasive circulating pancreatic cells [57]. Notably, these studies align closely with the clinical observations for some patients with tumors of very small sizes, which present advanced disease and very short survival of less than 2 years [23].

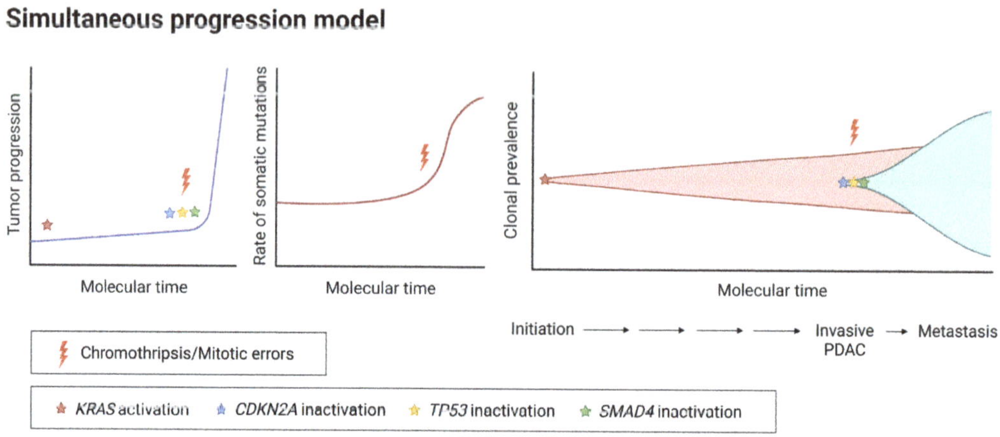

Figure 2. Schematic representation of the simultaneous progression model of PDAC evolution. Tumor progression, accumulation rate of somatic mutations, and clonal prevalence over molecular time, are represented.

2.3. *Are the Two Models Mutually Exclusive or Complementary?*

As Theodosius Dobzhansky wrote, "nothing in biology makes sense except in the light of evolution" and molecular oncology is no exception [60]. Neo-Darwinian cancer evolution models typically assume the sequential acquisition of mutations over time [61], such as the classical progression model of PDAC evolution. In regard to this concept, microevolution involves smaller, incremental genetic changes that occur over time, contributing to the gradual adaptation and diversification of tumor cells within the microenvironment [44,61]. Yet, the classical model of accumulating genetic changes under selective pressure does not fully explain the diverse evolutionary trajectories of pancreatic cancer.

Alternatively, tumor macroevolution in PDAC can be driven by single "catastrophic" events that cause complex genomic rearrangements involving numerous genetic loci [49,50,62]. Macroevolution refers to large-scale evolutionary changes, such as quick and synchronous chromosomal losses and gains, that lead to significant alterations in tumor biology and behavior [61–63]. These combined genomic alterations accelerate tumor evolution far more dramatically than any single type of event alone, creating unstable tumors with aggressive behavior [18,63]. According to the simultaneous model, evidence supports that numerous genomic aberrations can occur in short bursts due to chromosomal instability, breakage–fusion–bridge cycles, chromoplexy, and chromothripsis [62].

Despite the impact of macroevolutionary events, clinical and experimental evidence consistently underscores the "four usual genes" as those subject to the most recurrent alterations in pancreatic cancer patients. This is also supported by the low level of difference among the driver genes and low genetic heterogeneity observed between primary tumors and metastases [19]. However, a more realistic scenario posits an interplay between point mutations in the driver genes, associated with gradual progression, accompanied by the large-scale genomic events, described in the simultaneous model (Figure 3). Early mutations, such as the ones in KRAS, could appear as a point consequence, while the later inactivation or activation of tumor suppressor genes or oncogenes, respectively, could be due to rearrangements of chromosomes and structural variants (SVs) (Figure 3) [64].

Alternatively, oncogenic variants of TP53 and CDKN2A in combination with KRAS can themselves cause chromosomal instability and rearrangements [46,65]. Indeed oncogene-induced DNA replication stress is often responsible for the presence of genomic instability [66]. Additional polyploidization and genome doubling can add up, providing fertile ground for the acquisition of further mutations and increased genomic gains of already amplified loci [64,66,67].

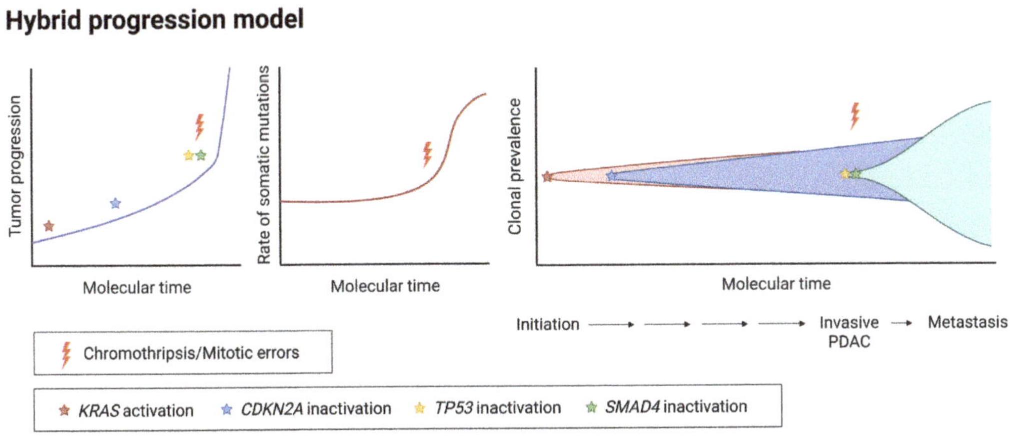

Figure 3. Schematic representation of a hybrid model of PDAC evolution, which combines features of the gradual and simultaneous progression process. Tumor progression, accumulation rate of somatic mutations, and clonal prevalence over molecular time, are represented. Adapted from [47].

Thus, tumor cells may alternate between prolonged phases of relative mutational stability and short periods of rapid evolutionary change [68] (Figure 3). Therefore, the two models discussed above are not mutually exclusive, but rather coexist in a hybrid way among tumor populations, contributing to the heterogeneity and adaptability of the tumors (Figure 3).

3. KRAS Allelic Frequencies in PDAC

3.1. Tumor Progression and Metastasis

In agreement with the accelerated tumor development process that we have described, several studies have reported complex genomic rearrangements in a significant percentage of human PDACs [49,51,63]. More specifically, Notta et al. and Mueller et al. reported a high frequency of chromothripsis and copy number alteration events in primary tumors that progressed to advanced disease [49,52]. Not surprisingly, in both studies, among the genetic regions with losses and gains were KRAS and the usual tumor suppressor genes [49,52]. Interestingly, increased genetic dosage of oncogenic KRAS was detected in both human and mouse PDACs [51,52]. Different mutant KRAS gene dosage states were found, including focal gains, arm-level gains, and copy-neutral loss of the wild-type allele [52]. Chan-Seng-Yue et al. described that apart from the increase in the copy number of the mutant KRAS, the complete loss of the wild-type KRAS gene is also a common event (Figure 4) [51]. Therefore, KRAS allelic imbalances in PDAC are defined by the loss of the wt allele (loss of heterozygosity, LOH), amplifications of the mutant, or both (Figure 4) [51,52]. The degree of this imbalance in PDAC is considered a continuum and is often categorized in minor and major imbalances (Figure 4) [51].

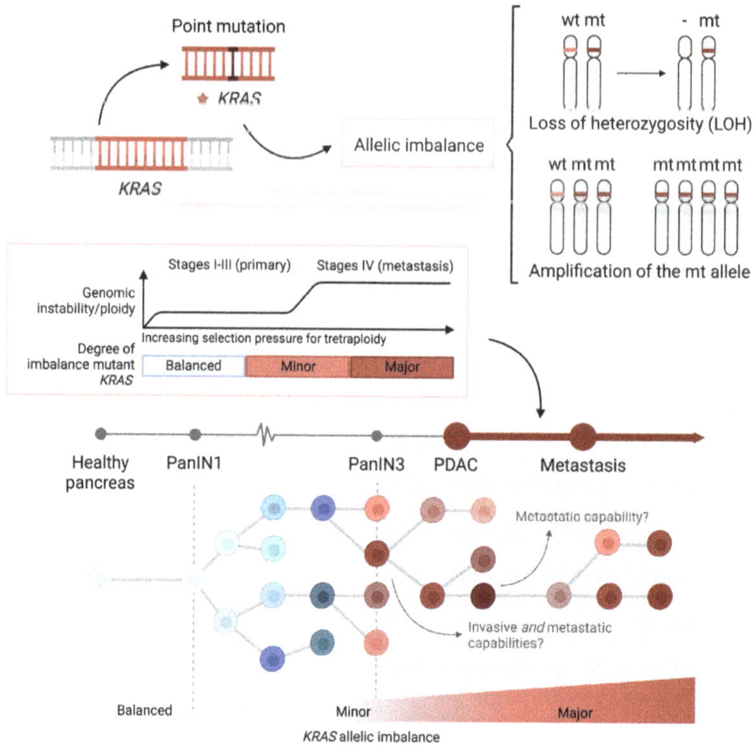

Figure 4. Schematic representation of KRAS allelic imbalances and clonal evolution during tumor progression, from precursor lesions until PDAC and metastasis.

Interestingly, it was previously shown that ADM and PanINs display some of the typical genetic alterations of PDAC. These include aneuploidy, heterozygous mutations, and a low rate of loss of heterozygosity (LOH) of the tumor suppressors, TP53 and CDKN2A [69]. Equally important, Mueller et al. recently observed that human PanINs (hPanINs) also display allelic imbalances in the KRAS oncogene. More specifically, KRAS mutant allele frequencies higher than 50% were detected in more than half of the examined hPanIN1a and hPanIN2. However, it is estimated that the actual frequency of the mutant copies is higher due to the contamination of healthy tissue in the microdissected region [52]. This proves the presence of amplifications in mutant KRAS already, during the early development of pancreatic cancer.

As the increased allelic frequency of mutant KRAS is also detected in low-grade PanINs and KRAS amplification takes place early in tumorigenesis, it seems that increased mutant KRAS dosage provides a clonal advantage (Figure 4) [52]. However, further changes in KRAS allelic states in PDAC during the evolution of tumors might follow distinct routes that are accompanied by the inactivation of tumor suppressors [70]. It is estimated that the loss of CDKN2A precedes mutant KRAS amplification [51,52]. According to the interplay between macroevolution and punctuated events during PDAC progression, an initial preneoplastic diploid phase precedes genomic instability and copy number alterations. Thus, KRAS point mutations are then followed by large structural rearrangements that are responsible for losses of the wt KRAS allele and other key driver genes. Next, genome doubling is another key molecular event that amplifies existing KRAS gains, leading to major allelic imbalances [51]. These genomic events drive rapid tumor progression and metastasis (Figure 4). Indeed, major KRAS allelic imbalances are detected in a considerable fraction of advanced-stage tumors and hepatic metastases [51,52].

Undoubtedly, allelic changes of KRAS are a frequent event in PDAC [71]. Studies involving genetic mouse models have demonstrated their impact in tumor growth and the metastatic capacity of pancreatic cells [51,52,72,73]. Not surprisingly, these functional studies have validated the oncosuppressive role of wt KRAS, as previously suggested [74]. However, the different cellular processes and molecular pathways mediated by the two alleles are still unknown. In other types of cancer that include colorectal, lung, and leukemia, KRAS allelic imbalances are also linked to tumor initiation, cell proliferation, invasion, and metastasis [72,75,76]. Yet, further studies are needed to explore how both the absence of the wt allele and increased dosage of oncogenic KRAS affect the interaction of tumor cells with the stromal and inflammatory microenvironment. This could shed light on the molecular mechanisms of the high metastatic capacity that characterizes these tumors.

3.2. *Response to Therapy*

3.2.1. Prognostic Value and Standard Chemotherapy

Overall, the tumor-specific allelic ratio and dosage of mutant KRAS influence PDAC biology and disease progression [71]. Indeed, it has been strongly demonstrated that PDAC tumors with genomic gains of KRAS mutant alleles have generally worse survival rates [51,77,78]. However, even if the increased mutant allele dosage is a notable predictor of poor survival, it has not shown strong prognostic value across the disease stages [78]. Similarly, the loss of the wild-type allele, often accompanied with copy number gain of mutant KRAS, does not seem to be representative of the overall survival of patients [78]. Thus, so far, the allelic status of KRAS is not included in the decision-making in terms of the therapeutic options and strategies for either resectable or advanced pancreatic cancer. Moreover, there is no evidence on whether tumors with distinct KRAS allelic frequencies may display a different response to the available cytotoxic chemotherapy. It is also worth mentioning that, among the primary resectable cases, the prevalence of patients that receive neoadjuvant therapy is low [78]. Therefore, KRAS allelic status could be a candidate biomarker for the decision of neo-adjuvant treatment or primary resection. Related studies can open up a line of research on personalized medicine to improve the clinical picture of the disease.

3.2.2. Targeted Therapies and Oncogenic Signaling

Lennerz and Stenzinger proposed in 2015 that both the allelic frequencies and genomic dosage of mutated KRAS affect PDAC biology [79]. Later on, transcriptomic analysis showed that PDAC cells with an increased KRAS dosage display differences in the downstream oncogenic signaling cascades of the Mitogen-activated Protein Kinase (MAPK) pathway, axon guidance, and Phosphoinositide 3-Kinase (PI3K), compared to cells with a balanced allelic status (Figure 5) [52]. Equally important, additional evidence has validated the function of wild-type KRAS as a tumor suppressor, antagonizing the mutant allele in KRAS dimerization and other biochemical properties [52,72,80]. In most of these studies, it was clearly demonstrated that the downstream signaling of the MAPK pathway is affected [52,72]. Indeed, pharmacological inhibition of MEK with the use of Trametinib and Selumetinib sensitized tumor growth in PDAC and lung mouse models [72,73]. On the contrary, the restoration of wt KRAS expression in human PDAC cells with LOH significantly attenuated the malignancy of tumor cells, confirming the tumor suppressive role of the wt allele [81]. Yan et al. also identified the HIPPO signaling pathway to be positively regulated by wt KRAS in PDAC cells [81]. However, other relevant genes and signaling cascades are also involved in the tumor-suppressive function of wt KRAS and need to be further investigated (Figure 5). Undoubtedly, the presence of the wt allele in tumors should be considered for the design of targeted therapeutic approaches. KRAS allele-specific therapies are now possible due to the recent development of allele-specific KRAS inhibitors [8]. Thus, further studies are necessary to target the broad KRAS signaling network with different combination approaches.

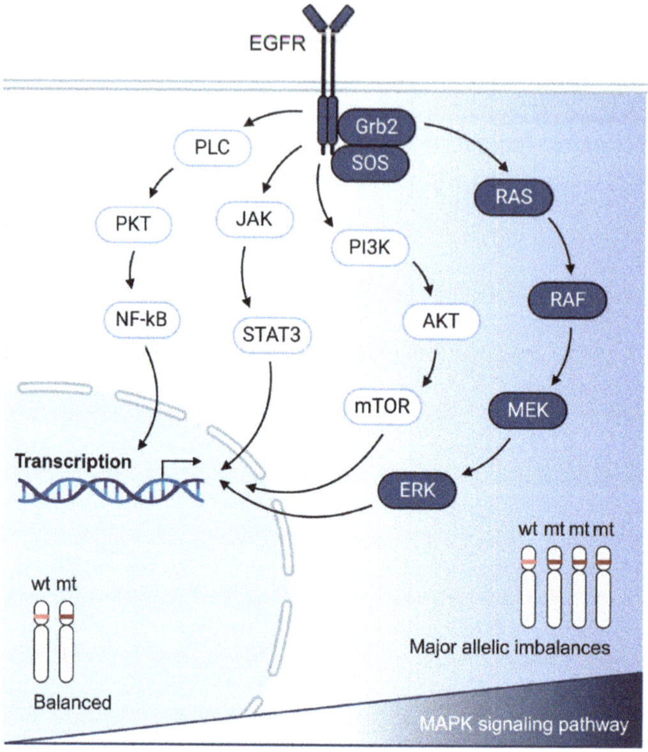

Figure 5. Graphical representation of the RAS signaling network. Increased MAPK signaling is represented in darker blue and is associated with major allelic imbalances and the increased dosage of oncogenic KRAS.

3.2.3. Targeted Therapies Against KRAS wt Tumors

While activating mutations in KRAS are present in the majority of PDAC cases, a significant subset of patients (approximately 10%) includes tumors with wild-type (wt) KRAS [14]. This subgroup of PDACs that harbor wt KRAS alleles has historically been overlooked in terms of targeted therapies, due to the absence of mutations in the KRAS oncogene. However, recent research has highlighted that wt KRAS PDAC tumors still undergo significant genetic alterations that activate alternative signaling pathways and drive tumor progression [82,83]. Among these genetic alterations, NRG1 fusions have emerged as a key event in the pathogenesis of wt KRAS PDAC. These fusions produce chimeric proteins that activate oncogenic signaling mediated by ERBB receptors and promote tumor cell proliferation and metastasis [84]. This suggests that PDAC patients with wt KRAS and NRG1 fusions can benefit from the inhibition of the ERBB signaling pathway [84]. Therefore, the identification of such fusions through molecular diagnostics can guide personalized treatment strategies, and include the use of ERBB inhibitors alone or in combination [84]. Indeed, a clinical study recently demonstrated the efficacy of Zenocotuzumab in NRG1 fusion–positive solid cancers [85].

Furthermore, other prevalent alterations of wt KRAS tumors include fusions, amplifications, and mutations in Receptor Tyrosine Kinases (RTKs), and genes involved in the MAPK and PI3K pathways, such as BRAF, FGFR, ERBB2, ALK, MET, and NF1 [14,82]. Multiple mutations are also often detected within DNA damage repair genes, as well as regulators of chromatin remodeling and the cell cycle [82]. From a therapeutic perspective, the identification of these frequent oncogenic fusion events together with targetable mutations, among patients with wt KRAS tumors, has presented an important opportunity to improve genome-guided treatment options for this subgroup of PDAC patients [9,86].

3.3. Challenges in Experimental Methodology

Despite the impact of KRAS as an oncogene, studies on KRAS allelic imbalances did not appear until very recently [51,52,76,78]. This is strongly attributed to the development of high-throughput sequencing methods, including whole genomic, exomic, and transcriptomic sequencing. Due to the molecular biology of the KRAS oncogene and the commonly detected point mutations on codon 12, it is often challenging to distinguish between the wt and mutant allele for the study of allelic ratios. Various protocols for allele-specific real-time PCR with customized probes have been developed to assess allelic abundancies of KRAS [87–90]. In regard to this concept, digital droplet allele-specific PCR (ddPCR) has increased the sensitivity of signal detection in samples with low concentrations of DNA substrates [91]. Taking it a step further, multiplex ddPCR with mutant- and wt-specific hydrolysis probes enables the detection and assessment of the fractional abundance of KRAS mutations in a cost-effective manner [92].

Next Generation Sequencing methods (NGS) have made it possible to assess KRAS allelic frequencies in many cohorts of pancreatic tumors. However, it should also be taken into account that most of these studies have assessed allelic imbalances using bulk technologies. This underlines that single-cell sequencing methods are needed to further explore intratumoral allelic imbalances [93]. In regard to this concept, technologies such as single-nucleus DNA sequencing have demonstrated that there is a large correlation with the pseudobulk variant allele frequency of KRAS [94]. Still, single-cell genomic sequencing is often unable to separate KRAS mt and wt alleles. Undoubtedly, research on KRAS biology in the context of allelic imbalances is highly challenging and will require further advanced molecular techniques and carefully designed functional studies.

Another major challenge in the research of pancreatic cancer includes the presence of the dense tumor microenvironment [95]. The two most distinguishing features are the robust desmoplasia of tumors, and the non-malignant components that constitute almost 80% of the tumor mass [96]. Briefly, the tumor microenvironment of PDAC is characterized by peritumoral desmoplastic stroma, which consists of a complex array of cellular components surrounded by extracellular matrix (collagens, hyaluronic acid, integrins, proteoglycans, and glycoproteins) [97]. The predominant cell types are cancer-associated fibroblasts, tumor-associated macrophages, regulatory T cells, and other cytotoxic, infiltrating immune cells, as well as endothelial cells [98]. This extensive microenvironment makes the study of the genomic characteristics of tumor epithelial cells very challenging. Common preanalytical approaches to exclusively isolate tumor cells include Laser Capture Microdissection (LCM) and cell-separating technologies (FACS, MACS) [99–101]. Other post-sequencing tumor cell enrichment methods include the development of bioinformatic tools that enable to assess the tumor cellularity of samples and, more importantly, to distinguish normal cell types in the tumor microenvironment from malignant cells [102–104].

4. Conclusions

At the time of diagnosis, only 20% of PDACs are localized, while more than 80% are detected metastasized to the liver and lung or spread to regional lymph nodes. Yet, most research on pancreatic cancer has focused on primary and early-stage resectable tumors, given the availability of these samples. Even so, genomic studies have revealed high mutational conservation between PDAC primary and metastatic tumors, with simple somatic variants to be closely concordant in untreated and advanced-disease patients. Thus, understanding the progression from primary PDACs to metastasis requires the extensive study of intratumoral heterogeneity with single-cell analyses [105].

In most cancers, complex genomic rearrangements that alter large genomic regions occur frequently during tumor evolution [47,49]. These genomic events, such as chromothripsis, chromoplexy, and genome doubling, give rise to copy number changes and genetically unstable tumors, defined by a large number of structural variants [13,51]. As Connon and Gallinger discuss, the PDAC evolutionary models and timing of progression can be gradual and stepwise or rapid and punctuated [47]. However, structural mutations and copy number alterations in the punctuated progression model can be difficult to interpret due to the large number of genes that are amplified or lost during these regional events [47]. Indeed, the simultaneous loss of genes, such as CDKN2A or SMAD4, and copy number gains of the transcription factor GATA6, or the KRAS oncogene, are frequently detected [51]. How these simultaneous gains and losses of multiple driver genes affect disease progression, organ-specific dissemination, as well as the response to standard chemotherapy, remains to be studied.

Various recent studies have shown that changes in KRAS allelic frequencies are important both for tumor progression, as well as major biological and clinical PDAC features [51,52,73,78]. Yet, there is still a long way to go. We and others have observed intratumoral gradients in KRAS allelic frequencies in a mosaic fashion (Figure 4) [51]. Thus, as it has been previously explained, more comprehensive analyses of KRAS copy number changes are needed in pure tumor cell populations and single-cell studies [53]. Considering the possible outgrowth of more aggressive minor clones, further studies are also necessary to uncover how subclones with major KRAS allelic imbalances obtain invasive capacity to metastasize. These observations increase the need for optimized KRAS molecular analyses and functional approaches, in order to study the antagonizing properties of mutant and wild-type KRAS alleles in regard to the KRAS signaling network.

Thus far, treatment with standard cytotoxic drugs is largely ineffective, especially in locally advanced and metastatic disease, highlighting the need for targeted therapies [3]. Currently, KRAS allelic imbalances are not taken into consideration for the design of therapeutic strategies, even though they alter tumor biology. Therefore, it is necessary to study how they define the response not only to existing chemotherapy, but also to new targeted approaches. Stratifying PDACs according to the KRAS allelic status can help define therapeutic vulnerabilities in groups of tumors with different allelic frequencies. Undoubtedly, PDAC is one of the malignant diseases most urgently in need of more personalized treatment, as well as the development of predictive signatures [106]. In regard to this concept of personalized therapy, the KRAS allelic status should also be included in the genetic characterization of tumors, as it may impact prognosis and the therapeutic sensitivity of patients.

Author Contributions: Manuscript preparation, V.L. and B.R.-P.; review and editing, V.L.; supervision, C.G. All authors have read and agreed to the published version of the manuscript.

Funding: This work was supported by the Cris Cancer Foundation and the CIBERONC research network (CB21/12/00121).

Conflicts of Interest: The authors declare that there are no conflicts of interest.

References

1. Siegel, R.L.; Giaquinto, A.N.; Jemal, A. Cancer statistics, 2024. *CA Cancer J. Clin.* **2024**, *74*, 12–49. [PubMed]
2. Sung, H.; Ferlay, J.; Siegel, R.L.; Laversanne, M.; Soerjomataram, I.; Jemal, A.; Bray, F. Global Cancer Statistics 2020: GLOBOCAN Estimates of Incidence and Mortality Worldwide for 36 Cancers in 185 Countries. *CA Cancer J. Clin.* **2021**, *71*, 209–249. [PubMed]
3. Nevala-Plagemann, C.; Hidalgo, M.; Garrido-Laguna, I. From state-of-the-art treatments to novel therapies for advanced-stage pancreatic cancer. *Nat. Rev. Clin. Oncol.* **2020**, *17*, 108–123. [CrossRef] [PubMed]
4. Gbolahan, O.B.; Tong, Y.; Sehdev, A.; O'Neil, B.; Shahda, S. Overall survival of patients with recurrent pancreatic cancer treated with systemic therapy: A retrospective study. *BMC Cancer* **2019**, *19*, 468.
5. Conroy, T.; Hammel, P.; Hebbar, M.; Ben Abdelghani, M.; Wei, A.C.; Raoul, J.-L.; Choné, L.; Francois, E.; Artru, P.; Biagi, J.J.; et al. FOLFIRINOX or Gemcitabine as Adjuvant Therapy for Pancreatic Cancer. *N. Engl. J. Med.* **2018**, *379*, 2395–2406. [CrossRef]
6. Conroy, T.; Desseigne, F.; Ychou, M.; Bouché, O.; Guimbaud, R.; Bécouarn, Y.; Adenis, A.; Raoul, J.-L.; Gourgou-Bourgade, S.; de La Fouchardière, C.; et al. FOLFIRINOX versus gemcitabine for metastatic pancreatic cancer. *N. Engl. J. Med.* **2011**, *364*, 1817–1825.
7. Von Hoff, D.D.; Ervin, T.; Arena, F.P.; Chiorean, E.G.; Infante, J.; Moore, M.; Seay, T.; Tjulandin, S.A.; Ma, W.W.; Saleh, M.N.; et al. Increased survival in pancreatic cancer with nab-paclitaxel plus gemcitabine. *N. Engl. J. Med.* **2013**, *369*, 1691–1703.
8. Hu, Z.I.; O'Reilly, E.M. Therapeutic developments in pancreatic cancer. *Nat. Rev. Gastroenterol. Hepatol.* **2024**, *21*, 7–24. [CrossRef]
9. Turpin, A.; Neuzillet, C.; Colle, E.; Dusetti, N.; Nicolle, R.; Cros, J.; De Mestier, L.; Bachet, J.-B.; Hammel, P. Therapeutic advances in metastatic pancreatic cancer: A focus on targeted therapies. *Ther. Adv. Med. Oncol.* **2022**, *14*, 17588359221118019.
10. Leroux, C.; Konstantinidou, G. Targeted Therapies for Pancreatic Cancer: Overview of Current Treatments and New Opportunities for Personalized Oncology. *Cancers* **2021**, *13*, 799. [CrossRef]
11. Hosein, A.N.; Dougan, S.K.; Aguirre, A.J.; Maitra, A. Translational advances in pancreatic ductal adenocarcinoma therapy. *Nat. Cancer* **2022**, *3*, 272–286. [PubMed]
12. Witkiewicz, A.K.; McMillan, E.A.; Balaji, U.; Baek, G.; Lin, W.-C.; Mansour, J.; Mollaee, M.; Wagner, K.-U.; Koduru, P.; Yopp, A.; et al. Whole-exome sequencing of pancreatic cancer defines genetic diversity and therapeutic targets. *Nat. Commun.* **2015**, *6*, 6744. [PubMed]
13. Waddell, N.; Pajic, M.; Patch, A.-M.; Chang, D.K.; Kassahn, K.S.; Bailey, P.; Johns, A.L.; Miller, D.; Nones, K.; Quek, K.; et al. Whole genomes redefine the mutational landscape of pancreatic cancer. *Nature* **2015**, *518*, 495–501. [PubMed]
14. Cancer Genome Atlas Research Network. Integrated Genomic Characterization of Pancreatic Ductal Adenocarcinoma. *Cancer Cell* **2017**, *32*, 185–203.e13.
15. Aguirre, A.J.; Nowak, J.A.; Camarda, N.D.; Moffitt, R.A.; Ghazani, A.A.; Hazar-Rethinam, M.; Raghavan, S.; Kim, J.; Brais, L.K.; Ragon, D.; et al. Real-time Genomic Characterization of Advanced Pancreatic Cancer to Enable Precision Medicine. *Cancer Discov.* **2018**, *8*, 1096–1111. [CrossRef]
16. Kleeff, J.; Korc, M.; Apte, M.; La Vecchia, C.; Johnson, C.D.; Biankin, A.V.; Neale, R.E.; Tempero, M.; Tuveson, D.A.; Hruban, R.H.; et al. Pancreatic cancer. *Nat. Rev. Dis. Primers* **2016**, *2*, 16022. [CrossRef]
17. Li, O.; Li, L.; Sheng, Y.; Ke, K.; Wu, J.; Mou, Y.; Liu, M.; Jin, W. Biological characteristics of pancreatic ductal adenocarcinoma: Initiation to malignancy, intracellular to extracellular. *Cancer Lett.* **2023**, *574*, 216391. [PubMed]
18. Baker, T.M.; Waise, S.; Tarabichi, M.; van Loo, P. Aneuploidy and complex genomic rearrangements in cancer evolution. *Nat. Cancer* **2024**, *5*, 228–239.
19. Makohon-Moore, A.P.; Zhang, M.; Reiter, J.G.; Bozic, I.; Allen, B.; Kundu, D.; Chatterjee, K.; Wong, F.; Jiao, Y.; Kohutek, Z.A.; et al. Limited heterogeneity of known driver gene mutations among the metastases of individual patients with pancreatic cancer. *Nat. Genet.* **2017**, *49*, 358–366.
20. Hruban, R.H.; Goggins, M.; Parsons, J.; Kern, S.E. Progression Model for Pancreatic Cancer. *Clin. Cancer Res.* **2000**, *6*, 2969–2972.
21. Maitra, A.; Hruban, R.H. Pancreatic cancer. *Annu. Rev. Pathol.* **2008**, *3*, 157–188. [PubMed]
22. Kim, J.Y.; Hong, S.M. Precursor Lesions of Pancreatic Cancer. *Oncol. Res. Treat.* **2018**, *41*, 603–610. [PubMed]
23. Real, F.X. A "catastrophic hypothesis" for pancreas cancer progression. *Gastroenterology* **2003**, *124*, 1958–1964. [CrossRef] [PubMed]
24. Real, F.X.; Cibrián-Uhalte, E.; Martinelli, P. Pancreatic cancer development and progression: Remodeling the model. *Gastroenterology* **2008**, *135*, 724–728.
25. Hruban, R.H.; Maitra, A.; Kern, S.E.; Goggins, M. Precursors to pancreatic cancer. *Gastroenterol. Clin. N. Am.* **2007**, *36*, 831–849.
26. Noë, M.; Niknafs, N.; Fischer, C.G.; Hackeng, W.M.; Beleva Guthrie, V.; Hosoda, W.; Debeljak, M.; Papp, E.; Adleff, V.; White, J.R.; et al. Genomic characterization of malignant progression in neoplastic pancreatic cysts. *Nat. Commun.* **2020**, *11*, 4085. [CrossRef] [PubMed]

27. Brune, K.; Abe, T.; Canto, M.; O'Malley, L.; Klein, A.P.; Maitra, A.; Adsay, N.V.; Fishman, E.K.; Cameron, J.L.; Yeo, C.J.; et al. Multifocal Neoplastic Precursor Lesions Associated With Lobular Atrophy of the Pancreas in Patients Having a Strong Family History of Pancreatic Cancer. *Am. J. Surg. Pathol.* **2006**, *30*, 1067–1076.
28. Hruban, R.H.; Adsay, N.V.; Albores-Saavedra, J.; Compton, C.; Garrett, E.S.; Goodman, S.N.; Kern, S.E.; Klimstra, D.S.; Klöppel, G.; Longnecker, D.S.; et al. Pancreatic intraepithelial neoplasia: A new nomenclature and classification system for pancreatic duct lesions. *Am. J. Surg. Pathol.* **2001**, *25*, 579–586. [CrossRef]
29. Marstrand-Daucé, L.; Lorenzo, D.; Chassac, A.; Nicole, P.; Couvelard, A.; Haumaitre, C. Acinar-to-Ductal Metaplasia (ADM): On the Road to Pancreatic Intraepithelial Neoplasia (PanIN) and Pancreatic Cancer. *Int. J. Mol. Sci.* **2023**, *24*, 9946. [CrossRef]
30. Hingorani, S.R.; Petricoin, E.F.; Maitra, A.; Rajapakse, V.; King, C.; Jacobetz, M.A.; Ross, S.; Conrads, T.P.; Veenstra, T.D.; Hitt, B.A.; et al. Preinvasive and invasive ductal pancreatic cancer and its early detection in the mouse. *Cancer Cell* **2003**, *4*, 437–450. [CrossRef]
31. Guerra, C.; Mijimolle, N.; Dhawahir, A.; Dubus, P.; Barradas, M.; Serrano, M.; Campuzano, V.; Barbacid, M. Tumor induction by an endogenous K-ras oncogene is highly dependent on cellular context. *Cancer Cell* **2003**, *4*, 111–120. [CrossRef]
32. Habbe, N.; Shi, G.; Meguid, R.A.; Fendrich, V.; Esni, F.; Chen, H.; Feldmann, G.; Stoffers, D.A.; Konieczny, S.F.; Leach, S.D.; et al. Spontaneous induction of murine pancreatic intraepithelial neoplasia (mPanIN) by acinar cell targeting of oncogenic Kras in adult mice. *Proc. Natl. Acad. Sci. USA* **2008**, *105*, 18913–18918. [CrossRef] [PubMed]
33. Kopp, J.L.; von Figura, G.; Mayes, E.; Liu, F.-F.; Dubois, C.L.; Morris, J.P.; Pan, F.C.; Akiyama, H.; Wright, C.V.E.; Jensen, K.; et al. Identification of Sox9-dependent acinar-to-ductal reprogramming as the principal mechanism for initiation of pancreatic ductal adenocarcinoma. *Cancer Cell* **2012**, *22*, 737–750. [CrossRef] [PubMed]
34. Wei, D.; Wang, L.; Yan, Y.; Jia, Z.; Gagea, M.; Li, Z.; Zuo, X.; Kong, X.; Huang, S.; Xie, K. KLF4 Is Essential for Induction of Cellular Identity Change and Acinar-to-Ductal Reprogramming during Early Pancreatic Carcinogenesis. *Cancer Cell* **2016**, *29*, 324–338. [CrossRef] [PubMed]
35. Carpenter, E.S.; Elhossiny, A.M.; Kadiyala, P.; Li, J.; McGue, J.; Griffith, B.D.; Zhang, Y.; Edwards, J.; Nelson, S.; Lima, F.; et al. Analysis of Donor Pancreata Defines the Transcriptomic Signature and Microenvironment of Early Neoplastic Lesions. *Cancer Discov.* **2023**, *13*, 1324–1345. [CrossRef]
36. De La O, J.P.; Murtaugh, L.C. Notch and Kras in pancreatic cancer: At the crossroads of mutation, differentiation and signaling. *Cell Cycle* **2009**, *8*, 1860–1864. [CrossRef]
37. Morris, J.P.; Cano, D.A.; Sekine, S.; Wang, S.C.; Hebrok, M. Beta-catenin blocks Kras-dependent reprogramming of acini into pancreatic cancer precursor lesions in mice. *J. Clin. Investig.* **2010**, *120*, 508–520. [CrossRef]
38. Hruban, R.H.; Takaori, K.; Klimstra, D.S.; Adsay, N.V.; Albores-Saavedra, J.; Biankin, A.V.; Biankin, S.A.; Compton, C.; Fukushima, N.; Furukawa, T.; et al. An illustrated consensus on the classification of pan-creatic intraepithelial neoplasia and intraductal papillary mucinous neoplasms. *Am. J. Surg. Pathol* **2004**, *28*, 977–987. [CrossRef]
39. Sipos, B.; Frank, S.; Gress, T.; Hahn, S.; Klöppel, G. Pancreatic intraepithelial neoplasia revisited and updated. *Pancreatology* **2009**, *9*, 45–54.
40. Jaffee, E.M.; Hruban, R.H.; Canto, M.; Kern, S.E. Focus on pancreas cancer. *Cancer Cell* **2002**, *2*, 25–28. [CrossRef]
41. Kanda, M.; Matthaei, H.; Wu, J.; Hong, S.M.; Yu, J.; Borges, M.; Hruban, R.H.; Maitra, A.; Kinzler, K.; Vogelstein, B.; et al. Presence of somatic mutations in most early-stage pancreatic intraepithelial neoplasia. *Gastroentrology* **2012**, *142*, 730–733.
42. Braxton, A.M.; Kiemen, A.L.; Grahn, M.P.; Forjaz, A.; Parksong, J.; Mahesh Babu, J.; Lai, J.; Zheng, L.; Niknafs, N.; Jiang, L.; et al. 3D genomic mapping reveals multifocality of human pancreatic precancers. *Nature* **2024**, *629*, 679–687. [PubMed]
43. Wilentz, R.E.; Geradts, J.; Maynard, R.; Offerhaus, G.J.; Kang, M.; Goggins, M.; Yeo, C.J.; Kern, S.E.; Hruban, R.H. Inactivation of the p16 (INK4A) tumor-suppressor gene in pancreatic duct lesions: Loss of intranuclear expression. *Cancer Res.* **1998**, *58*, 4740–4744.
44. Yachida, S.; Iacobuzio-Donahue, C.A. Evolution and dynamics of pancreatic cancer progression. *Oncogene* **2013**, *32*, 5253–5260. [CrossRef]
45. Hosoda, W.; Chianchiano, P.; Griffin, J.F.; Pittman, M.E.; Brosens, L.A.; Noë, M.; Yu, J.; Shindo, K.; Suenaga, M.; Rezaee, N.; et al. Genetic analyses of isolated high-grade pancreatic intraepithelial neoplasia (HG-PanIN) reveal paucity of alterations in TP53 and SMAD4. *J. Pathol.* **2017**, *242*, 16–23. [CrossRef]
46. Hingorani, S.R.; Wang, L.; Multani, A.S.; Combs, C.; Deramaudt, T.B.; Hruban, R.H.; Rustgi, A.K.; Chang, S.; Tuveson, D.A. Trp53R172H and KrasG12D cooperate to promote chromosomal instability and widely metastatic pancreatic ductal adenocarcinoma in mice. *Cancer Cell* **2005**, *7*, 469–483. [CrossRef]
47. Connor, A.A.; Gallinger, S. Pancreatic cancer evolution and heterogeneity: Integrating omics and clinical data. *Nat. Rev. Cancer* **2022**, *22*, 131–142.
48. Makohon-Moore, A.P. Emerging and extensive clonal evolution in the pancreas. *Trends Cancer* **2024**, *10*, 669–670. [PubMed]

49. Notta, F.; Chan-Seng-Yue, M.; Lemire, M.; Li, Y.; Wilson, G.W.; Connor, A.A.; Denroche, R.E.; Liang, S.-B.; Brown, A.M.K.; Kim, J.C.; et al. A renewed model of pancreatic cancer evolution based on genomic rearrangement patterns. *Nature* **2016**, *538*, 378–382. [CrossRef]
50. Notta, F.; Hahn, S.A.; Real, F.X. A genetic roadmap of pancreatic cancer: Still evolving. *Gut* **2017**, *66*, 2170–2178.
51. Chan-Seng-Yue, M.; Kim, J.C.; Wilson, G.W.; Ng, K.; Figueroa, E.F.; O'Kane, G.M.; Connor, A.A.; Denroche, R.E.; Grant, R.C.; McLeod, J.; et al. Transcription phenotypes of pancreatic cancer are driven by genomic events during tumor evolution. *Nat. Genet.* **2020**, *52*, 231–240.
52. Mueller, S.; Engleitner, T.; Maresch, R.; Zukowska, M.; Lange, S.; Kaltenbacher, T.; Konukiewitz, B.; Öllinger, R.; Zwiebel, M.; Strong, A.; et al. Evolutionary routes and KRAS dosage define pancreatic cancer phenotypes. *Nature* **2018**, *554*, 62–68. [PubMed]
53. Real, F.X.; De Andrés, M.P. Mutant Kras Dosage and Chromothripsis: The Right Ingredients for a Pancreatic Cancer Catastrophe. *Trends Cancer* **2018**, *4*, 399–401. [CrossRef]
54. Murphy, S.J.; Hart, S.N.; Lima, J.F.; Kipp, B.R.; Klebig, M.; Winters, J.L.; Szabo, C.; Zhang, L.; Eckloff, B.W.; Petersen, G.M.; et al. Genetic alterations associated with progression from pancreatic intraepithelial neoplasia to invasive pancreatic tumor. *Gastroenterology* **2013**, *145*, 1098–1109.e1. [PubMed]
55. Makohon-Moore, A.; Iacobuzio-Donahue, C.A. Pancreatic cancer biology and genetics from an evolutionary perspective. *Nat. Rev. Cancer* **2016**, *16*, 553–565.
56. Yachida, S.; Jones, S.; Bozic, I.; Antal, T.; Leary, R.; Fu, B.; Kamiyama, M.; Hruban, R.H.; Eshleman, J.R.; Nowak, M.A.; et al. Distant metastasis occurs late during the genetic evolution of pancreatic cancer. *Nature* **2010**, *467*, 1114–1117. [PubMed]
57. Rhim, A.D.; Mirek, E.T.; Aiello, N.M.; Maitra, A.; Bailey, J.M.; McAllister, F.; Reichert, M.; Beatty, G.L.; Rustgi, A.K.; Vonderheide, R.H.; et al. EMT and dissemination precede pancreatic tumor formation. *Cell* **2012**, *148*, 349–361.
58. Haeno, H.; Gonen, M.; Davis, M.B.; Herman, J.M.; Iacobuzio-Donahue, C.A.; Michor, F. Computational modeling of pancreatic cancer reveals kinetics of metastasis suggesting optimum treatment strategies. *Cell* **2012**, *148*, 362–375.
59. Tuveson, D.A.; Neoptolemos, J.P. Understanding metastasis in pancreatic cancer: A call for new clinical approaches. *Cell* **2012**, *148*, 21–23.
60. Dobzhansky, T. Nothing in Biology Makes Sense except in the Light of Evolution. *Am. Biol. Teach.* **1973**, *35*, 125–129.
61. Vendramin, R.; Litchfield, K.; Swanton, C. Cancer evolution: Darwin and beyond. *EMBO J.* **2021**, *40*, e108389. [CrossRef] [PubMed]
62. Turajlic, S.; Sottoriva, A.; Graham, T.; Swanton, C. Resolving genetic heterogeneity in cancer. *Nat. Rev. Gen.* **2019**, *20*, 404–416. [CrossRef]
63. Murphy, S.J.; Hart, S.N.; Halling, G.C.; Johnson, S.H.; Smadbeck, J.B.; Drucker, T.; Lima, J.F.; Rohakhtar, F.R.; Harris, F.R.; Kosari, F.; et al. Integrated Genomic Analysis of Pancreatic Ductal Adenocarcinomas Reveals Genomic Rearrangement Events as Significant Drivers of Disease. *Cancer Res.* **2016**, *76*, 749–761. [CrossRef] [PubMed]
64. Campbell, P.J.; Yachida, S.; Mudie, L.J.; Stephens, P.J.; Pleasance, E.D.; Stebbings, L.A.; Morsberger, L.A.; Latimer, C.; McLaren, S.; Lin, M.-L.; et al. The patterns and dynamics of genomic instability in metastatic pancreatic cancer. *Nature* **2010**, *467*, 1109–1113. [CrossRef]
65. Qiu, W.; Sahin, F.; Iacobuzio-Donahue, C.A.; Garcia-Carracedo, D.; Wang, W.M.; Kuo, C.-Y.; Chen, D.; Arking, D.E.; Lowy, A.M.; Hruban, R.H.; et al. Disruption of p16 and activation of Kras in pancreas increase ductal adenocarcinoma formation and metastasis in vivo. *Oncotarget* **2011**, *2*, 862–873. [CrossRef] [PubMed]
66. Negrini, S.; Gorgoulis, V.G.; Halazonetis, T.D. Genomic instability--an evolving hallmark of cancer. *Nat. Rev. Mol. Cell Biol.* **2010**, *11*, 220–228. [CrossRef]
67. Bielski, C.M.; Zehir, A.; Penson, A.V.; Donoghue, M.T.A.; Chatila, W.; Armenia, J.; Chang, M.T.; Schram, A.M.; Jonsson, P.; Bandlamudi, C.; et al. Genome doubling shapes the evolution and prognosis of advanced cancers. *Nat. Genet.* **2018**, *50*, 1189–1195. [CrossRef]
68. Hayashi, A.; Hong, J.; Iacobuzio-Donahue, C.A. The pancreatic cancer genome revisited. *Nat. Rev. Gastroenterol. Hepatol.* **2021**, *18*, 469–481. [CrossRef]
69. Baumgart, M.; Werther, M.; Bockholt, A.; Scheurer, M.; Rüschoff, J.; Dietmaier, W.; Ghadimi, B.M.; Heinmöller, E. Genomic instability at both the base pair level and the chromosomal level is detectable in earliest PanIN lesions in tissues of chronic pancreatitis. *Pancreas* **2010**, *39*, 1093–1103. [CrossRef]
70. Tsanov, K.M.; Barriga, F.M.; Ho, Y.-J.; Alonso-Curbelo, D.; Livshits, G.; Koche, R.P.; Baslan, T.; Simon, J.; Tian, S.; Wuest, A.N.; et al. Metastatic site influences driver gene function in pancreatic cancer. *bioRxiv* **2024**, 585402.
71. Krasinskas, A.M.; Moser, A.J.; Saka, B.; Adsay, N.V.; Chiosea, S.I. KRAS mutant allele-specific imbalance is associated with worse prognosis in pancreatic cancer and progression to undifferentiated carcinoma of the pancreas. *Mod. Pathol. Off. J. USA Can. Acad. Pathol. Inc.* **2013**, *26*, 1346–1354.
72. Ambrogio, C.; Köhler, J.; Zhou, Z.-W.; Wang, H.; Paranal, R.; Li, J.; Capelletti, M.; Caffarra, C.; Li, S.; Lv, Q.; et al. KRAS Dimerization Impacts MEK Inhibitor Sensitivity and Oncogenic Activity of Mutant KRAS. *Cell* **2018**, *172*, 857–868.e15. [PubMed]

73. Fey, S.K.; Najumudeen, A.K.; Watt, D.M.; Millett, L.M.; Ford, C.A.; Gilroy, K.; Simpson, R.J.; McLay, K.; Upstill-Goddard, R Chang, D.; et al. KRAS Loss of Heterozygosity Promotes MAPK-Dependent Pancreatic Ductal Adenocarcinoma Initiation an Induces Therapeutic Sensitivity to MEK Inhibition. *Cancer Res.* **2024**, *85*, 251–262.
74. Kent, O.A. Increased mutant KRAS gene dosage drives pancreatic cancer progression: Evidence for wild-type KRAS as a tumc suppressor? *Hepatobiliary Surg. Nutr.* **2018**, *7*, 403–405. [CrossRef]
75. Najumudeen, A.K.; Fey, S.K.; Millett, L.M.; Ford, C.A.; Gilroy, K.; Gunduz, N.; Ridgway, R.A.; Anderson, E.; Strathdee, D.; Clar W.; et al. KRAS allelic imbalance drives tumour initiation yet suppresses metastasis in colorectal cancer in vivo. *Nat. Commu* **2024**, *15*, 100.
76. Burgess, M.R.; Hwang, E.; Mroue, R.; Bielski, C.M.; Wandler, A.M.; Huang, B.J.; Firestone, A.J.; Young, A.; Lacap, J.A Crocker, L.; et al. KRAS Allelic Imbalance Enhances Fitness and Modulates MAP Kinase Dependence in Cancer. *Cell* **201** *168*, 817–829.e15. [CrossRef]
77. Bielski, C.M.; Donoghue, M.T.A.; Gadiya, M.; Hanrahan, A.J.; Won, H.H.; Chang, M.T.; Jonsson, P.; Penson, A.V.; Gorelick, A Harris, C.; et al. Widespread Selection for Oncogenic Mutant Allele Imbalance in Cancer. *Cancer Cell* **2018**, *34*, 852–862.e4.
78. Varghese, A.M.; Perry, M.A.; Chou, J.F.; Nandakumar, S.; Muldoon, D.; Erakky, A.; Zucker, A.; Fong, C.; Mehine, M.; Nguyen, B et al. Clinicogenomic landscape of pancreatic adenocarcinoma identifies KRAS mutant dosage as prognostic of overall surviva *Nat. Med.* **2025**, *31*, 466–477.
79. Lennerz, J.K.; Stenzinger, A. Allelic ratio of KRAS mutations in pancreatic cancer. *Oncologist* **2015**, *20*, e8–e9.
80. Cheng, D.K.; Oni, T.E.; Thalappillil, J.S.; Park, Y.; Ting, H.C.; Alagesan, B.; Prasad, N.V.; Addison, K.; Rivera, K.D.; Pappin, D.J et al. Oncogenic KRAS engages an RSK1/NF1 pathway to inhibit wild-type RAS signaling in pancreatic cancer. *Proc. Natl. Aca Sci. USA* **2021**, *118*, e2016904118.
81. Yan, H.; Yu, C.C.; Fine, S.A.; Youssof, A.L.; Yang, Y.R.; Yan, J.; Karg, D.C.; Cheung, E.C.; Friedman, R.A.; Ying, H.; et al. Los of the wild-type KRAS allele promotes pancreatic cancer progression through functional activation of YAP1. *Oncogene* **202** *40*, 6759–6771. [CrossRef] [PubMed]
82. Philip, P.A.; Azar, I.; Xiu, J.; Hall, M.J.; Hendifar, A.E.; Lou, E.; Hwang, J.J.; Gong, J.; Feldman, R.; Ellis, M.; et al. Molecula Characterization of KRAS Wild-type Tumors in Patients with Pancreatic Adenocarcinoma. *Clin. Cancer Res. Off. J. Am. Asso Cancer Res.* **2022**, *28*, 2704–2714. [CrossRef]
83. Topham, J.T.; Tsang, E.S.; Karasinska, J.M.; Metcalfe, A.; Ali, H.; Kalloger, S.E.; Csizmok, V.; Williamson, L.M.; Titmuss, E Nielsen, K.; et al. Integrative analysis of KRAS wildtype metastatic pancreatic ductal adenocarcinoma reveals mutation an expression-based similarities to cholangiocarcinoma. *Nat. Commun.* **2022**, *13*, 5941. [CrossRef] [PubMed]
84. Heining, C.; Horak, P.; Uhrig, S.; Codo, P.L.; Klink, B.; Hutter, B.; Fröhlich, M.; Bonekamp, D.; Richter, D.; Steiger, K.; et al. NRG Fusions in KRAS Wild-Type Pancreatic Cancer. *Cancer Discov.* **2018**, *8*, 1087–1095. [CrossRef]
85. Schram, A.M.; Goto, K.; Kim, D.-W.; Macarulla, T.; Hollebecque, A.; O'Reilly, E.M.; Ou, S.-H.I.; Rodon, J.; Rha, S.Y.; Nishino, K et al. Efficacy of Zenocutuzumab in NRG1 Fusion-Positive Cancer. *N. Engl. J. Med.* **2025**, *392*, 566–576. [CrossRef]
86. Hu, H.-F.; Ye, Z.; Qin, Y.; Xu, X.-W.; Yu, X.-J.; Zhuo, Q.-F.; Ji, S.-R. Mutations in key driver genes of pancreatic cancer: Molecularl targeted therapies and other clinical implications. *Acta Pharmacol. Sin.* **2021**, *42*, 1725–1741. [CrossRef]
87. Horikoshi, T.; Lenz, H.J.; Danenberg, K.; Koch, O.M.; Bertino, J.R.; Danenberg, P.V. Quantitative determination of the ra tio of mutated to normal ras genes in the blood of leukemia patients by allele-specific PCR. *Leuk. Res.* **1994**, *18*, 693–70 [CrossRef] [PubMed]
88. Modrek, B.; Ge, L.; Pandita, A.; Lin, E.; Mohan, S.; Yue, P.; Guerrero, S.; Lin, W.M.; Pham, T.; Modrusan, Z.; et al. Oncogeni activating mutations are associated with local copy gain. *Mol. Cancer Res. MCR* **2009**, *7*, 1244–1252. [CrossRef]
89. Lang, A.H.; Drexel, H.; Geller-Rhomberg, S.; Stark, N.; Winder, T.; Geiger, K.; Muendlein, A. Optimized allele-specific real-tim PCR assays for the detection of common mutations in KRAS and BRAF. *J. Mol. Diagn. JMD* **2011**, *13*, 23–28. [CrossRef]
90. Chubarov, A.S.; Oscorbin, I.P.; Filipenko, M.L.; Lomzov, A.A.; Pyshnyi, D.V. Allele-Specific PCR for KRAS Mutation Detectio Using Phosphoryl Guanidine Modified Primers. *Diagnostics* **2020**, *10*, 872. [CrossRef]
91. Denis, J.A.; Patroni, A.; Guillerm, E.; Pépin, D.; Benali-Furet, N.; Wechsler, J.; Manceau, G.; Bernard, M.; Coulet, F.; Larsen, A.K et al. Droplet digital PCR of circulating tumor cells from colorectal cancer patients can predict KRAS mutations before surger *Mol. Oncol.* **2016**, *10*, 1221–1231. [PubMed]
92. Alcaide, M.; Cheung, M.; Bushell, K.; Arthur, S.E.; Wong, H.-L.; Karasinska, J.; Renouf, D.; Schaeffer, D.F.; McNamara, S.; D Tertre, M.C.; et al. A Novel Multiplex Droplet Digital PCR Assay to Identify and Quantify KRAS Mutations in Clinical Specimen *J. Mol. Diagn. JMD* **2019**, *21*, 214–227.
93. Nam, A.S.; Chaligne, R.; Landau, D.A. Integrating genetic and non-genetic determinants of cancer evolution by single-ce multi-omics. *Nat. Rev. Genet.* **2021**, *22*, 3–18. [CrossRef]
94. Zhang, H.; Karnoub, E.-R.; Umeda, S.; Chaligné, R.; Masilionis, I.; McIntyre, C.A.; Sashittal, P.; Hayashi, A.; Zucker, A Mullen, K.; et al. Application of high-throughput single-nucleus DNA sequencing in pancreatic cancer. *Nat. Commun.* **202** *14*, 749. [PubMed]

95. Halbrook, C.J.; Lyssiotis, C.A.; Di Pasca Magliano, M.; Maitra, A. Pancreatic cancer: Advances and challenges. *Cell* **2023**, *186*, 1729–1754. [PubMed]
96. Neesse, A.; Michl, P.; Frese, K.K.; Feig, C.; Cook, N.; Jacobetz, M.A.; Lolkema, M.P.; Buchholz, M.; Olive, K.P.; Gress, T.M.; et al. Stromal biology and therapy in pancreatic cancer. *Gut* **2011**, *60*, 861–868.
97. Liot, S.; Balas, J.; Aubert, A.; Prigent, L.; Mercier-Gouy, P.; Verrier, B.; Bertolino, P.; Hennino, A.; Valcourt, U.; Lambert, E. Stroma Involvement in Pancreatic Ductal Adenocarcinoma: An Overview Focusing on Extracellular Matrix Proteins. *Front. Immunol.* **2021**, *12*, 612271.
98. Truong, L.-H.; Pauklin, S. Pancreatic Cancer Microenvironment and Cellular Composition: Current Understandings and Therapeutic Approaches. *Cancers* **2021**, *13*, 5028. [CrossRef]
99. Fend, F.; Raffeld, M. Laser capture microdissection in pathology. *J. Clin. Pathol.* **2000**, *53*, 666–672.
100. Boyd, Z.S.; Raja, R.; Johnson, S.; Eberhard, D.A.; Lackner, M.R. A tumor sorting protocol that enables enrichment of pancreatic adenocarcinoma cells and facilitation of genetic analyses. *J. Mol. Diagn. JMD* **2009**, *11*, 290–297.
101. Burdziak, C.; Alonso-Curbelo, D.; Walle, T.; Reyes, J.; Barriga, F.M.; Haviv, D.; Xie, Y.; Zhao, Z.; Zhao, C.J.; Chen, H.-A.; et al. Epigenetic plasticity cooperates with cell-cell interactions to direct pancreatic tumorigenesis. *Science* **2023**, *380*, eadd5327. [PubMed]
102. Cibulskis, K.; Lawrence, M.S.; Carter, S.L.; Sivachenko, A.; Jaffe, D.; Sougnez, C.; Gabriel, S.; Meyerson, M.; Lander, E.S.; Getz, G. Sensitive detection of somatic point mutations in impure and heterogeneous cancer samples. *Nat. Biotechnol.* **2013**, *31*, 213–219.
103. Puleo, F.; Nicolle, R.; Blum, Y.; Cros, J.; Marisa, L.; Demetter, P.; Quertinmont, E.; Svrcek, M.; Elarouci, N.; Iovanna, J.; et al. Stratification of Pancreatic Ductal Adenocarcinomas Based on Tumor and Microenvironment Features. *Gastroenterology* **2018**, *155*, 1999–2013.e3. [CrossRef] [PubMed]
104. Gao, R.; Bai, S.; Henderson, Y.C.; Lin, Y.; Schalck, A.; Yan, Y.; Kumar, T.; Hu, M.; Sei, E.; Davis, A.; et al. Delineating copy number and clonal substructure in human tumors from single-cell transcriptomes. *Nat. Biotechnol.* **2021**, *39*, 599–608. [PubMed]
105. Swanton, C. Intratumor heterogeneity: Evolution through space and time. *Cancer Res.* **2012**, *72*, 4875–4882. [CrossRef]
106. Iovanna, J.; Dusetti, N. Speeding towards individualized treatment for pancreatic cancer by taking an alternative road. *Cancer Lett.* **2017**, *410*, 63–67. [CrossRef]

MDPI AG
Grosspeteranlage 5
4052 Basel
Switzerland
Tel.: +41 61 683 77 34

Cancers Editorial Office
E-mail: cancers@mdpi.com
www.mdpi.com/journal/cancers

www.ingramcontent.com/pod-product-compliance
Lightning Source LLC
Chambersburg PA
CBHW062041220926
51850CB00010B/2626
* 9 7 8 3 7 2 5 8 6 9 4 2 8 *